Japanese/American Technological Innovation
The Influence of Cultural Differences on Japanese and American Innovation in Advanced Materials

Japanese/American Technological Innovation
The Influence of Cultural Differences on Japanese and American Innovation in Advanced Materials

Proceedings of the symposium on Japanese/American Technological Innovation held December 1990 at the University of Arizona, Tucson, Arizona.

Edited by

W. David Kingery
Professor of Materials Sciences and Engineering
Professor of Anthropology
University of Arizona
Tucson, Arizona

Conference Supported by
U.S. Air Force Office of Scientific Research
NGK Spark Plug Company
Sumitomo Metals Company
University of Arizona College of Engineering
University of Arizona Program on Culture, Science and Technology

Elsevier
New York • Amsterdam • London • Tokyo

No responsibility is assumed by the publisher for any injury and/or damage
to persons or property as a matter of products liability, negligence or otherwise,
or from any use or operation of any methods, products, instructions, or ideas
contained herein.

Elsevier Science Publishing Company, Inc.
655 Avenue of the Americas, New York, New York, 10010

Sole distributors outside the United States and Canada:

Elsevier Science Publishers B.V.
P.O. Box 211, 1000 AE Amsterdam, The Netherlands

© 1991 by Elsevier Science Publishing Company, Inc.

This book has been registered with the Copyright Clearance Center, Inc.
For further information, please contact the Copyright Clearance Center, Inc.,
Salem, Massachusetts.

All inquiries regarding copyrighted material in this publication, other than
reproduction through the Copyright Clearance Center, Inc., should be directed to:
Rights and Permissions Department, Elsevier Science Publishing Company, Inc.,
655 Avenue of the Americas, New York, New York 10010. FAX 212-633-3977.

This book is printed on acid-free paper.

ISBN 0-444-01633-3

Current printing (last digit)
10 9 8 7 6 5 4 3 2 1

Manufactured in the United States of America

CONTENTS

dedication vii

preface ix

participants xi

INTRODUCTION xv

INNOVATION IN CONTEXT 1
    Japan and the United States: A Cultural Comparison
        Eric Poncelet 3
    Japanese and American Approaches to Technological
        Innovation: Cultural Influences
        Eric Poncelet 23
    Innovation in the Arts: Japanese and American Approaches
        Jacques Maquet 35
    Models of Innovation and Their Policy Consequences
        Stephen J. Kline 43
    Some Conjectures about Innovation in Nascent and
        Infant Advanced Material Technologies
        W. David Kingery 59
    A Comparison of Japanese and U.S. High-Technology Transfer
        Practices
        Robert S. Cutler 67
    The Japanese Research Environment - A Student's View
        from Within
        Mary I. Buckett 81
    Japanese Research Students in the United States
        Richard C. Bradt 91
    Cross-Cultural Studies of Management and Organizational
        Behavior
        Toshimasa Kii 93
    Japanese Rural Household Organization and Patterns of
        Scientific/Industrial Group Management
        Robert McC. Netting 99
    Japanese Culture and Innovation
        Shin-Pei Matsuda 105

HIGH TEMPERATURE OXIDE SUPERCONDUCTORS 111

LOW PRESSURE DIAMOND SYNTHESIS 135

SILICON NITRIDE STRUCTURAL CERAMICS 161

MANAGEMENT OF INNOVATION 207

COMMENTARIES 231
    Smithsonian Horizons
        Robert McC. Adams 233
    Cultural Influences on Technological Development in the
        United States and Japan: An Archaeologist's Musings
            Michael Brian Schiffer 235
    Strains in the American Industry-Science-Education Triad
        Robert McC. Adams 243
    The Distinction Between Science and Engineering
        Bryan Pfaffenberger 249
    The Management of Knowledge and the Role of Culture
        in the R/D Enterprise
            Rustum Roy 253
    An Ethnographic Approach to the Conference:
        An Anthropological Perspective
            Eric Poncelet 259
    Conference Discussions
        W. David Kingery 263

SUMMARY 283

AUTHOR INDEX 287

SUBJECT INDEX 289

This volume is dedicated

to

Donald R. Ulrich, 1936-1990

    Dr. Donald R. Ulrich, deputy director of the Chemical and Atmospheric Directorate of the Air Force Office of Scientific Research, was a major contributor to the organization and support of this conference. The Air Force programs he developed and managed were focused on the development of chemistry-based processing methods to achieve superior performance of advanced materials, on the development of novel multifunctional materials, on the exploration of materials with greatly improved optical and mechanical behavior, and on the development of understanding of a broad spectrum of phenomena in materials. His programs have resulted in advanced materials innovations important for both defense and commercial applications. Don was a strong proponent of international cooperation in science and engineering, and put in place programs which involved participants from France, Japan and the United Kingdom as well as the United States. His recognition of the importance of innovations as the ultimate goal of research, the importance of the internationalization of technology, and the critical inter-relations between culture and technology were typical of his farsighted approach to mission-oriented research. His creative leadership will be sorely missed.

PREFACE

First of all, I should like to appreciate the support of the United States Air Force Office of Scientific Research, the Sumitomo Metals Company, NGK Spark Plug Company, the University of Arizona College of Engineering and Program on Culture, Science and Technology. The Air Force Office of Scientific Research support was a result of several conversations with Dr. Donald R. Ulrich, to whom this volume is dedicated. His vision for the Air Force included the development of an increasingly effective management of innovation. He himself managed a remarkably productive program of research and development on chemistry and materials. He recognized that differences in the Japanese and American cultures affect the ability of the Air Force to tap into Japanese technological capabilities, affect relationships which are developing between the U.S. and Japanese forces and are important for long range Defense Department planning and for the development of an industrial base in the United States. It was his hope and expectation that this conference would be a contribution toward identifying and perhaps alleviating cultural differences which affect international cooperation.

Second, I should like to express my appreciation to the participants who were willing to steal a week from their busy schedules to participate at the meeting. Most of the hundreds of books and articles about innovation in Japan and the United States have been anecdotal reports written by traveling university presidents, corporate executives, management consultants and professional journalists. Like all Americans, they know that Japanese innovations have led to high quality and low cost autos, cameras and consumer electronics. All agree that America is strong in basic science and Japan is strong in innovative products. Anecdotes are provided as supporting evidence along with statistics that contrast American and Japanese performance. Our meeting was different in that most of our discussants came from the ranks of the doers — innovative research workers and managers of innovative research who have years of accumulated personal experience. In addition, both the Japanese and American technical participants had substantial interactions with their foreign counterparts in corporate, university and national laboratories; in a real sense, it was a meeting of colleagues. The topic of the meeting was restricted to three types of advanced materials, each of which is in a precommercial or early commercial stage of development. This permitted markedly more focussed discussions.

I should like to express my special appreciation to the group anthropologists--R. McC. Adams, T. Kii, W. Longacre, J. Maquet, R. Netting, B. Pfaffenberger, E. Poncelet, M. Schiffer and T. Weaver — who ventured with enthusiasm into the new field of the anthropology of technology. They accepted to attend with some reluctance since none are specialists in the field. At the conference they provided a depth of knowledge and sensitivity to social and cultural issues that proved stimulating and was essential to the meeting's success.

The conference assembled with an informal reception on Sunday evening and continued to a Friday closing. Its business could have been concluded in three days of nonstop meetings, but there was general agreement that achievement of real communication, as opposed to presentations, between members of very different peer groups required some time to develop. The atmosphere and effectiveness of interactions noticeably changed (for the better) as the week progressed. The participants clearly benefitted from both the formal conference sessions and relaxed informal discussions. This report on the conference is not a tidy sequential narrative of the meeting, but rather a presentation of many different points of view in a format that I hope conveys the image that technological innovation is a process rather than an action — a process that brings together many different strands of activity, all of which are influenced by the social and cultural context in which they are embedded.

Finally, I should like to thank Eric Poncelet for his contributions as research assistant and colleague in bringing together much of the background information, proposing and preparing an essay on the ethnography of a Japanese-American conference, and helping edit this conference report. Transcripts of many conference discussions and typescript for these proceedings were prepared by Dorothyanne Peltz.

W. David Kingery

PARTICIPANTS

Adams, Dr. Robert McC., Secretary
Smithsonian Institution
Washington, DC 20560

Angus, Prof. John C.
Dept. of Chemical Eng.
Case Western Reserve Univ.
Cleveland, Ohio 44106

Bradt, Dr. Richard C.
Dean, Mackay School of Mines
University of Nevada-Reno
Reno, NV 89557-0047

Buckett, Ms. Mary I.
Dept. MSE
Northwestern University
2145 Sheridan Rd.
Evanston, IL 60208

Button, Dr. Daniel P.
Dupont Japan Tech. Ctr.
12F-Elec., Dupont Tower
Shin-Nikko Bldg.
10-1, Toranomon 2-chome
Minato-ku Tokyo 105 Japan

Chiang, Prof. Yet-Ming
M.I.T.
Bldg. 13, Rm 4090
Materials Science and Engineering
77 Massachusetts Ave.
Cambridge, MA 02139

Cutler, Dr. Robert S.
12306 Captain Smith's Court
Potomac, MD 20854

Elder, Dr. J. Tait
1256 A Highway 61 E
Two Harbors, MN 55515

Jackson, Prof. Kenneth
Arizona Materials Laboratory
4715 E. Fort Lowell
Tucson, AZ 85712

Kamo, Dr. Mutsukazu
Mukizaishitsu Kenkyusho
National Institute for Research in
  Inorganic Materials
101 Namiki, Tsukuba, 305 Japan

Katz, Dr. R. Nathan
Norton Prof. of MSE
Worcester Polytechnic Institute
100 Institute Rd.
Worcester, MA 01690

Kii, Dr. Toshimasa
Dept. of Sociology
Georgia State Univ.
University Plaza
Atlanta, Ga 30303

Kingery, Prof. W. David
Materials Science & Engineering
University of Arizona
338E Mines Bldg. 12
Tucson, AZ 85721

Kitazawa, Prof. Koichi
University of Tokyo
Dept. of Industrial Chemistry
Dept. of Superconductivity
Hongo, 7-3-1
Bunkyo-ku, Tokyo, 113, Japan

Kline, Prof. Stephen J.
Mechanical Eng. Dept.
Stanford Univ.
Stanford, CA 94305

Kobayashi, Prof. Kazuo
Dept. of MSE
Faculty of Engineering
Nagasaki University
1-14, Bunkyo-cho, Nagasaki-shi
Nagasaki 852, Japan

Kritchevsky, Dr. Gina, Director
Corporate New Business Development
Donnelly Corp.
4545 E. Fort Lowell, Suite 301
Tucson, AZ 85712

Larbalestier, Dr. David M.
University of Wisconsin
909 Engingeering Research Bldg.
1500 Johnson Drive
Madison, Wisconsin 53706

Longacre, Dr. Wm.
Anthropology Dept.
University of Arizona
Anthropology 210
Tucson, Az 85721

Manak, Dr. Rita C.
Director, Office of Technology Transfer
University of Arizona
OALS
Box 5, 845 N. Park Ave.
Tucson AZ 85719

Maquet, Dr. Jacques
University of California
Department of Anthropology
408 Hilgard Ave.
Los Angeles, CA 90024-1553

Matsuda, Dr. Shin-Pei
Director
Superconductor Research Center
Hitachi Research Laboratory
Hitachi, Ltd.
Hitachi, Ibaraki, Japan 319-12

McLean, Dr. Arthur F.
Ford Motor Co. (Retired)
6225 N. Camino Almonte
Tucson, AZ 85718

Morgan, Dr. Peter E.D.
Rockwell International
1049 Camino Dos Rios
Thousand Oaks, CA 91360

Moriyoshi, Dr. Yusuke
Mukizaishitsu Kenkyusho
National Institute for Research in
  Inorganic Materials
101 Namiki, Tsukuba
305 Japan

Netting, Prof. Robert McC.
Anthropology Department
University of Arizona
Anthropology 404
Tucson, AZ 85721

Ogren, Dr. John
TRW
Bldg. O-1, Room 2281
1-Space Park
Redondo Beach, CA 90278

Pfaffenberger, Prof. Brian
University of Virginia
107 Woodstock Drive
Charlottesville, VA 22901

Poncelet, Mr. Eric
Department of Anthropology
University of Arizona
Tucson, AZ 85721

Richerson, Dr. David W.
Ceramatec, Inc.
2425 South 900 West
Salt Lake City, Utah 84119

Roy, Dr. Rustum
The Pennsylvania State Univ.
Materials Research Laboratory
University Park, PA 16802

Sato, Mr. Yasushi
Sumitomo Metals Co.
Department of MSE
University of Arizona
Tucson, AZ 85721

Schiffer, Prof. Michael Brian
Anthropology Department
University of Arizona
Anthropology 314A
Tucson, AZ 85721

Smerdon, Dr. Ernest T
Dean, College of Engineering & Mines
University of Arizona
Civil Engineering 100
Tucson, AZ 85721

Swalin, Prof. Richard
Arizona Materials Laboratory
4715 E. Fort Lowell
Tucson, AZ 85712

Tajima, Dr. Yo
NGK Spark Plug Co. Ltd
14-18 Takatsuji-cho, Mizuho-iku
Nagoya, 467 Japan

Uhlmann, Prof. D.R.
University of Arizona
Head, Department of MSE
4715 E. Fort Lowell
Tucson, AZ 85712

Watanabe, Dr. Seiichi
Director, Research Ctr.
Sony Corp.
174 Fujitsuka-cho Hodogaya-ku
Yokohama 240 Japan

Weaver, Prof. Thomas
Anthropology Department
University of Arizona
Anthropology 225
Tucson, AZ 85721

Wolsky, Dr. Alan M.
Energy & Environmental Systems
Argonne National Laboratory
9700 S. Cass Ave.
Argonne, IL 60439

Yokoi, Mr. Hitoshi
NGK Spark Plug Co.
Department of MSE
University of Arizona
109 Mines Bldg. #12
Tucson, AZ 85721

INTRODUCTION

This volume reports on a week-long conference of about forty people. Roughly half were scientists-engineers who have been and are successful researchers, managers and innovators playing leadership roles in materials research and development; these participants were equally divided between Japanese and Americans. In order to have a fairly narrow and focussed discussion all are working on only three types of advanced materials which are in the pre-commercial, nascent or infant commercial stages of development. Another quarter of the participants were active or former successful managers of materials development and/or serious scholars studying innovation and technology transfer. The final quarter of participants consisted of anthropologists and sociologists, most of whom had little or no background in technology or technological innovation. In a very real sense the conference was an experiment in an emerging field that I shall call the anthropology of technology.

Until about twenty years ago the history of technology was dominated by detailed narrative descriptions which were generally positivist and reductionist. Historians of technology refer to this as an internalist view of technology that is seen as somehow independent of the surrounding society. It was a widely held view that technology was responsible for social change rather than vice versa — technological determinism. Progress was taken as inevitable and more often than not the mechanism of progress was seen as a linear progression from pure science to applied science (equated with technology) to engineering and manufacturing. This is a process that can be directed and controlled by following a progression from research to development to production and marketing. In the last two or three decades a "new" and more realistic history of technology has developed which focuses on technological change and views technology as a system that involves social organization, management behavior, human perceptions, environment, culture, economics and politics as important formative factors. It contends that technological systems must be seen in the social and cultural context in which they are constructed. The engineering community is beginning to recognize and accept this. The topic of the 1990 National Academy of Engineering Annual Meeting Symposium was "Engineering as a Social Enterprise".

Stephen Kline in a following chapter "Models of Innovation and Their Policy Consequences" describes how and why we must be concerned with the nature of the socio-technical systems in which technology is embodied and that such systems are extremely complex. Innovation may involve changes in not only products and processes of manufacture, but also of the social arrangements of the system, fiscal and legal matters, the socio-technical systems of distribution and use and the overall system as a whole. It necessarily follows that the linear model of research-development-production-marketing cannot possibly be correct. Kline proposes a more complex paradigm for innovation that involves a number of different feedback loops.

In discussing technological innovation in some depth, we are somewhat on the horns of a dilemma. We need the participation of the scientific and engineering craftsmen who are out there doing the innovation and have firsthand knowledge of what's actually happening. But their model for such a discussion tends to be one of technological determinism and an almost inevitable sequence of events that we know to be false or misleading. We have addressed this problem by bringing the technologists and managers together with anthropologists who have little background in technical detail but a presumed sensitivity to the social and cultural contexts of technological activities and socio-technological change. In a further restriction on the scope of our endeavor, the topic is limited to influences of cultural differences — specifically between Japan and the United States — on technological innovation. We have chosen to focus on Japan vis-à-vis the United States for two reasons: first there are distinct cultural differences which many books and articles have associated with relative success in science, technology and corporate management; second, Japan and America are world leaders in the field of advanced materials.

Culture refers to the framework of behavior, beliefs and customs common to a society, the structure within which events and behavior are interpreted, the values and expectations within which the world is ordered. Japan and America have national cultures but there are also a variety of subcultures associated with smaller social organizations such as universities, national laboratories and corporations within which technological innovation occurs. Cultures are shaped by a variety of influences and change over time. The national cultures of both Japan and the United States are continually changing and sometimes thought to be converging. In the United States, the changing role of women has led to revolutionary changes in the family and the workplace. Managers of American national laboratories and many companies are actively attempting to modify their laboratory or corporate culture. Japanese women are rarely seen in technological and managerial roles, but that seems to be changing. Traditionally, Japanese salarymen have lifetime employment and work is the most important activity in their life. But there is some fraction of the generation under age 30, the "Shinjinrui", who are not afraid to change jobs, who want to avoid the "three k's" — kiken (dangerous), kitsui (hard) and kitanai (messy). Shinjinrui are said to care more about their pleasures than their work. Cultural contexts are deeply embedded and difficult to alter, but we must understand that our conference was populated mostly by participants in the 35-65 year age group. It is that strata of our societies, leaders of this and coming decades, who are discussing and being discussed.

Advanced materials are a particularly appropriate area of technology for a meeting such as this. Materials are an enabling technology that make possible improved performance of devices critical to the improved performance of more complex systems. As such, these materials contribute more to high-value-added manufacturing than their own cost or market value. Almost a century ago, development of electrical lighting systems was dependent upon the development of new materials for the filaments of incandescent bulbs as a critical component of much larger electrical systems that included generating stations, transformers, power lines and new financial institutions. There were improved carbon filaments, bulbs with incandescent solid electrolyte Nernst glowers and a variety of refractory metals. In the end, drawn-wire tungsten allowed the development of brighter, longer-life incandescent lighting that made the whole system more efficient and effective. The same can be said for ceramic ferrite magnetic memory devices which transformed the development of early computer systems. All industries and all systems are limited by the performance of materials as components. If we had lighter metals . . . . . if we had tougher ceramics . . . . . if we had higher temperature superconductors . . . . . .

In mature industries, such as binoculars, cameras, clocks and watches, hand calculators, microwave ovens, motorcycles, semiconductors, tape recorders, televisions receivers and VCRs amongst others, it is clear that Japanese innovations have led the world and enabled Japan to capture a significant market share. Dozens of analyses and interpretations have been published. Technological innovations have been accompanied by the development of a Japanese manufacturing system which combines high quality with low cost of production. This has been achieved by flexible manufacturing schedules with relatively short production runs of several closely related products. Manufacturing concerns in Japan have considered the development and well-being of human resources as a principle objective. This has led to workers cross-trained over a range of skills, effective participation and acceptance of responsibility at all levels of the hierarchy, and an organizational approach of cooperative work groups which range across a variety of related disciplines and include technicians and craftsmen as well as engineers. There is an emphasis on seeking consensus, open communication and decision-making as close as possible to the point of implementation. Finally, there are strong communications and interactions with customers and suppliers that provide rapid feedback into design, production and engineering.

This conference has been concerned with earlier stages of product development and particularly the invention of new products and processes and their development in nascent and infant industries. At this stage of the innovation process, there has been much less comparison of Japanese and American industries. The nucleation and growth of new industries may well be the critical step in the development of new high-value-added manufacturing. It is often stated that the United States is the source of major new concepts and products for which the Japanese devise more effective manufacturing processes. That reductionist picture is one which requires careful examination. In order to focus our discussions, we have selected three advanced materials that are widely considered to be harbingers of the future. In 1986 the discovery and confirmation of high temperature oxide superconductors unleashed extensive international research which has led to these materials being on the threshold of commercial innovation. Synthetic diamonds formed at low pressures have been undergoing technological development for about ten years and commercial products are now being manufactured on a limited scale. Silicon nitride structural ceramics have been actively pursued for almost forty years and are now in regular production in what may be described as an infant industry. Taken as a group, these advanced materials provide insights into the progression from invention to nascent commercialization to infant industry.

Superconductivity was discovered by Heike Kamerlingh Onnes, a physicist at the University of Leyden, in 1911. He found that the electrical resistance of mercury disappears completely at temperatures below 4.2° Kelvin. Onnes received a Nobel prize in 1913. In the superconducting state, an electrical current persists undiminished for weeks. During the 1950s and 1960s, there were hundreds of new superconductors discovered and a transition temperature of 23.2°K for Nb_3Ge was achieved in 1973. With expensive liquid helium refrigeration, high magnetic fields can be developed with these materials that are required for nuclear magnetic resonance medical imaging and will be required for the Superconducting Super Collider (SSC) project. In 1986 Johannes Bednorz and Karl Müller discovered a new oxide superconductor, barium lanthanum copper oxide, with a transition temperature of about 35°K. They received a Nobel prize in 1989. Within a year of their announcement, new oxide superconductors were discovered having transition temperatures greater than the boiling point of nitrogen, 77°K. These discoveries opened the possibility of devices operating with relatively inexpensive liquid nitrogen refrigeration. At a March 1987 meeting of the American Physical Society, a session on high temperature superconductivity was jammed with more than a thousand scientists. The full potential of oxide superconductors is a long way off, but an international race to transform discovery into devices and products is in full swing.

Almost a hundred years ago, the synthesis of diamonds was claimed by Moissan, but it was not until 1955 that Bundy, Hall, Strong, and Wentorf at the General Electric Co. announced successful diamond synthesis at high temperatures and intensely high pressures. This is now a commercial manufacturing process. At about the same time William G. Eversole at Union Carbide successfully synthesized diamond from the vapor phase at low pressures. After a lull of more than two decades, Russian work (Spitzyn and Bouilov and Derjaguin) published in 1981 and Japanese work (Kamo, Sato, Matsumoto and Setaka) published in 1983 showed that crystalline diamond could be grown on a non-diamond substrate from a gaseous mixture of hydrogen and methane. This presaged the likelihood of commercial innovation. Diamond is the hardest substance known, the best heat conductor, transparent over a wide spectrum and with attractive electrical properties for a number of applications. In December 1990, *Science* magazine (p. 1640) described the vapor synthesis of diamond to be a "glittering prize for materials science." Wear resistant coatings for cutting tools, heat sinks for electrical components, special windows and tweeters in stereo speakers are commercial products in a nascent industry.

Silicon nitride is a material combining low density, high hardness and high strength at high temperatures together with a resistance to oxidation that allows it to replace or extend the performance of metal parts operating at or subject to high temperatures. The anticipation of being able to construct high temperature more efficient gas turbines led to the initiation of a serious development program in Britain during the 1950s. This was taken up again in an ambitious series of demonstration engines during the 1970s. The U.S. Department of Defense funded a program with Ford which led to the demonstration that an automotive ceramic gas turbine engine was feasible; the Garrett Turbine Engine Company operated a turboprop engine containing silicon nitride parts; Kyocero Corporation in Japan operated a three cylinder silicon nitride diesel automotive engine. During this decade, there were continuing improvements in compositions, processes, properties and design with brittle materials. In the late 1970s, commercial silicon nitride cutting tools came to the market and in the 1980s, silicon nitride ball bearings, automotive diesel engine glow plugs and swirl chambers in rocker arm wear pads were introduced by Mazda, Isuzu, Toyota and Mitsubishi. More complex parts, high performance turbocharger rotors, were first introduced into commercial production in 1985. The high cost of materials and reliable processing must be substantially lowered for large scale cost-competitive replacement of metals by silicon nitride structural ceramics. At present the total annual sales of this infant industry are about $200 million.

Comparisons of Japanese and American invention and innovation is not a novel activity. More than ten years ago, Constance Holden (*Science, 210,* p. 751, Nov. 1980) reported in a House of Representatives report: "We believe that the Japan rate of industrial progress and stated economic goals should be as shocking to Americans as was Sputnik". She went on to bring out most of the generalizations that have been many times elaborated. It is our hope that a more narrow focus on advanced materials, on nascent industries and on the influences of culture will provide an immediacy and depth to our discussions of national differences and the rates of innovation with new products. We all appreciate that there is a great deal of diversity in both nations in the size, complexity and wealth of organizations which generate innovations. By focusing on nascent industries and considering separately the contributions of different innovating organizations such as national laboratories, universities, large corporations and small entrepreneurial firms, we may hope to relate residual differences in innovation styles with cultural variation. We may ask whether the effectiveness of different innovating units reflect culturally based priorities and whether national preferences exist for innovation in nascent or mature industries.

As we examine the record, we find that some Japanese and American innovations have occurred nearly simultaneously. In Japan the locus of research innovations has been more effective at national laboratories than in the United States. In contrast it seems that more innovations have come from universities in the United States than occurs in Japan. We may ask whether these and other similarities and differences in the nature, rate and direction of innovations in advanced materials can be explained, wholly or in part, as a result of cultural differences, social organization, corporate structure, management style, the cost of capital, government policies or pure chance. Addressing these questions requires considering innovation as a socio-technological *process* embedded in a cultural and social context.

When we question how culture and social context affect innovation, we may inquire as to the effectiveness of communication between designers, engineers and users. We may ask if the record shows that Americans are better at bold new theories and Japanese better at incremental improvements. If so, can we attribute this to different educational systems as has often been suggested? Americans generally believe that design by committee is catastrophic while Japanese seem much more comfortable with that process. Does this suggest that the style of Japanese and American innovation differ rather than the content. That is, can there be group consensus innovation in Japan which is as effective as directed team innovation in the United States? We may ask what accounts for the success of Japanese national laboratories. Again, in addressing cultural content, we may ask whether advanced material innovations occur in the same mode as agricultural innovations or as innovations in mature industries. We may ask if laboratory, village, corporate and household mobilization of labor follow the same patterns. Two major differences of Japanese and American life are the strong "vertical" social structure in Japan and Japanese group consensus decision-making versus American directed team approach. We may ask how these cultural differences affect invention and innovation. It is by considering and answering some of these questions that we may add context to our understanding of innovation in Japan and the United States.

To address these sorts of questions, the conference included several presentations and extended discussions of general and particular contexts of technological innovation. Some of these are brought together in the next section of this report. As background material for the participants, Eric Poncelet prepared a report on a Cultural Comparison Contrasting Japan and the United States and another report on Cultural Influences on Japanese and American Approaches to Technological Innovation. These review a large literature in a way that brought all participants up to a level playing field. From the Japanese point of view Shin-Pei Matsuda of Hitachi Research Laboratory presented his impression of cultural influences on Japan. Jacques Maquet of UCLA presented a stimulating anthropological approach toward Japanese and American approaches to innovation in the arts. Focusing on the particular case of cultural basis for innovation styles in Japan and the United States and the implication of these for competitiveness. Stephen Kline who is professor of Mechanical Engineering and also of Values Technology Science and Society at Stanford University provided a detailed evaluation of the nature of technology, the concept of innovation, the failures of a simple linear model which is often assumed and the necessity that it be replaced with a more complex model in order to understand the nature of innovation. His report emphasizes the socio-technical nature of industry and technology and the necessity to look at it as a complex system. With these models in mind, he was able to discuss innovation in Japan and its relationship to Japanese cultural norms. The conference is mostly concerned with invention, nascent industries and infant industries; David Kingery presented some conjectures about innovation in advanced materials within these contexts.

There were four additional contributions that provided the conference with a framework for thinking about innovation. Robert Cutler provided a discussion of Japanese and American technology transfer practices. Mary Buckett, a graduate student who spent time in Japan, provided a student's view of the Japanese research environment. Toshimasa Kii, a sociologist with experience advising American and Japanese companies about the management of technology, described differences of management and organizational behavior that are important in establishing effective communications. Finally, Robert Netting, a distinguished cultural anthropologist provided an anthropological perspective on Household Agricultural Technology and the models it suggested for the behavior of modern technological innovation.

The conference then discussed specific histories, the first of which was High Temperature Oxide Superconductors. This discussion was led on the Japanese side by Prof. Koichi Kitazawa, a distinguished researcher at the University of Tokyo with an international reputation in oxide superconductor studies. Dr. Shin-Pei Matsuda, director of Hitachi's Superconductor Research Center, provided another Japanese point of view. On the American side Dr. David Larbalestier, Director of the Applied Superconductor Laboratory at the University of Wisconsin, Prof. Rustum Roy of Pennsylvania State University and Dr. Alan Wolsky of the Argonne National Laboratory provided different viewpoints on the technology and its innovations.

On the subject of Low Pressure Diamond Synthesis, one of the pioneers in the field and a continuing contributor to the technology, Prof. John Angus of Case Western Reserve University, Prof. Rustum Roy of Pennsylvania State University and Dr. John Ogren of TRW provided different viewpoints on the American experience. On the Japanese side Dr. Mutsukazu Kamo and Dr. Yusuke Moriyoshi provided input from the NIRIM laboratory in Tsukuba City where many of the technical innovations were developed.

In the discussion of silicon nitride structural ceramics Japanese innovations were represented by Kazuo Kobayashi of Nagasaki University and by Dr. Yo Tajima of NGK Spark Plug Company which is one of the world's leading manufacturers of silicon nitride material. On the American side David Richerson now at Ceramatec but previously an active researcher at Norton Company and at Garrett Turbine Company was discussion leader. Dr. R. N. Katz, now at Worchester Polytechnic Institute but formerly director of the AMRAC Program on ceramics for gas turbines, Dr. Arthur McLean former manager of the Ford Motor Company program on the advanced turbines and, from a different viewpoint, Dr. Richard Bradt, now dean of the Mackay School of Mines at the University of Nevada-Reno, participated in discussions.

A number of people with technical backgrounds participated in discussions of the management of innovation. Dr. Daniel P. Button, Dupont Japan Technical Center, was able to discuss Dupont's experience both in Japan and the United States. Dr. J. Tait Elder was a venture leader at 3M Company and is now a management consultant. Prof. Kazuo Kobayashi now at Nagasaki University was, until last year, director of a MITI research laboratory in Kyusho. Dr. Gina Kritchevsky is director of Corporate New Business Development of the Donnelly Corp. Arthur McLean was manager of the Ford Motor Company research program on ceramic automobile gas turbine materials. Prof. Richard Swalin, now at the University of Arizona, is the former vice president for research at Allied Chemical Corp. Dr. Seiichi Watanabe is Director of the Central Research Center of Sony Corp. There was thus present a rich collection of innovation managers familiar with general management concerns and having a close involvement with technological innovation.

Along with the specialists and managers in discussing the role of culture, we were fortunate to have a distinguished group of anthropologists and sociologists who actively participated in the discussions. These included Robert McC. Adams, Secretary of the Smithsonian Institution, Toshimasa Kii, a sociologist familiar with Japanese and American technology, Dr. William Longacre, head of the Anthropology Department at the University of Arizona, Dr. Jacques Maquet, professor of anthropology at UCLA, Prof. Robert Netting, cultural anthropologist at the University of Arizona, Prof. Brian Pfaffenberger, anthropologist and historian of technology at the University of Virginia, Dr. Michael Schiffer, professor of archaeology at the University of Arizona, and Prof. Thomas Weaver, professor of cultural anthropology at the University of Arizona. In addition to these anthropologists, there were a number of participants who had close interactions with the relationships between universities, government, national laboratories and industrial corporations including Prof. Richard Bradt of the University of Nevada-Reno, Dr. Robert Cutler, formerly with the NSF, Prof. Kenneth Jackson, formerly with Bell Laboratories and now at the University of Arizona, Prof. Koichi Kitazawa at the University of Tokyo, Prof. Kazuo Kobayashi now at Nagasaki University but formerly director of a MITI laboratory, Dr. Rustum Roy a leading materials scientist who is also one of the founders of the STS movement in the United States, Prof. Donald R. Uhlmann, head of the Department of Materials Science and Engineering at the University of Arizona, Dr. Alan Wolsky, director of Energy and Environmental Systems at the Argonne National Laboratory, and Dr. Peter Morgan, Rockwell International who spent six months as a visiting researcher at the Hitachi Research Laboratory in Japan.

For intense topical discussions, the conference divided into working groups focussed on high temperature oxide superconductors, low pressure diamond synthesis, silicon nitride structural ceramics and management of innovation. Each of these working groups summarized its discussions and presented them to the conference participants meeting as a whole. Four sections of the report are devoted to the discussion of these working groups.

Finally, there were insightful commentaries presented by all of the participants during the meeting and some of these have been collected together as a separate section of this report. With these we have included an informal report of oral discussions, based on our notes of the meeting. In attempting to summarize the conference proceedings, we are faced with the problem that the influence of culture and social context on technological innovation is an inherently messy subject. For instance, there has been in recent years a national furor for technology in Japan, a technological enthusiasm that recalls the context of American technology at the turn of the century. In *American Genesis* (Viking Penguin, New York, 1989, p. 1), Thomas P. Hughes recalls an image presented by historian and literary critic Perry Miller of Americans who "flung themselves into the technological torrent, how they shouted with glee in the midst of the cataract, and cried to each other as they went headlong down the chute that here was their destiny...." (Perry Miller, *The American Scholar,* xxxi, 51-69, 1961-62). This pretty well describes Japan of the 1970's and 1980s. In this same more recent period, the national consensus about technology in America, if any, was rather one of deep concern over Three Mile Island, the Challenger disaster, holes in the ozone layer, global heating and so forth. I am convinced that these different national attitudes deeply affect the ways in which the science and engineering communities, national laboratories, universities, CEO's, corporate managements, government agencies, legislators and ordinary citizens approach nascent technology and high-value-added new products. But I must admit that I have no way of quantitatively measuring what may be one of the most important aspects of the cultural context within which these developments are embedded.

Japanese/American Technological Innovation
The Influence of Cultural Differences on Japanese
and American Innovation in Advanced Materials

INNOVATION IN CONTEXT

During the conference several participants prepared and presented discussions of topics affecting the role of Japanese/American culture on technological innovation in advanced materials. In this section we have collected a number of those presentations which range from very general discussion of Japanese and American culture to very personal reports of individual experiences and perceptions of innovation in context.

JAPAN AND THE UNITED STATES: A CULTURAL COMPARISON

Eric Poncelet, University of Arizona

Introduction

There are two general schools of thought which may be considered in regards to the analysis of Japanese and American approaches to technological innovation. The first, known as the Universal or "Organizational" approach, stresses the universality of organizational principles and has been heavily pursued by economists and business administration scholars. An example of this school and one which focuses on managerial factors in specific is the work of William Ouchi (1981) on "Theory Z".[1] The second school, known as the Cultural/Historical approach, considers approaches to technological innovation to be outgrowths of cultural tradition as molded by historical experience. This second school of thought, where technological innovation is examined within the context of culture and society, will be the focus of our discussion.

The purpose of this paper is to examine Japanese and American cultures in the areas which influence their respective approaches to technological innovation. The discussion is broken down into three main and interrelated sections: 1) general cultural characteristics; 2) social structure; and 3) ecological, economic and political histories. In each section, characteristics from Japan and the U.S. will be compared and contrasted.

Before commencing the comparison, a definition of the word "culture" is in order. The term "culture", as used here refers to:

...the patterns of behavior and belief common to members of a society. It is the rules for understanding and generating customary behavior. Culture includes beliefs, norms, values, assumptions, expectations, and plans for action. It is the framework with which people see the world around them, interpret events and behavior, and react to their perceived reality (Spradley & Rynkiewich 1975).

Culture is learned and shared by members of the same society. It develops as a way of understanding and coping with a particular environment, and it changes continuously over time. Finally, culture influences and is, in turn, influenced by many aspects of human existence including religion and myth, language and communication, the creative arts, social structure, education, the environment, economics and politics.

This type of large-scale cultural comparison, or "national character" study, always runs the risk of succumbing to misleading stereotypes. For instance, almost every trait, good and bad, has at some time been attributed to the American and Japanese peoples. Numerous explanations have been advanced to account for the attributed qualities to the point that social scientists have spent much time and energy simply debating the validity of the concept of "national character". Consequently, it is important to recognize that the descriptions of cultural patterns presented in this paper are not meant to be absolutes but merely representative of the dominant forms found in each society. We believe that this broad-based approach is quite useful in serving our ultimate goal of relating Japanese and American cultural influences to technological innovation.

General Cultural Characteristics

The outline for this first section on general cultural characteristics is broken down into four parts: a discussion of values and attitudes, religions and mythologies, language and communication, and approaches to art. The last three parts may be considered as symbols or direct manifestations of culture.

Values and Attitudes

Values are conceptions of what is desirable. They serve as criteria for judgment, preference and choice. They are not individual goals or activities but are instead the rules by which goals are selected and activities chosen. Attitudes are the enactment of these values, and both of them change over time. Many traditional values in Japan and the U.S. have carried on to the present while others have weakened. When you consider the effects of these changes in addition to the great complexity of cultural systems, it is not surprising that contradictions and conflicts exist in the value systems and attitudes of Japan and the U.S. However, these contradictions and conflicts are somewhat resolved by their overall integration into society.

The discussion below turns now to some fundamental differences in Japanese and American values and attitudes. While the total list of differences is extensive, we have narrowed them down to the areas we find most pertinent to each country's approach to technological innovation. These areas include different values/attitudes towards: individuals and groups, confrontation and conformity, self-reliance and dependence, motivation, competition, responsibility, mobility, change, freedom, personal relations, work, temporal orientation, and science and technology.

Individuals and Groups

In traditional Japan, individualism was not held in especially high regard. The development of ego control (i.e. the capacity to act according to one's own judgments and beliefs) was inhibited from infancy, and personality, ambitions, and feelings were submerged in the interest of the family unit. This emphasis on standardization, uniformity and conformity inhibited the development of individual initiative, self-expression and realization, and personal responsibility (in the western sense). In society, it was considered brash for an individual to make decisions or even to simply urge the acceptance of an individual opinion. The individual Japanese was not an autonomous whole but a fraction of the whole. It was the group that was the primary point of reference. Group welfare and security was considered more important than individual welfare and autonomy. However, as individuals and group were fused into one, it was impossible for individuals to belong to more than one group. In more recent times, individual values have become more increasingly prized so long as they do not lead to nihilistic self-centeredness or selfishness.

Individualism in the U.S. is more highly valued, and the point of reference is the self instead of the group. According to American individualism, each person has within himself the right to make his own decisions, develop his own opinions, solve his own problems, have his own things, learn to view the world from his own perspective, and make it or break it in life on the basis of his own judgments. The emphasis on individualism is perhaps the most influential American value of all because it permeates all of the other values. Group affiliation is also encouraged, but this applies primarily to voluntary organizations and multiple-group affiliation, structures not commonly found in Japan.

Confrontation and Conformity

Confrontation is generally avoided in Japan, and because confrontation often necessitates avenging one's honor, indirection is heavily used to preserve harmony. This desire for harmony is strong and ultimately results in an emphasis on conformity. However, this conformity is not simply passive obedience. It also includes conformity to changing situations and contributes to the competition between groups.

When faced with a problem, Americans like to get to its source, and confrontation has traditionally been a highly valued means of accomplishing this. Americans also tend to be unwilling to compromise over "principles". Compromising solely for the purpose of maintaining group harmony is considered to be equivalent to selling out one's principles.

Self-Reliance and Dependence

Along the lines of suppressing individualism, the Japanese have long preferred dependence over self-reliance. The Japanese do not actually stress being dependent. Rather, mutual dependence is simply taken for granted. This desire for belongingness has become a necessary basis for establishing identity, and it is fueled by the fear of being left alone or ostracized by the group.

In line with "rugged" individualism, self-reliance is also highly valued in the U.S. The virtuous man in the U.S. is one who "owes nothing to any man". However, this emphasis on self-reliance is accompanied by a fear of dependence, itself considered as weak or immature, which ultimately creates insecurity due to the denial of the importance of others. This insecurity manifests itself in the lack of permanency in relationships, jobs, and families.

Motivation

Motivation in Japan has been traditionally based on the normative values of indebtedness, loyalty, and social and moral obligations (especially to one's family). Achievement, as it is oriented toward merit acquired by individual contributions to the goals of the group, has also been a powerful motivator. More recently, personal and family happiness, economic pursuits and profit, leisure time, and consumerism have increased in importance as sources of motivation.

For Americans, individual externalized achievement is the dominant motivating factor although ascription also exists as a variation. There is also a national confidence in effort. The basic belief is that one only has to try to succeed. Failure is attributed to the lack of will or effort on the part of the individual. Other sources of motivation include upward mobility, increased social status, a drive for security, and profit. This American success ethic is essentially a simplified product of the Protestant ethic where individual success has replaced spiritual salvation.

Competition

In Japan, individual competition is generally avoided. Competition between groups, on the other hand, is highly valued as a means of encouraging unanimity of effort with a group.

In the U.S., competition, be it individual, within groups, or between groups, is highly valued as a means for achieving success.

Responsibility

In Japan, responsibility is diffused in the group. Each group member shares the responsibility of every other member.

In the U.S., responsibility tends to be compartmentalized and is individually maintained.

Mobility

Mobility in Japan, especially as it affects the permanence within a group, has never been highly valued. Emphasis has been placed instead on maintaining one's personal relations. This value draws its roots from the importance of sedentism to the village production of rice.

In the U.S., a high value is placed on mobility. Indeed, mobility has proven to be a phenomenally successful economic strategy for Americans in general going all the way back to the expansion of the American frontier.

Change

In Japan, emphasis has long been placed on the supremacy of custom. A traditional fear of isolation has led to a deeply rooted conservatism which dislikes change and avoids initiatives. Of late, and riding the coat-tails of recent economic success, risk-taking has begun to rise in value.

In the U.S., there is a strong impetus to discard old ways and adopt new ones. However, long term political and economic stability have resulted in an increased impetus for Americans to rest upon their laurels.

Freedom

The Japanese traditional ideology of life may be described as a resignation to the "irresistible" realities. One does not find imbedded within this ideology the same notion of freedom found in the West. Instead, life is conducted without independent thought and is accompanied by a deferment to the status quo. Maximum individual freedom is obtained when very young (a time of no shame) or very old (a time when obligations to society are small or none).

Americans treat freedom as an inalienable right. For Americans, freedom means that they will be subject to minimal external constraint in pursuing their desires. This also implies a freedom to deviate from the norm. In the U.S., maximum individual freedom is obtained as an adult.

Personal Relations

Personal relations in Japan are characterized by both a deep-felt need to be nurtured in a sympathetic, understanding and harmonious environment and an outward fear of incurring or creating unnecessary obligations. Hence arises the Japanese preference for the protection provided by a group. Friendships within the group, when they do occur, are marked by high emotional content and typically extend into the private lives of

Personal relations in the U.S. are less characterized by a fear of adversarial relations than in Japan. They tend to be numerous but lacking in permanence and depth. Hence, they are distinguished by more frequent changes in friendships and group memberships. On the whole, Americans do not feel the pressure of obligations or reciprocity more common in Japan.

the actors. New forms of personal relations, however, suggest a change in the Japanese lifestyle toward individuation and a decrease in the sense of social obligation and reciprocity.

Approach to Work

The traditional Japanese approach to work is based on an ethic characterized by perseverance, discipline, and curiosity. However, the will and desire to pursue this ethic comes from obligations arising from the individual workers' decision to take on the work, not from coercion from above. Also supporting this work ethic is occupational training emphasizing the development of spiritual strengths such as composure, endurance, acceptance, social responsibility, and self-reflection. In recent times, hard work is not considered as much of a virtue as before. More of an emphasis is now being placed on leisure and the comforts and pleasure of today's consumer-oriented society. However, high value is still placed on producing quality products.

The traditional American approach to work is heavily influenced by the Protestant ethic which equates hard work and the accumulation of material wealth with virtuousness. Nevertheless, Americans still commonly regard work as "toil and trouble", something that a person must do to survive but not necessarily enjoy. Work is distinguished from play, but both are approached with the same sense of seriousness.

Temporal Orientation

Traditionally, temporal orientation was focused on the past because it was from the past that Japan's distinctive national character emerged. Currently, with Japan's rise to economic power, temporal orientation has shifted to the future, especially where technological issues are involved. What has not changed though is the high value placed on duration and endurance.

Temporal orientation in the U.S. is primarily directed toward the future, if not the immediate future. An emphasis is placed on immediate gratification. Americans equate time with money, place an accent on youthfulness, and maintain the attitude that one can always improve on the present.

Science and Technology

Science, for the Japanese, refers to applied science and is not generally perceived as being separate from technology (or manufacturing for that matter). However, the Japanese do emphasize the importance of technology over pure science. Traditionally, the Japanese attitude toward technology has been one where mastery was the goal and imitativeness, as a step toward mastery, was regarded as a virtue. This emphasis is

In the U.S., science and technology are seen as separate, and of the two, science (and specifically the natural sciences) is given a higher status. Science is believed to be the key to all improvement, and there is a constant drive in the U.S. to continue creating new things. Technology also plays an important part in this but only in a supporting role to science. This emphasis on science is evident in the high number of Nobel prizes in the natural

evident in the superiority of Japanese manufacturing processes. Finally, technology and technological change are seen as being closely related to culture and cultural change.

sciences garnered by the U.S. since 1900. Finally, both science and technology are generally perceived as being independent of culture.

Religion and Mythology

Religions are involved with man's attempts to gain some control over unpredictable events. They begin where science and technology leave off and play an important role in determining world view. As religions and myths are influenced by values and attitudes, so do they influence them in return.

Religion in Japan emerges basically from three non-exclusive sources: Shinto, Buddhism, and Confucianism. Shintoism is the oldest religion in Japan and is often regarded as the indigenous religion. It is based on vague conceptions that everything in nature is imbued with some degree of divinity. Buddhism was brought over to Japan from China (via Korea) in the sixth century A.D. and has since become the largest religion in Japan. Buddhism brings with it the belief in the transcendental and a sense of interconnectedness with the world. It also introduces a deep sense of impermanence, the concept of life after death, and compassion for all living things. Confucianism, also imported from China, is both a religion and a moral philosophy. It stresses the importance of ancestor worship and filial piety and places an emphasis on obedience, loyalty, and conformity to the social order. The assimilation of these three different religions into Japanese culture is very complete. The Japanese have a "this-worldly" cast of mind, and no concept of God exists abstractly or separate from the human world.

Religion in the U.S. differs from religion in Japan in that it is both a spiritual philosophy and a social ethic. The religious beliefs and practices are highly concerned with general morality in such areas as family relations, sexual customs, and civic responsibilities. The dominant form of religion is Christianity, and the dominant types are Protestantism and Catholicism. Judaism also plays a major role. While all of these religions are monotheistic, they see themselves as separate and distinct. Religion is not a particularly unifying institution in American life, and the spirit of the country is rather secular and rationalistic.

Mythology in Japan is closely related to the Shinto religion. The "origin myth" holds that Japan was created by the gods and goddesses and later ruled by a god and his divine attendants. The Japanese people are considered to be the direct descendants of these divine attendants. Consequently, both the land and the people of Japan are held to be divine. The Japanese also have a vision for "utopia". While the word itself means "no place", the Japanese model for it is often thought to be some distant idealized country that must be emulated or surpassed.

American mythology has been a powerful influence in the minds of Americans, and of all the American myths, none has been stronger than that of the loner moving west across the broad expanses of land. This is the myth of the frontier and the pioneering spirit, the heroic romantic cowboy and the trailblazing mountain man. This pioneering spirit stresses such values as individualism, self-reliance, autonomy, mobility, courage, self-actualization, personal growth and humanitarianism.

Religious and mythological distinctions between Japan and the U.S. have influenced cultural differences in the following areas: individual vs. group emphasis, linear vs. cyclical orientation, relations between church and state, and science and technology.

Individual vs. Group Emphasis

In Japan, devotion is not usually directed toward something transcendental

Christianity is a religion of the indi-

but toward something connected with the group. There is no belief in a single deity, but there is a heavy emphasis on obedience, loyalty, and conformity to the social order.

vidual. There is only one God, and salvation is achieved on an individual level.

Linear vs. Cyclical Orientation

In Japan, there is more of a cyclical orientation to life and the passage of time.

Judeo-Christian religions follow a linear orientation to life and the passage of time. Both are sequential and non-iterative.

Relations between Church and State

Japanese religions have long played an influential role in Japanese politics. A lack of separation between church and state is inherent in Confucian ethics. From Shintoism arises the belief in the divine ancestry of the emperor line and the association of the state with moral consciousness. Shintoism has played an important role historically in unifying the nation and bolstering nationalism.

In the U.S., church and state are more strictly separated. While support from the church is deemed to be politically important, the church and the state often find themselves in somewhat adversarial positions.

Mythology, Science and Technology

In Japan, society is considered to be "perfectible". The mastery of technology has played an important role in pursuing idealized or utopian states.

In the U.S., the myth of the western frontier has been replaced recently by the frontier of science. This is clearly demonstrated by the role played by the U.S. space program over the past thirty years.

Language and Communication

Communication is the process of transmitting thoughts and ideas from one mind to another, and language is a primary means for accomplishing this. Other forms of communication in addition to language include proxemics, kinesics, and writing. Language is of interest here because it acts as a mirror for the culture in which it develops; it reflects the dominant values and attitudes and in turn influences them. Japanese and American cultural differences are influenced in regards to the following areas: the communicative role played by language, the relationship of language to social structure, and the level of directness/indirectness with which language is applied.

Role of Language

In Japan, language is only one of many means of communication. As there is a strong belief that verbal language is not necessarily the best medium for enhancing human understanding, non-verbal communication plays an important role. Today, over-reliance on verbal language is still considered to be indicative of abruptness, immaturity, and possibly dishonesty.

The U.S. follows a tradition which assumes that language is "the" means of communication. A thought unspoken is considered to be useless.

Relation to Social Structure

Language in Japan constitutes a strong and deeply imbedded verbal status system. Communication between persons must include specific language which defines the status relationship between them. Communication must account for differences in rank and gender. Little equality exists in communication.

In the U.S., language is not as closely tied to status or social structure. Some forms of language (e.g. slang) are even used as expressions of affiliation and social conformity in order to transcend social stratification.

Directness and Indirectness

The Japanese are very sensitive about the use of incorrect and inappropriate language. They must be language-conscious at all times in order to avoid making errors or statements which might cause insult or violate norms of gender and social hierarchy. Overassertiveness is generally avoided. Consequently, their speech patterns are characterized by indirectness and ambiguousness and reflect their inherent desire to avoid conflict-causing situations.

For Americans, direct, clear, and effective speech is highly valued and considered to be practically an art form. Speaking one's mind and "telling it like it is" is encouraged, and less regard is paid to the effects of one's speech on others.

Art

Art, like language and communication, is also expressive of national character. Japanese and American cultural differences are revealed by the manner in which each country's art is intended to relate to ordinary, everyday life.

Approach to Art

Japanese art is seen only in relation to the realities and emotions of ordinary life. It is not an independent creation in a realm all its own but an instrument for the beautification of the whole surrounding in which it is placed. Both fine art and decorative art are intended for use in everyday life. They are expressions of a cultural inclination toward refining and polishing the sensuous world of the here and now. Much of Japanese everyday life is itself regarded as art, whether this be a tea ceremony or a military technique.

American art follows the European model which considers art to be the expression of an independent life of its own, a life which transcends ordinary human existence. American art deals with abstract life and is performed primarily for its own sake. Life is represented not as "being" but as a "search" for something else. Even when it is performed as a means of self-expression, American art is not intended for practical use to the same extent as Japanese art.

Social Structure

The second main section of this discussion concerns social structure. Social structure, as the term is utilized here, refers to the interrelated parts and roles in which a populace is ordered and assigned. It also assumes that the component institutions and regularized activities of society function in such as a way as to maintain system equilibrium. Social structure both derives from and derives cultural activity.

This section on social structure will be broken down into four parts. The first part will discuss the key features of Japanese and American social structure. The next two parts will describe two major components of social structure: the family and industrial organization. The final part will discuss education systems as they support and maintain the social structure.

Key Features of Social Structure

Chie Nakane (1970) has described two main and inter-related features critical to the description of Japanese and American social structure: group/individual orientation, and vertical/horizontal structure. These will be discussed in relation to their influences on frame vs. attribute emphasis, occupational systems, and structural relations.

Group/Individual Orientation

In line with Japanese values, group orientation takes priority over individual orientation in Japanese social structure. Nakane believes that all of Japanese social organization actually stems from this group consciousness and orientation. Groups in Japan are held together by two means: a natural feeling of solidarity between group members, and internal organization. The solidarity between group members is emotionally based and carries with it an "us" versus "them" attitude. It becomes difficult for the group members to transcend the group and act individually. Japanese internal organization is characterized by the spirit of familism. Social groups demand exclusive allegiance from their members, and individuals must be primarily absorbed in the group from which they derive a livelihood. People feel that the will of the organization will grow naturally out of their conformity.

U.S. society fosters individualistic tendencies rather than group-orientedness in American social structure. An individual may be a member of many groups and will be inclined to give preference to his private interest over that of any of the groups. The individual will also be more self-sufficient or independent when it comes to personal development and need satisfaction.

Vertical/Horizontal Structure

In Japan, there is a predominant emphasis on vertical rather than horizontal structure orientation, and a strong departmentalism constructed along the vertical tie is latent in all social groups. Vertical structure shapes not only attitudes and behavior but overshadows everything else including character, personality, profession, ability, and accomplishment. This vertical structure is most prevalent in larger, older, and more stable groups.

U.S. social structure is based on two basic but antithetical principles: 1) the principle of unequal status and of superior/inferior ranking, and 2) the principle of equality. In other words, U.S. social structure functions not only on the vertical tie but on the horizontal tie as well.

Emphasis on Frame or Attribute

Nakane has described the criterion for group formation in Japan as stressing situational position in a particular frame (e.g. locality) rather than universal attribute (e.g. membership). For instance, a designer of steam turbines would consider himself an employee of Toshiba instead of a mechanical engineer in general. This indicates an emphasis of group structure over group function.

American group formation is based on functioning by attribute rather than by situational position in a particular frame. Hence, we find the formation of professional societies in the U.S. Along the horizontal tie, function takes priority over structure.

Occupational Systems

Japan has no firmly defined occupation system, and groups do not tend to have clearly marked divisions of labor. The group enjoys the efforts of the individual, but the individual roles of each member are not clearly determined. Reasons for this lack of horizontal structure stem from a lack of contact with similar people outside of one's group and the tendency of building social groups by frame.

U.S. social structure is characterized by a strong division of labor by function and sharply defined groups.

Structural Relations

Relationships in Japan take place along the vertical tie. These vertical relationships are emotional and become the actuating principle in creating cohesion among group members. No two individuals are equal. One single relationship exists between individuals or groups or none at all. Thus one can understand the important Japanese custom of exchanging name cards in order to define rank. Finally, Japanese vertical relations are supported by the spirit of familism. The leader ties the vertical group together and enters with his subordinates into relations of mutual dependence.

In the U.S., relations extend along the horizontal tie and across groups. Horizontal structure is connected to the principles of equality in that it provides people with a sense of self-respect.

Family Structure

The first major component of social structure to be discussed is family structure. The family is the basic unit of social structure, and the dominant form of family structure present in both Japan and the U.S. is currently the nuclear family (only one married couple). However, it is by comparing the American nuclear family with the traditional Japanese stem family household that one may best understand the differences in Japanese and American culture. The traditional Japanese family structure is still at the root of Japanese social structure today. Differences in Japanese and American family structures will be explored regarding structural forms, and family relations.

Structural Forms

Traditional Japanese family structure revolved around the traditional household and was linked with the concept of "ie". "Ie" refers to a corporate residential group and is an example of group forming criteria based on frame. It is characterized by the following: in extended family (i.e., more than one married couple) with hierarchical relations; obedience and filial piety from the family members toward the head in return for security; solidarity between the family members; and a deep sense of importance in carrying on the family line. Males were ranked higher than females, the old ranked higher than the young, and it was generally the eldest son who carried on the family line and succeeded as the family head.

Contemporary U.S. family structure is characterized by the nuclear family of husband, wife and children. The parents typically have few children and old people and unmarried adults generally live apart from their kin. The nuclear family operates as a single, and sometimes isolated, independent unit.

Family Relations

The central core relationship within the traditional Japanese family was the parent-child rather than the husband-wife. The household head was primarily concerned with the household as a whole, not the individuals. A strong sense of obedience and solidarity among family members remains today.

The range of American kinship is narrower than in Japan, and relations tend to be more fluid. Americans typically do not want to get too tied down with obligations to family and will often rely on voluntary associations of common interest rather than strong kinship ties. Family relations such as marriage tend to be characterized by considerable latitude for variation in terms of roles and relationships.

Industrial Organization

The second major component of social structure is industrial organization. Industrial organization is also heavily influenced by cultural values and attitudes. The discussion below will compare Japanese and American industrial organizations in the following areas: general structure, rank, employment system, recruiting, labor mobility, motivation to work, labor relations, and industry-government relations.

General Structure

Japanese industrial organization is heavily marked by vertical orientation and an emphasis on group formation. The corporate group is based on frame. The corporate structure also has its roots in the "ie" and the family household structure. The factory owner is like the head of a household, new employees are like new family members, and the employer-employee relationship is like that of the father-child. There is no firmly defined

The general structure of U.S. industrial organization is horizontally oriented with priority given to attribute. More emphasis is placed on the individual than on the group, and there is a clear division of labor.

occupational system, division of labor, or individual roles. Finally, high value is placed on the harmonious integration of the work members.

Rank

Rank in Japan is established primarily by duration of service and age (i.e. seniority) rather than ability. The institutionalization of rank is more prevalent in larger firms than in smaller ones, and rank based on merit is becoming more common, especially in firms involved with technological innovation.

Rank in U.S. industrial organizations is more heavily influenced by merit than in Japan although seniority also plays an important role.

Employment Systems

Japanese industrial organization in the larger and more stable firms is characterized by a lifetime employment system. Under this system, both the employer and employee assume that the employment relationship is permanent, that the company will not lay off or discharge the employee, and that the employee will not change to another employer during his career. In effect, the Japanese employment system rewards tenure instead of short term performance, and commitment and loyalty are demanded in return for high job security.

In the U.S., there is less insurance of employment tenure and a greater influence of market forces in the allocation on the labor supply. The U.S. employment system places a higher reward on short-term performance.

Recruiting

Japanese firms typically recruit directly from schools, and new recruits enter through the bottom of the age ranking instead of through an open labor market. New recruits are not hired for specific jobs. Instead, they are hired based on the assumption that the labor will be required sometime in the future and that in-house training will provide additional skills as needed.

The recruiting system in the U.S. is based on buying ready-made labor as it is required rather than purchasing future potential, unshaped labor.

Labor Mobility

Intra-firm mobility is heavily encouraged, but inter-firm mobility is small (especially among larger firms). This is related to the stable and secure nature of Japanese social structure and the difficulties involved with leaving one's former group to join a new one.

Both inter-firm and intra-firm mobility are high in the U.S. There is less stigma attached with leaving one company for the purposes of joining another.

Motivation to Work

Motivation of the Japanese workforce comes from a number of different areas. First, as the company operates as an extended family, it employs the entire person instead of just the labor and takes great interest in its employees' well-being both inside and outside of work. The company is a place for its employees to participate in social activities. As a result, personal success and company success become inextricably interconnected. Second, motivation arises from a perceived sense of collective social responsibility and obligations to groups both inside and outside of the company. A third motivating force is direct compensation, and wages are based on individual and family need as well as on age and length of service.

The motivation to work in U.S. industrial organizations comes primarily from a contractual sense of responsibility to the employer and work groups even though economic relations tend to be more depersonalized. However, motivation to work is also largely oriented toward the individual. Compensation in U.S. industrial organizations is influenced by a preference for straight, contribution-based wages and for limited company involvement in personal aspects of workers' lives. Wages are paid according to work performed as opposed to individual need, and workers seem to be interested in their firms primarily as a source of income.

Labor Relations

Labor relations in Japan are characterized by the following: an emphasis on company membership rather than occupation or skill identification; few craft or skill unions; union membership generally undifferentiated by job; priority given to human resources; a human relations management approach which places greater importance on moral character than on technical proficiency; face-to-face relations with little reliance placed on written contracts; and relatively little resistance to new technologies.

Labor relations in the U.S. are characterized by the following: an emphasis on occupation and skills rather than company membership; craft and skill unions marked by job differentiation; priority given to mechanical rather than human resources; a management approach which treats workers as components of a manufacturing process; a dependence on written contracts; and a greater history of resistance to new technologies.

Industry-Government Relations

Government ministries are highly regarded by the Japanese. Industry has a close relationship with the Japanese Ministry of International Trade and Industry (MITI) often involving the exchange of workers.

In the U.S., a "laissez-faire" philosophy is still preferred. American businesses and government agencies tend to have more adversarial relations than in Japan.

Education

The final section in this comparison of Japanese and American social structure concerns education. Education is important to social structure, and to values and attitudes as well, because it is through the educational system that much of this information is taught, maintained and propagated. In short, schools train people to fit into their cultures. Differences in the Japanese and American education systems are pertinent in the following areas: the roles of educational institutions, the influence on creative thought, the emphasized fields of study, and the status of study abroad.

Role of Educational Institutions

The main role of Japanese schools is to impart basic educational skills upon their students. This is accomplished predominately through the use of rote learning. The primary function of the pre-university school system is to prepare students for the college entrance exam, the exam which largely determines which students will attend which universities (and ultimately get what jobs). University curriculums are less structured than in the U.S., but this does not seem to interfere with the primary role of universities, that of placing its students into the workforce.

U.S. schools also serve to impart basic educational skills upon their students (though the results are currently somewhat less successful than in Japan). Where U.S. schools differ from Japanese schools is in their greater emphasis on independent thought and "Socratic-style" exchange. This continues on into the universities where there is a greater focus on the development of critical thinking skills. More research opportunities are made available to students in U.S. universities than in Japanese universities, and U.S. universities play a stronger role in scientific research.

Influence on Creative Thought

The rigid format of the Japanese college entrance exam can cause students to restrict their intellectual breadth, concentrate on following guidelines at the expense of originality, eliminate extracurricular activities, and neglect their social development. The Japanese educational system has been criticized for failing to adequately encourage independent and creative thought.

In the U.S., students are more encouraged to question the system and display independent and critical thought. This is considered to be vital for the continuation of democracy.

Emphasis of Study

The Japanese educational system places a greater emphasis on the field of engineering (due to its perceived economic importance) than in the U.S.

The American educational system gives higher status to the fields of medicine, law, business, and science then one finds in Japan.

Study Abroad

In Japan, it is greatly encouraged and prestigious to study abroad (especially in Western universities).

In the U.S., study abroad is not as highly valued.

Ecology, Economics and Politics

The third and final main section of this paper concerns the exploration and comparison of Japanese and American ecological, economic, and political histories. Each country's ecology, economy, and political structure has influenced and been influenced by culture, and this section will take a brief look at some of the changes that have taken place over time. The purpose of this section is to provide additional perspective to the above discussions of culture and social structure.

Ecology

Japanese and American ecologies will be briefly compared in regards to the following areas: geography, natural resources, population, foreign influences, and the role of technology.

Geography

Japan covers approximately the same area as the state of California, but most of the land is mountainous, leaving only 15% of the total area arable.

The U.S. has over twenty times the area of Japan.

Natural Resources

Japan suffers from a lack of natural resources.

The U.S. is rich in minerals and soil resources.

Population

Japan's population is approximately 120 million, and the population density exceeds 300/km². This density figure increases rapidly when one excludes the large areas of sparsely inhabited mountainous regions.

The U.S. population is approximately double that of Japan, and its population density is an order of magnitude smaller.

Foreign Influence

Since the eighth century A.D., the Japanese archipelago has maintained a relatively isolated position characterized by a rather small inflow and outflow of immigrants. The foreign influences which did enter Japan (e.g. Buddhism, governmental structure, and language from China) were successfully integrated into Japan's own culture.

American demography has been profoundly influenced by migration, both from other countries to the U.S. and within the U.S. The U.S. has been greatly affected by its role as a cultural "melting pot."

Role of Technology

Prior to the Meiji Restoration, Japan was characterized by a historical lack of focus on science and technology. This was due in part to Japan's isolation during feudal times, hindrances to the flow and exchange of new theories and findings, the fact that Japanese technology focused primarily on implements for home and agriculture rather than on tools for production, and the absence of an atmosphere conducive to free inquiry. Since then, borrowing technology from the West in conjunction with minor indigenous technological innovations has become the dominant strategy for acquiring and maintaining its current economic position. Japan's technological success has been

Much of the early U.S. technology was imported from Europe, but after a while, the U.S. became the world leader in the export of technology. It is interesting to note that these large exports of domestic technology have not been matched by significant imports of foreign technology. This raises questions in regards to the receptivity in the U.S. to external technological innovation.

accomplished with few if any accompanying major discoveries of a magnitude capable of changing the basic direction of technological development.

Economics

Japanese and American economics will be compared in the following areas: history, influence of foreign investment, internal vs. external markets, and the role of the military sector.

History

During the Tokugawa period (1600-1868), the Japanese economy was driven by agriculture, but cottage industry was becoming more and more common. In the following Meiji period (1868-1912), government policies promoted industrialization, increased production, national wealth, and military strength. In the years following World War II, Japan's post-war economic development was heavily influenced by Japan's ties to the U.S. In the 1960s, with the acquisition of advanced technology and management methods from the West, Japan became an open economy. Today, Japan is challenging the U.S. for the role as the world's economic leader.

The U.S. economy started on the path to industrialism by following in the footsteps of the European Industrial Revolution. By the early 1900s, the U.S. had taken over as the world's dominant economic power, developing highly efficient systems of mass production and "scientific" management.

Foreign Investment in Each Country

Japan's industrial development occurred with little help from foreign capital, only imported technology and internal education.

The U.S.'s economic development was partly financed by European foreign investment.

Internal vs. External Markets

During Japan's economic rise in industrialism, low purchasing power in combination with high production necessitated that the Japanese search for and establish markets abroad. Japan also needed to establish external trade ties for the purposes of importing raw materials and foreign technologies.

The U.S. economy has always been primarily focused on an extremely large and diverse internal market.

Role of the Military Sector

During the Meiji Restoration, increased military expenditures were a powerful factor serving to speed up industrialization. Since World War II, military factors have played a minor role in the economy.

Traditionally, U.S. economic growth has been assisted by a heavy emphasis in the military sector.

Politics
Japanese and American politics will be compared in relation to stability and change over the past couple of centuries.

History: Stability and Change

Since the mid-1800s, the Japanese political system has undergone a series of dramatic changes. The Tokugawa feudal regime and its policy of isolationism was succeeded by the restoration of the Emperor in 1868 and an embarkment on a path of imperialism. After World War II and the democratic reform of the political system, Japan dropped its military goals and pursued the path of economic development.

The U.S. has pursued a political path leading from colonization by various European countries to an active imperialism of its own. However, its democratic political structure has remained remarkably stable for over the past two centuries.

Summary

The comparison of general Japanese and American cultural and social characteristics put forward above is abbreviated at the risk of being "stereotypic" and incomplete. It is important to remember though that the comparisons presented are relative. Their intent is to shed light on some of the cultural differences which may influence Japanese and American approaches to technological innovation.

It is also apparent that change is inherent in the Japanese and American cultural systems and that in many ways these systems are in the process of converging. Nevertheless, in order to better understand current cultural values, attitudes and social structures, it is also important to understand the sources from which they arise.

Footnotes

[1] In his 1980 book entitled *Theory Z*, William Ouchi formulated a theory in which he distinguished A-like organizations (hierarchical bureaucracies) from Z-like organizations ("industrial clans"). He claimed that Z-like firms tended to produce happier, more involved, and, hence, more productive workers. According to Ouchi, these clans operate by socializing each member completely so that they merge individual and organizational goals, thereby creating a motivational force. Ouchi concluded that Z-like organizations (typical of Japanese firms) were more effective than A-like organizations (typical of American firms) because they were able to coordinate workers' efforts more efficiently and enhance workers' emotional well-being, both of which result in higher productivity (Kaplan & Ziegler 1985).

Bibliography

Abegglen, J.
　1973　Management and Worker: The Japanese Solution. Kodansha International Ltd. Tokyo, Japan.
Albert, Ethel M.
　1970　Conflict and Change in American Values. In The Character of Americans, M. McGiffert (ed). The Dorsey Press. Homewood, Illinois.
Arensberg, Conrad M. and Arthur H. Niehoff
　1975　American Cultural Values. In The Nacirema: Readings on American Culture, J. P. Spradley and M. A. Rynkiewich (eds.). Little, Brown and Company, Inc.
Befu, Harumi
　1971　Japan: An Anthropological Introduction. Harper & Row, Publishers, Inc.
Benedict, Ruth
　1946　The Chrysanthemum and the Sword: Patterns of Japanese Culture. Boston, Mass.
Billington, Ray Allen
　1974　America's Frontier Heritage. University of New Mexico Press. Albuquerque, New Mexico.
Cleaver, Charles Grinnell
　1976　Japanese and Americans: Cultural Parallels and Paradoxes. University of Minnesota Press. Minneapolis.

Cole, R. E.
   1979 Work, Mobility, and Participation: A Comparative Study of American and Japanese Industry. University of California Press.
Davis, David B.
   1968 Ten-Gallon Hero. In The American Experience: Approaches to the Study of the United States, Hennig Cohen (ed.). Houghton Mifflin Company. Boston, Mass.
De Mente, Boye
   1981 Japanese Manners & Ethics in Business. Phoenix Books Publishers. Phoenix.
De Mente, Boye and Fred Thomas Perry
   1968 The Japanese as Consumers: Asia's First Great Mass Market. John Weatherhill, Inc. New York.
Fukutake, Tadashi
   1962 Man and Society in Japan. The University of Tokyo Press. Tokyo.
   1974 Japanese Society Today. University of Tokyo Press. Tokyo.
   1982 The Japanese Social Structure: Its Evolution in the Modern Century. University of Tokyo Press. Tokyo.
Gorer, Geoffrey
   1964 The American People: A Study in National Character. W. W. Norton & Company, Inc. New York, New York.
Graham, Saxon
   1957 American Culture: An Analysis of Its Development and Present Characteristics. Harper & Brothers, Publishers. New York, New York.
Grossberg, Kenneth A.
   1981 Japan Today, K. A. Grossberg (ed.). Institute for the Study of Human Issues, Inc. Philadelphia, Pennsylvania.
Hague, John A.
   1979 America and the Post-Affluent Revolution. In American Character and Culture in a Changing World: Some Twentieth-Century Perspectives, John A. Hague (ed.). Greenwood Press. Westport, Connecticut.
Hall, Edward T. and Mildred R. Hall
   1987 Hidden Differences: Doing Business with the Japanese. Anchor Press/Doubleday. Garden City, New York.
Haring, Douglas G.
   1956 Japanese National Character: Cultural Anthropology, Psychoanalysis, and History. In Personal Character and Cultural Milieu, D. G. Haring (ed.). Syracuse University Press.
Hasegawa, Nyozedan
   1965 The Japanese Character: A Cultural Profile. Greenwood Press.
Helvoort, Ernest van
   1979 The Japanese Working Man: What Choice? What Reward? University of British Columbia Press. Vancouver.
Henry, Jules
   1963 Culture Against Man. New York: Random House.
   1975 Golden Rule Days: American Schoolrooms. In The Nacirema: Readings on American Culture, J. P. Spradley and M. A. Rynkiewich (eds.). Little, Brown and Company, Inc.
Hsu, Francis L. K.
   1975 American Core Value and National Character. In The Nacirema: Readings on American Culture, J. P. Spradley and M. A. Rynkiewich (eds.). Little, Brown and Company, Inc.
Ishida, Takeshi
   1971 Japanese Society. Random House, Inc. New York.
Kaplan, David and Charles A. Ziegler
   1985 Clans, Hierarchies and Social Control: An Anthropologist's Commentary on Theory Z. Human Organization Vol. 44, No. 1.
Kato, Shuichi
   1986 The Sources of Contemporary Japanese Culture. In Cultural Tradition in Japan Today, by Nippon Steel Corporation. Japan.
Khruslov, Georgyi
   1973 National Characteristics Reflected in Japanese Language. In Japan and the Japanese, Compiled by the Mainichi Newspapers. Japan Publications, Inc. San Francisco.
Lebra, Takie S. and William P. Lebra
   1986 Japanese Culture and Behavior, T. S. Lebra and W. P. Lebra (eds.). University of Hawaii Press. Honolulu, Hawaii.

Lee, Everett L.
  1968 The Turner Thesis Re-examined. In The American Experience: Approaches to the Study of the United States, Hennig Cohen (ed.). Houghton Mifflin Company. Boston, Mass.
London, Herbert I. and Albert L. Weeks
  1981 Myths That Rule America. University Press of America, Inc. Washington, D.C.
Mead, Margaret
  1942 And Keep Your Powder Dry. William Morrow and Company. New York, New York.
Minami, Hiroshi
  1986 The Japanese Mind. In Cultural Tradition in Japan Today, by Nippon Steel Corporation. Japan.
Miyazaki, Isamu
  1986 Japan's Economic Dynamism: Values in Tradition. In Cultural Tradition in Japan Today, by Nippon Steel Corporation. Japan.
Morton, W. Scott
  1970 Japan: Its History and Culture. Thomas Y. Crowell Company. New York, New York.
Murphy, Robert F.
  1971 The Dialectics of Social Life: Alarms and Excursions in Anthropological Theory. Basic Books, Inc. Publishers. New York.
Nakane, Chie
  1970 Japanese Society. University of California Press. Berkeley and Los Angeles, CA.
Nishio, Kanji
  1986 Education in Japan. In Cultural Tradition in Japan Today, by Nippon Steel Corporation. Japan.
Olsen, Edward A.
  1978 Japan: Economic Growth, Resource Scarcity, and Environmental Constraints. Westview Press. Boulder, Colorado.
Ouchi, William G.
  1981 Theory Z: How American Business can meet the Japanese Challenge. Addison-Wesley Publishing Company.
Ozaki, Robert S.
  1978 The Japanese: A Cultural Portrait. Charles E. Tuttle Company. Japan.
Potter, David M.
  1970a Economic Abundance and the Formation of American Character. In The Character of Americans, M. McGiffert (ed). The Dorsey Press. Homewood, Illinois.
  1970b The Quest for the National Character. In The Character of Americans, M. McGiffert (ed). The Dorsey Press. Homewood, Illinois.
Rohlen, Thomas P.
  1974 For Harmony and Strength: Japanese White-collar Organization in Anthropological Perspective. University of California Press.
  1986 "Spiritual Education" in a Japanese Bank. In Japanese Culture and Behavior, T. S. Lebra and W. P. Lebra (eds.). University of Hawaii Press. Honolulu, Hawaii.
Sakamoto, Yoshikazu
  1981 Frameworks of Japanese Identity. In Japan Today, K. A. Grossberg (ed.). Institute for the Study of Human Issues, Inc. Philadelphia, Pennsylvania.
Schneider, David M. and George C. Homans
  1975 Kinship Terminology and the American Kinship System. In The Nacirema: Readings on American Culture, J. P. Spradley and M. A. Rynkiewich (eds.). Little, Brown and Company, Inc.
Smith, Robert J.
  1983 Japanese Society: Tradition, Self, and the Social Order. Cambridge University Press. Cambridge.
Sours, Martin H.
  1982 The Influence of Japanese Culture on the Japanese Management System. In Japanese Management: Cultural and Environmental Considerations, S. M. Lee and G. Schwendiman (eds.). Praeger Publishers.
Spradley, James P. and Michael A. Rynkiewich
  1975 The Nacirema: Readings on American Culture, J. P. Spradley and M. A. Rynkiewich (eds.). Little, Brown and Company, Inc.
Stewart, Edward C.
  1972 American Cultural Patterns: A Cross-Cultural Perspective. Intercultural Network, Inc. LaGrange Park, Illinois.
Sumiya, Mikio
  1986 The Japanese Work Ethic. In Cultural Tradition in Japan Today, by Nippon Steel Corporation. Japan.
Vogel, Ezra F.
  1979 Japan as Number One: Lessons for America. Harvard University Press. Cambridge, Mass.

Warner, W. Lloyd
    1962 American Life: Dream and Reality. The University of Chicago Press. Chicago, Illinois.
Watanabe, Teresa
    1990 Cookie-Cutter Education. Los Angeles Times, June 24.
Whitehill, A. M. and Shinichi Takezawa
    1968 The Other Worker: A Comparative Study of Industry Relations in the U.S. and Japan. East-West Center Press. Honolulu, Hawaii.
Williams, Robin, M. Jr.
    1970 Changing Value Orientations and Beliefs on the American Scene. In The Character of Americans, M. McGiffert (ed). The Dorsey Press. Homewood, Illinois.
Yamada, Keiji
    1986 Traditional Characteristics of Japanese Technology. In Cultural Tradition in Japan Today, by Nippon Steel Corporation. Japan.
Yamazaki, Masakazu
    1981 Social Intercourse in Japanese Society. In Japan Today, K. A. Grossberg (ed.). Institute for the Study of Human Issues, Inc. Philadelphia, Pennsylvania.
Yanase, Mutsuo
    1986 Japanese Scientists: As Seen through the Achievements of Kumagusu Minakata. In Cultural Tradition in Japan Today, by Nippon Steel Corporation. Japan.
Yasuda, Takeshi
    1986 Communication in Modern Japan. In Cultural Tradition in Japan Today, by Nippon Steel Corporation. Japan.
Zelinsky, Wilbur
    1973 The Cultural Geography of the United States. Prentice-Hall, Inc. Englewood Cliffs, New Jersey.

JAPANESE AND AMERICAN APPROACHES TO TECHNOLOGICAL INNOVATION: CULTURAL INFLUENCES

Eric Poncelet, University of Arizona

The purpose of this paper is to describe some of the major differences between Japanese and American approaches to technological innovation and to suggest some of the cultural characteristics influencing these differences. The discussion will proceed in two parts. First, a description will be presented concerning what is meant by innovation and the innovation process. This will be followed by a comparison of the different approaches to technological innovation and brief listings of pertinent cultural influences. This discussion suffers somewhat from being over-generalized, but it does reveal some important distinctions between the two countries.

Innovation and the Innovation Process

Economists have conventionally defined an innovation as "the introduction of a new (or improved) product into the market". This paper assumes a broader definition and chooses to adopt a version suggested by Stephen Kline. According to Kline (1989; 1990), an innovation is any change in the social and technical systems of design, manufacture, distribution, and/or use which provides an improvement in cost, quality and/or match to customer requirements. This broader definition is instructive because it acknowledges that innovations may arise in the areas of manufacturing processes, human resource management, marketing, distribution, and use as well as in the area of products.

Technological innovation is performed by humans and takes place in what we call the "innovation process." There are various ways of looking at this process. One model, which Kline (1990) has called the "linear model", has been conventionally adopted in the U.S. since the end of World War II. It is not so much a model which Americans work from as it is a belief about the way innovations are produced. According to this model innovation takes place in four distinct and sequential phases: a research phase, a development phase, a production phase, and a marketing phase. Research is considered to be the initiating step and the source of all innovations, and no feedback role is built into the system. This linear model has been used as a justification for doing basic science research in the U.S.

A second model for the innovation process is one based on Kline's (1989; 1990) "chain-link model". The general process starts with a market finding phase followed by design, production, marketing and distribution, and use phases. It differs from the linear model in a number of ways: there are multiple paths from which innovations may arise and many forms of feedback; research is not normally considered to be the initiating step (in fact, research occurs in and contributes to all phases in the innovation process) and the primary source of innovations is now held to be stored knowledge and technological paradigms. This model more closely corresponds to the Japanese perception of the innovation process.

For purposes of discussion, two other general phase descriptions will also be used to describe the innovation process: the invention phase and the implementation phase. The invention phase refers to the idea generation, research, and development of a product or process. The implementation phase refers to everything after the invention phase which is required to bring the product or process to commercial use.

Japanese and American Approaches to Technological Innovation

This section compares Japanese and American approaches to technological innovation in twelve different areas: the innovation process, industrial organization, human resources, worker productivity, business strategy, customer focus, decision-making processes, utilization of external resources, role of government, role of universities, occupational status by field, and creativity.

1. *The Innovation Process*

Comparison:

General Strategy

Japan has followed a general strategy of emphasizing incremental process and product improvements based on existing technology and knowledge. The focus has been on the implementation phase (especially manufacturing) with the goal of producing high quality products at low costs. This strategy works best in maturing and mature industries.

The U.S. tends to follow dual strategies emphasizing, on the one hand, product breakthroughs based on research, and on the other hand, controlling customer preferences through marketing techniques. There is a strong focus on both the invention and marketing phases. The goal is developing and selling new products and creating demand for existing ones. The breakthrough strategy works best in nascent industries where scientific discoveries play a key role; the marketing strategy is adapted to mature industries.

Differences in Resource Allocation

Japanese firms tend to allot more of their resources to the middle part of the innovation process (i.e., process engineering and manufacturing facilities) than do their U.S. counterparts. Japanese firms allot approximately twice as much of their total innovation costs to manufacturing equipment and facilities as American firms (Mansfield 1988b).

U.S. firms tend to allot more of their resources to the latter part of the innovation process (i.e. marketing) than do Japanese firms. American firms allot approximately twice as much of their total innovation costs to marketing start-up as Japanese firms (Mansfield 1988b).

Functioning of the Innovation Process

Japanese firms tend to treat the innovation process as a continuing, cyclical, iterative process. In this sense, functioning of the innovation process in Japan may be likened to the operation of a rugby team--each individual works as part of a team with a common goal. There is a strong interface between the different phases (e.g. good coordination between product design and engineering), and feedback plays an important role in all of the phases throughout the process. This feedback is evident in the manner in which the

American firms tend to treat the innovation process as a linear, sequential process which begins with research and ends with the marketing of a product. Along these lines, the innovation process in the U.S. operates as a relay race team-- each group performs its task and then passes on the baton. Strong coordination between the different phases is often lacking (e.g. between R&D and manufacturing), and this encourages the actors in each phase to operate as separate, functionally specialized groups. Feedback also

users of R&D (e.g. production, marketing, customers) play a greater role in shaping R&D programs.

tends to be weak. Future R&D in U.S. firms tends to be more influenced by previous R&D projects rather than by user feedback.

Actors

In Japan, the actors in the innovation process operate as part of group teams. Individual members and the groups have multi-functional and overlapping roles. A team often follows an innovation through the entire process. Multi-functional teamwork tends to be most beneficial to the rate of innovativeness when it occurs during the implementation phase.

The actors in U.S. firms are primarily individuals or teams with strong individual leaders. The individuals have specialized roles and are compartmentalized. When one individual's task is complete, the innovation moves on to the next person. Individual contributions tend to be most beneficial to the rate of innovativeness when they occur during the invention phase.

Role of Science and Technology

Japan pays more attention to technological knowledge than to scientific knowledge. Innovations tend to involve "learning by doing" and experimentation more so than theories and interpretations. Research is ultimately performed with manufacturing in mind.

In the U.S., technology is seen predominantly as the application of scientific knowledge, and science is considered to be the primary source of knowledge. Hence, there is an emphasis on theory in the innovation process. Research is performed more with a focus on future research than on manufacturing.

Sources of Technology

Japan is distinct in its efficient and effective use of external ideas and technologies. Japanese firms have demonstrated significant cost and time advantages over the U.S. when it comes to putting out commercial products and processes based on foreign technologies (Mansfield 1988b).

The U.S. has imported much less technology from external sources than Japan and preferred to rely primarily on internal ideas and technologies, but this is in the process of changing.

General Results

Japanese firms tend to focus on innovation throughout the innovation process. In general, they are currently more successful than U.S. firms in actualizing the timely production of higher quality, lower cost products and processes.

U.S. firms have been successful in producing breakthrough innovations. However, much of the effort of U.S. firms has been on improving financial performance, soliciting government assurance, seeking stability, and reducing risk instead of improving their performance in the implementation phase of the innovation process.

Cultural Influences affecting the Innovation Process:[1]
- Values/attitudes toward groups vs. individuals.
- Values/attitudes concerning technology vs. science.

- Values/attitudes toward change (existing technology vs. new science).
- Values/attitudes toward shared vs. compartmentalized responsibilities.
- Sources of motivation (social obligations vs. externalized achievement).
- Influence of religion (cyclical vs. linear world views).
- Influence of art (art as intended for everyday use vs. art for art's sake).
- Influence of education (emphasis on technology vs. emphasis on science).
- History of foreign influences.

2. Industrial Organization

Comparison:

Structure

The structure of Japanese industrial organization tends to be more flexible and unified within firms and hierarchical between firms. Internal structure is made up of multi-functional and less specialized groups.

The structure of U.S. industrial organizations tends to be more hierarchical within firms and horizontal or equal between firms. Internal structure is more compartmentalized, functionally specialized, and individual oriented.

Relationships Within Firms and Between Firms

In Japan, relationships within and between firms are long-term and characterized by dependence. Families of companies exist which operate in a way similar to the traditional extended family. A closer (i.e. more permanent and stable) working relationship exists between customers and suppliers. Trust and dependability are preferred over lower cost.

In the U.S., relationships within and between firms are short-term and characterized by independence. The choice of suppliers changes more frequently and is heavily dependent on lowest cost. The U.S. relationships most similar to the Japanese style is the corporate culture which has developed in the military-industrial complex.

Information Flow

Japanese firms and industries are characterized by rapid and efficient flow of information along vertical ties. Good feedback exists between departments and between companies. Technology transfer is also rapid and almost always interpersonal.

U.S. firms are marked by less effective communication between departments and between companies. The lines of communication tend to be longer and less personal than in Japan.

Adaptability

Japanese bureaucracy is more flexible, organic (i.e. priority is given to participants' behavior), and adaptable to changing structures or functions. This type of bureaucracy is more effective under uncertain conditions.

U.S. bureaucracy is more rigid, mechanistic (i.e. priority is given to the formal, bureaucratically programmed structure), and slow to adapt to change. This type of bureaucracy is most effective under more stable and relatively certain conditions.

Employment Patterns

Japanese employment patterns are characterized by longer term employment (especially for larger stable firms where lifetime employment exists), promotability largely dependent on age and rank, high in-house training, and more intra-firm mobility.

U.S. employment patterns are characterized by more short-term employment, promotions related to individual performance, and increased dependence on economic conditions.

Cultural Influences affecting Industrial Organization:
- Social structure (vertical vs. horizontal).
- Family structure and relations.
- Values/attitudes implicit in social and family structures:
  - Dependency vs. self-reliance.
  - Competition (intra-firm vs. inter firm).
  - Mobility and commitment.
  - Communication (interpersonal vs. impersonal).

3. *Human Resources*

Comparison:

Management Approach

Japanese firms place a greater emphasis on human resources, and human relations are held to be more important relative to research or manufacturing results than in the U.S. Companies tend to operate as extended families in that they are more involved in and concerned with the personal lives of their employees. Intra-firm relations are typically not competitive.

In the U.S., firms place less emphasis on their human resources and hold human relations to be less important relative to research or manufacturing results than in Japan. Firms are also less involved in the personal or private lives of their employees. Intra-firm relations tend to be more competitive.

Results

Japanese labor relations are more stable than U.S. labor relations. Many fewer days are lost to labor disputes in Japan when compared to the U.S. There is also lower inter-firm mobility allowing Japanese firms to better reap the fruit of their in-house training programs.

Labor relations are less stable in the U.S. when compared to Japan. Employment is also characterized by higher inter-firm mobility, thus preventing much continuity within a workforce and requiring more constant training of new employees.

Cultural Influences affecting Human Resources:
- Values/attitudes toward mobility (low vs. high).
- Values/attitudes toward personal relations (preference for long-term, in-depth group involvement vs. short-term, multiple affiliations).
- Values/attitudes toward work (discipline and perseverance vs. "toil and trouble").
- Values/attitudes toward competition (low intra-group vs. high intra-group).
- Family structure (extended family and dependence vs. nuclear family and independence).

- Organization of labor relations (company membership vs. skill identification).

4. *Worker Productivity*

Comparison:

Worker Productivity

In general, Japanese industry maintains a higher level of worker productivity than does American industry. For example, Japanese plants turn out two to three more cars per worker than U.S. plants. Responsibility for quality control is assumed primarily by the workers themselves.

On the average, American industry has a lower level of worker productivity than in Japan. Quality controls fall primarily in the hands of inspectors rather than on the shoulders of the workers themselves.

Cultural Influences affecting Worker Productivity:
- Values/attitudes toward work (higher value of quality vs. lower value of quality).
- Sources of motivation (company success vs. individual success).
- Values/attitudes toward responsibility (group vs. individual).

5. *Business Strategy*

Comparison:

General Strategy

The primary goal of Japan's general business strategy is survival (i.e. to secure long-term market share). A primary strategy for securing this has been to focus on improving existing processes and products. There is also greater support for higher-risk, long-term projects.

The primary goal of the U.S.'s general business strategy is to maximize short-term profit. Consequently, U.S. companies tend to focus their innovative strategies on developing new products capable of securing immediate rewards. However, short-term gain is often incompatible with, and hence leads to diminished support for, high-risk, long-term projects.

Cultural Influences affecting Business Strategy:
- Temporal orientation (past vs. near future).
- Structure of industrial organizations (traditional, inter-dependent, extended family structure vs. independent, nuclear family structure).
- Economic histories (higher risk, less mature economy vs. lower risk, more mature economy).

6. *Customers Forces*

Comparison:

Focus on the Customer

In Japanese firms, high priority is given to serving and satisfying the customer. Close relationships are established

Satisfying the customer is also held in high regard by American firms, but there has also been the tendency in the past for

between suppliers and customers based on mutual trust. As a result, customers tend to have a greater input into the innovation process, and the resulting innovations are inherently geared towards meeting customer needs and desires.

American suppliers to try to dictate the needs of their customers. A case in point is the emphasis placed on advertising in the U.S. As a result, response to customer claims and requests tends to be slower and more reluctant than in Japan, and customer-supplier relations are often more adversarial. In the end, innovations are more independent of customer needs.

Cultural Influences affecting Business Strategy:
- Values/attitudes toward personal relations (long-term, harmonious relations vs. short-term, more adversarial relations).
- Social structure orientation: groups vs. individuals (customer and supplier as a team vs. customer and supplier as separate components).

7. *Decision-making Processes*

Comparison:

Management Approach

Japan's managerial approach to decision-making is based on consensus and cooperation by all affected personnel. Furthermore, responsibility is dispersed on company employees as a whole. This approach allows for increased involvement by the team players and good feedback to earlier phases of the innovation process.

The U.S. management approach to decision-making utilizes constructive conflict among existing functional divisions and individual decisions. Responsibility falls on individuals and at the top. This process is more rapid, but not all affected personnel participate.

Results

While the consensus decision-making in Japan tends to take longer, it has contributed to an effective implementation phase in the innovation process and a high level of quality control.

U.S. decisions are made more rapidly, but what often suffers is inter-phase communication and ultimately quality control. U.S. firms devote twice as much of their innovation investment to manufacturing start-up as Japanese firms (Mansfield 1988b).

Cultural Influences affecting Decision-Making Processes:
- Values/attitudes toward groups vs. individuals.
- Values/attitudes toward conformity vs. confrontation.
- Values/attitudes toward dependence vs. self-reliance.
- Values/attitudes toward responsibility (group vs. individual).

8. *Utilization of External Resources*

Comparison:

Use of Foreign Resources

Japan makes greater use of foreign resources for the purposes of technological innovation than does the U.S. Japan has been historically, and currently still is, a net importer of foreign technology. Foreign scientific and technological knowledge is brought in via extensive scanning and monitoring of publications, attendance of foreign conferences and trade fairs, the sending of researchers and students abroad, joint ventures, acquisition of foreign companies, and the relocation of R&D labs overseas. Japan has greater access to U.S. scientific and technological data bases than the U.S. has to those in Japan.

The U.S. has made less extensive use of foreign resources and has been marked by a seeming aversion to things "not made here". The U.S. has prided itself on being the largest exporter of technology and the largest producer of scientific and technological publications (Fuji Corp. 1983). The U.S. has not been quick to import external technologies, has sent fewer researchers and students abroad, and does not have an efficient system for bringing in scientific and technological information from the outside.

Imitation of Foreign Innovations

The Japanese have been more successful at imitating foreign technological innovations because they have been borrowing highly visible product innovations (i.e. technologies of a hardware nature).

The U.S. has not had great success in imitating Japanese innovations because of the difficulties inherent in borrowing subtle process innovations involving tacit knowledge from the implementation phase.

Cultural Influences affecting Utilization of External Resources:
- Values/attitudes concerning self-reliance.
- Historical role of technology.
- Economic history (Japanese targeting of foreign markets vs. U.S. dependence on domestic markets).

9. *Role of Government*

The most successful governmental policies aimed at increasing a (developed) country's competitiveness in innovation are those which create an environment in which companies can gain a competitive edge rather than those that involve government directly in the process (Porter 1990).

Comparison:

Relations between Industry and Government

The Japanese government utilizes indirect methods of spurring innovation. Japanese industry has a strong relationship with government in the form of the Ministry of International Trade and

American industry operates in a predominantly free-market system with little intervention from government. The U.S. government has refrained from targeting specific sectors of industrial R&D for sup-

Industry (MITI). Employees from industry are commonly sent to MITI laboratories to participate in the development of new technologies. Some of MITI's functions include: researching and investing in strategic technologies, making foreign scientific and technological developments readily available to Japanese industries, and controlling technology flow by exercising its right to examine and approve all technical alliances in the light of their ultimate benefit to the nation. In sum, closer relations exist between government research labs and industry in Japan than in the U.S., and government more successfully acts as an organizer and coordinator of private industry. However, industry still funds the bulk of R&D geared toward innovation.

port except in the case of the space and defense industries (where the U.S. government and industry have a very close relationship). The U.S. government does not have a direct counterpart to MITI, but it does fund a great deal of basic research. However, this government funding is directed toward science, leaving technological innovations to industry. In sum, weaker and often times adversarial relations exist between government and industry in the U.S. when compared with Japan. On the whole, the U.S. government plays a relatively minor role in stimulating U.S. innovative competitiveness.

Cultural Influences affecting the Role of Government:
- Values/attitudes toward group vs. individual (national policies vs. individual enterprise).
- Social structure: emphasis on frame (government and industry on the same team) vs. attribute (government and industry working separately and in competition).
- Communication (indirect vs. direct).

10. *Role of Universities*

Comparison:

Role of Universities

While both Japanese and American universities are involved in research and development, Japanese university R&D is more directed toward applied research and specific projects. In neither country do universities receive much R&D funding from industry, and Japanese universities receive less than American universities. Furthermore, in 1982, Japanese industries invested twice as much money in foreign universities as it did in domestic ones (National Research Council 1989). In sum, Japanese university research has played a minor role in the innovation process.

R&D in U.S. universities is directed toward general and basic science research. There is a closer relationship between industry research and university research in the U.S. than in Japan, but R&D funding from industry to the universities is still small. Nevertheless, university research, as it supplies industry, does fit well into the U.S.'s linear conception of the innovation process. The transfer of technology from universities to industry though is not always effective.

Cultural Influences affecting the Role of Universities:
- Social structure: emphasis on frame (specific research project) vs. attribute (general research topic).

11. *Occupational Status by Field*

Economic and innovative success depends on where a country's talented people choose to work.

Comparison:

Status of Occupations

The field of engineering is highly valued in Japan. In 1984, nearly six times as many undergraduate students and twice as many doctoral students received their degrees in engineering as in the natural sciences (National Research Council 1989).

In the U.S., the natural sciences are more highly valued than engineering, and science is seen as playing a more important role in relation to technology than manufacturing. In 1985, approximately one and one-half times as many undergraduate students and two and one-half times as many doctoral students received their degrees in the natural sciences rather than in engineering (National Research Council 1989). The fields of medicine, law, and business/accounting are also more highly valued in the U.S. than in Japan.

Cultural Influences affecting Occupational Status by Field:
- Values/attitudes toward science and technology.

12. *Creativity*

Comparison:

Management of Creativity

Japanese industry's strategy toward creativity is to stimulate it through the cross-fertilization of ideas within a project team. Individual independent creativity is less encouraged.

U.S. industry's strategy toward creativity is to devise a creative environment and leave individual professionals alone to create. Individual independent creations are highly encouraged and highly rewarded.

Cultural Influences affecting Creativity:
- Values/attitudes toward group conformity vs. individuality.
- Values/attitudes toward group dependence vs. self-reliance.
- Educational systems as they reflect cultural values (following guidelines vs. independent critical thought).

Footnote
[1] For a more complete and detailed description of these cultural descriptions, please refer to "Japan and the United States: A Cultural Comparison," by Eric Poncelet in this volume.

Bibliography

Abegglen, J.
  1973 Management and Worker: The Japanese Solution. Tokyo, Japan: Kodansha International Ltd.
Cole, R. E.
  1979 Work, Mobility, and Participation: A Comparative Study of American and Japanese Industry. University of California Press.
Drucker, Peter F.
  1990 The Emerging Theory of Manufacturing. Harvard Business Review, May-June.
Fuji Corporation
  1983 Japan Science and Technology Outlook. Based on Kagakugijutsu Hakusho, a white paper of the Science and Technology Agency. Japan: The Fuji Corporation.
Gee, Sherman
  1981 Technology Transfer, Innovation, and International Competitiveness. New York: John Wiley & Sons.
Gomory, Ralph E.
  1989 From the 'Ladder of Science' to the Product Development Cycle. Harvard Business Review, November-December.
Herbert, Evan
  1989 Japanese R&D in the United States. Research-Technology Management, November-December.
Hull, Frank and Koya Azumi
  1984 R&D in Japan versus the U.S.
  1989 Teamwork in Japanese and U.S. Labs. Research-Technology Management, November-December.
Johnson, Howard W. et al.
  1983 International Competition in Advance Technology: Decisions for America. A Consensus Statement Prepared by the Panel on Advanced Technology Competition and the Industrialized Allies. Washington, D.C.: National Academy Press.
Jorgenson, Dale W.
  1988 Technological Innovation and Productivity Change in Japan and the United States: Productivity and Economic Growth in Japan and the United States. AEA Papers and Proceedings, Vol. 78, No. 2.
Kaplan, David and Charles A. Ziegler
  1985 Clans, Hierarchies and Social Control: An Anthropologist's Commentary on Theory Z. Human Organization, Vol. 4, No. 1.
Kelly, Patrick and Melvin Kranzberg et al.
  1978 Technological Innovation: A Critical Review of Current Knowledge. P. Kelly and M. Kranzberg, eds. San Francisco Press, Inc.
Kline, Stephen Jay
  1989 Models of Innovation and their Policy Consequences. Report INN-4. Department of Mechanical Engineering, Stanford University.
  1990 Innovation Styles in Japan and the United States: Cultural Bases; Implications for Competitiveness. The 1989 Thurston Lecture. Report INN-3B. Department of Mechanical Engineering, Stanford University.
Lynn, Leonard H.
  1982 How Japan Innovates: A Comparison with the U.S. in the Case of Oxygen Steelmaking. Boulder, Colorado: Westview Press.
Mansfield, Edwin
  1988a Industrial R&D in Japan and the United States: A Comparative Study. AEA Papers and Proceedings, Vol. 78, No. 2.
  1988b Industrial Innovation in Japan and the United States. Science, vol. 241.
  1988c The Speed and Cost of Industrial Innovation in Japan and the United States: External vs. Internal Technology. Management Science, Vol. 34, No. 10.
Melcher, Arlyn J. and Bernard Arogyaswamy
  1982 Decision and Compensation Systems in the United States and Japan: Contrasting Approaches to Management. In Management by Japanese Systems, S. M. Lee and G. Schwendiman, eds. Praeger Publishers.
Murayama, Motofusa
  1982 A Comparative Analysis of U.S. and Japanese Management Systems. In Management by Japanese Systems, S. M. Lee and G. Schwendiman, eds. Praeger Publishers.
Nakane, Chie
  1970 Japanese Society. Berkeley: University of California Press.
Nanto, Dick Kazuyuki
  1982 Management, Japanese Style. In Management by Japanese Systems, S. M. Lee and G. Schwendiman, eds. Praeger Publishers.

National Research Council
   1989   The Working Environment for Research in U.S. and Japanese Universities: Contrasts and Commonalities. National Academy Press. Washington, D.C.
Poncelet, Eric
   1991 Japan and the United States: A Cultural Comparison. In Japanese/American Technological Innovation: The Influence of Cultural Differences on Japanese and American Innovation in Advanced Materials. W. David Kingery, ed. New York: Elsevier Science Publishing Co., Inc.
Porter, Michael E.
   1990 The Competitive Advantage of Nations. Harvard Business Review, March-April.
Ronstadt, Robert and Robert J. Kramer
   1982 Getting the Most out of Innovation Abroad. Harvard Business Review, March-April.
Rosenberg, Nathan and W. Edward Steinmueller
   1988 Why are American Such Poor Imitators? AEA Papers and Proceedings, Vol. 78, No. 2.
Susskind, Charles and Martha Zybkow
   1978 The Argument. In Technological Innovation: A Critical Review of Current Knowledge. P. Kelly and M. Kranzberg, eds. San Francisco Press, Inc.
Swords-Isherwood, Nuala
   1984 The Process of Innovation: A Study of Companies in Canada, the United States and the United Kingdom. London: British-North American Committee.
Whitehill, A. M. and Shinichi Takezawa
   1968 The Other Worker: A Comparative Study of Industry Relations in the U.S. and Japan. Honolulu, Hawaii: East-West Center Press.

INNOVATION IN THE ARTS: JAPANESE AND AMERICAN APPROACHES

Jacques Maquet, University of California at Los Angeles

It is only at the Renaissance that art emerged in the Western reality as a separate category of objects made just to be looked at. To prompt and sustain the beholder's admirative attention, the art object had to have, among other qualities, novelty. Artists were expected to make objects different from what had been made previously.

In the arts, an innovation is not, as in economics, the introduction of a new object into the market. It is a new configuration of forms, a new style that is socially recognized as different from the existing styles--for example, in architecture, the Gothic style was an innovation, different from the Romanesque style. Innovations are based on a new cluster of techniques and procedures. These techniques and procedures either were invented by the innovators (such as the vanishing-point perspective at the Renaissance, or oil painting in the 15th century), or had been used previously and put together in a new cluster by the innovators. Innovations are usually collective, originating in a group of artists, a school. Russian rayonism, German expressionism, American action painting are examples of innovative movements which presented new types of visual forms, and new ideas on what painting should achieve [1].

Creativity refers to the ability of an individual to generate something that is new to her or him. Invention and innovation refer to what is new to the society. If an isolated young girl discovers by herself the way to represent depth by drawing oblique lines on a two-dimensional surface, she has created something new to her but not to her society. She is highly creative but her technique of perspective is not an invention.

One

During the fourth quarter of the 19th century, impressionism was the first significant school in the Western-type painting of Japan. The Western-type painting, called *yôga* in Japanese, began before impressionism and the Meiji Restoration. It began in the 18th century, at the time of the "Dutch Studies"--studies of occidental books, mainly on applied sciences, brought by the Dutch traders in Nagasaki [2]. Shiba Kôkan (1747-1818) was a *yôga* painter in the Dutch line, during the Tokugawa period [3].

The distinction between *yôga*, the "occidental school," and *nihonga*, the "traditional school" [4], illustrates a typical pattern of Japanese cultural thinking: conflict avoidance by compartmentalization [5].

Interest in Western painting during the Edo period was not initiated by Japanese painters but by the shogunate government. As art historian Michiaki Kawakita wrote, "Western art, like Western languages and science, was being studied because it was considered necessary for defense and diplomacy as well as for the advancement of architecture and civil engineering" [6].

At the end of the Tokugawa period, the painter Kawakami Togai, was appointed director of the Department of Painting, at the Institute for the Study of Western Documents. He was also teaching Western painting at the Military Academy. Yet his own works were of the *nihonga* type and he did not consider Western painting as aesthetic [7].

His students were more enthusiastic about Western painting. What impressed them was its achievements in representing the world as the eye sees it by means of techniques such as perspective and modeling with light and dark. As painter Shiba Kôkan had commented, one century earlier, modeling made it possible "to distinguish a sphere from a circle" [8].

At the end of the 19th century and the beginning of the 20th, an impressionist school developed. The leading painters were Asai Chu (1856-1907) and Kuroda Seiki (1866-1924)--they had studied in France, the latter for nine years [9].

For these painters, the rendering of the visual impression of the present moment was the best way to represent the world as the eye sees it. Attention was paid to the luminosity of the reflected light, the contour lines marking the limits of shapes were attenuated or suppressed, the scale of values was fully used, a flexible composition was favored, and the easels were set up outdoors--hence its name in Japan, the Plein-Air school. For that generation of Japanese painters, impressionism was a better realism.

At the time impressionism was flourishing in Japan, a group of ten American painters--who took the name of "the Ten"--were also painting impressionist canvasses. Their first common exhibition was in 1895. Several of them had studied in Paris. Some were landscape painters, such as J. H. Twachtman (1853-1902) and Childe Hassam (1859-1935), others were genre painters like T. W. Dewing (1851-1938) [10].

Two

When considering the influence of culture on innovation in the arts, Japanese and American impressionisms are interesting as both were historically derived from French impressionism. They began as technological transfers. This did not prevent the movements to be innovative in Japanese and American societies, neither the painters to be creative. What did result from the national impressionisms in Japan and the States? To answer this question, let us briefly consider the source, French impressionism.

It began as a movement when a few painters, having been rejected by the official *Salon* in Paris, organized in 1874 their own private exhibition. The initial core group included Claude Monet (1840-1926), Jean Renoir (1841-1919), Alfred Sisley (1839-1899), and Frédéric Bazille (1841-1870). The picture of a harbor in a misty morning, by Monet, had the title "Impression: Sunrise." The name *impressionism* was coined by a critic, adopted by the painters, and used to denote their group that lasted as a school for at least twelve years (their last exhibition together was in 1886) [11].

As art historian E. H. Gombrich has put it, the 19th century was, in the visual arts, the time of a permanent revolution [12]. Impressionism was part of the revolution. The classical old masters painted their subjects as they knew the subjects were in the outside world: three-dimensional, sharply in focus, with local color. The impressionists wanted to paint the subjects as they were seen: permeated by the luminosity of the moment, their local color modified by incident and reflected light, and the middle- and background figures somewhat blurred.

Impressionists were also opposed to the romanticism of painters who expressed their own emotions and excitement about the scene they represented. They wanted to record the subject as visually perceived, not what they felt about it.

Impressionism, as the intellectual position we have just sketched out, was an innovation in the French artworld of the second half of the 19th century. At that time classicism and romanticism were not abstract systems but living forces represented by two remarkable and influential painters of the preceding generation, the conservative and academic Jean-Auguste Ingres (1780-1867) and the rebellious and emotional Eugène Delacroix (1798-1863). In the Paris of the 1860s and 70s, impressionism was recognized as new, even revolutionary by the art public.

The innovation was in the ideas and in the clustering of painting procedures. Were these pictorial techniques invented by the innovators? No, each had antecedents. Turner (1775-1851) had emphasized the rendition of atmospheric luminosity; Delacroix had blurred the borderlines of shapes; Rembrandt (1606-69) had been the master of chiaroscuro, the gradations of values from lightness to darkness; early photographers had framed their compositions with a new flexibility; and François Millet (1814-75) had painted outdoors, in the fields near Barbizon. Yet the combination of these previously known techniques in a strong cluster was a collective innovation. It was the basis for a new style.

Three

The Japanese impressionists, Asai, Kuroda and their colleagues imported the French impressionist system, ideas and practices, as a whole. They did it by going abroad, by learning and mastering the techniques.

For these 19th century *yôga* artists, the techniques to be mastered included not only the impressionist ones, but all of the Western antecedents of these techniques. Impressionism was, on the Japanese cultural scene, more novel than on the American or even the French one. As individuals, the Japanese impressionists were not imitators. Like the French disciples of Monet and Renoir, they had internalized the procedures and the spirit of the new style, and they painted original works within the impressionist paradigm.

Rapid and successful assimilation of a portion of a foreign culture has been a recognized pattern in Japanese history. From China, Japan adopted script and Buddhism, the ranking system of official titles, the space organization of a capital, and many other cultural items. As Ruth Benedict has put it, more than forty years ago, "it is difficult to find anywhere in the history of the world any other such successfully planned importation of civilization by a sovereign nation" [13].

As big as the imported portion may have been, there was always selection and interpretation of what was adopted. The list of titles given in China to officials who had passed state examinations was adopted in Japan. But the system was reinterpreted: in Japan, the titles were bestowed upon hereditary nobles and feudal lords [14]. Chinese *Ch'an* Buddhism was adopted in Japan, and reinterpreted into *Zen* which is distinctly different from it.

Something similar happened with the adoption of impressionism. The study of European art had been planned by the shogunate government, and later the imperial government, but some artists' interpretations were not planned. An article entitled "Green Sun," that has been called an impressionist manifesto, was published by Takamura Kôtarô in 1910. It began by this sentence: "I ask for an absolute freedom in the world of the arts. I am thus ready to acknowledge that the artist's personality [in the Japanese text, the German word *Persönlichkeit* was not translated] has a limitless authority. . . . If somebody sees as red what seems blue to me . . . my only concern will be to discern how he treats what he perceives as red" [15]. We are far from the realism of the disciples of Kawakami. Takamura had interpreted impressionism, and gone beyond it--first step in the direction of nonfigurative painting.

In America too, impressionism was adopted as a whole system during the last decade of the 19th century. As in Japan, the movement was innovative, and the individual painters creative. The Ten attempted to Americanize what was at that time perceived as an avant-garde French movement. Their effort was not vigorous, and as art historian Harold Osborne has put it, it was a "quietist and unadventurous impressionism" [16]. It did not, as the French and Japanese impressionisms did, evolve into other new and lively art trends.

Four

Japanese and American impressions brought change on their respective art scenes by the same processes, adoption and assimilation. Yet, the results were different, because the approaches were different.

In the fourth quarter of the 19th century, Japan was at the beginning of the Meiji Restoration, still very close to the Edo period during which the aesthetic concern was not focused on objects made, or selected, only to be looked at--what we now call art objects [17]. (The situation was the same in pre-Renaissance Europe.) The aesthetic quality--the formal excellence--was present, and perceived, in instrumental objects (religious buildings and rituals, artifacts to be used in political and military ceremonies, and in everyday life at the court and at home). Instrumental objects embodying formal perfection, in different degrees, were made by craftsmen [18]. This high esteem for the craftsmanship quality and the aesthetic aspect of artifacts was still dominant in early Meiji Japan.

The introduction of Western art and Western art movements was confined to the *yôga* compartment of the culture, a sort of foreign enclave. It was not incorporated in the mainstream culture; it was not threatening it. For the painters working in the *yôga* compartment, it was different. They were aware that someday the category of art in its international dimension would be included in the Japanese culture, and they were committed to achieve that. Meanwhile compartmentalization succeeded in assimilating and insulating foreign art and its potentially dangerous reinterpretations, such as the radical individualism of the Green Sun manifesto.

In the America of the fourth quarter of the 19th century, the cultural dependence on Europe was, for artists, a divisive question. Some, following the examples of the 18th century painters John Copley (1738-1815) and Benjamin West (1738-1820) identified themselves with the European artworld. The most Europe-oriented settled in Europe, as Copley and West had done in the 18th century. Mary Cassatt (1845-1926) joined the French impressionist group and lived in Paris.

Others wanted to develop styles of painting independent from the European movements. This was not easy as they could not rely upon a tradition of quality craftsmanship comparable to the Japanese one. The immigrants to the East coast of North America had not developed a high level of proficiency in that field. In addition, American artists were, as much as the European artists, descendants from the Renaissance. They did not have a separate ancestry. Finally, compartmentalization is not an American strategy: a new movement is necessarily in competition with the others.

The Ten recognized that they were adopting and assimilating a European movement. But, at the same time, they wanted to be independent from Europe. They were on the American scene, in competition with others and could not develop in an enclave. Their movement ended with the death of the last of them. It was not at the origin of other vital developments, new avant-gardes, as in Japan and France.

Five

About one century later, in the last decade of the 20th century, the situation has changed in Japan and America.

The art enclave in the Japanese culture has been included in the mainstream. As in all the urban contemporary societies, art (as distinct from aesthetic craftsmanship) is a recognized part of the culture, even if the appreciation of it is not the same for every segment of the urban population. A minority that includes mostly the better educated and the more affluent--but not all of them--has a serious interest in the contemporary visual arts. The rest knows about the arts of today, but is rather indifferent or puzzled.

In the past, traditional craftsmanship and the customary icons were easily assessed and appreciated by everybody. It is not the same with conceptual art, minimal art, and other trends called *zen.ei* (avant-garde).

The art of today in Japan is international in the sense that it is situated on the contemporary "one-world" art scene that is no longer fragmented by national boundaries. And it is Japanese in the sense that it is made by Japanese in Japan. It seems that for these post-Hiroshima generations, there is not any longer a problem of cultural identity in terms of either Japanese, or foreign.

An exhibition that traveled in North America in 1990-91 offers an example of an art at the same time Japanese and international. Organized by the Hara Museum of Contemporary Art and the Los Angeles County Museum of Art, it is called *A Primal Spirit* and presents the works of ten sculptors.

These ten sculptors did not constitute a group before the exhibition--"in fact, some of them had never met, nor heard of some of the others" [19]. Yet they have a common attitude to the materials they use: wood, fibers, stone, metals, minerals and fired clay. These materials are not just raw materials to be transformed, manipulated, and processed for making objects that conform to the artist's intention. They should be allowed to speak for themselves, as it were.

This attitude to materials derives from a movement, *mono-ha*, which emerged in the turbulent late 60s. *Mono-ha*, which means "school of things," proposed to let the things, or materials, appear "as they stood, bare and undisguised" [20]. The main proponent of *mono-ha*, Lee U-Fan, wrote in 1969, "we must learn to behold the world as it is without making a representation of it." To achieve this, the thing has to be presented in a way that makes its structure apparent; the structure is "a great intermediary that allows to see more clearly the world as it is, in all its gestures and aspects" [21].

In interviews conducted by the organizers of the exhibition, one of the sculptors, Koichi Ebizuka (b. 1951) told that most Japanese artists working today use a material "without attempting any dialogue with the material. . . . I can only use a material when I have stripped away the various levels of meaning and found its original form. . . . Every tree has an inherent, individualized expression that reflect such things as the place where it has lived" [22].

Another sculptor, Toshikatsu Endo (b. 1950) said that "when I work with fire or water, I see them not simply as material . . . I am not creating some figure through the use of fire; fire as a phenomenon is what I am interested in" [23].

From this first common attitude--giving priority to the material over the object--follows a second: recognizing the impermanent nature of the presented thing. The works exhibited by seven of the ten artists are installations that are dismantled when the exhibition is over. Even works that do not have to be dismantled, grow old, deteriorate, and decay. Kazuo Kenmochi (b. 1951) uses woods that have been rejected, the waste to be found on a construction site. By "picking this wood up at the final stage in its cycle, and by using it in my work, I allow it to stand . . . once more." This reprieve will come to an end also. In that case, concludes the artist, "I don't care if a work of mine is destroyed by the wind, or rots in the rain" [24].

A third position is that the artist, as human being, is not separated from, and opposed to nature. Trees, like humans, are unique. Kimio Tsuchiya (b. 1955) said, in an interview: "I do not feel that there is any great difference in the value of a human life compared with the life of a tree . . . The wood I use . . . cannot be used for construction or for traditional sculpture . . . it may have been beautiful when alive but, once cut, it has no commercial value . . . it is just plowed under by bulldozers to get it out of the way. Watching this, I feel as though the bulldozer is plowing into my own flesh. Wood is not just matter" [25].

Each human and each tree is unique. It does not mean that they are separate beings: they are fragments. "As a human being I do not constitute the center of the earth but exist only as a single point within nature's cycle," says Takamasa Kunyasu [26].

From this third common position--that the artist, as human being, is not separated and opposed to nature--follows the idea that the work is not, first of all, an expression of the artist's self. As Toshikatsu Endo said, "Anyone could make works like mine; they are just drawings of circles. Anyone can draw a perfect circle. . . . It is as though I am making the work because no one else is making it" [27].

Emiko Tokushige (b. 1939), the only female artist in the exhibition, said, "I am rather envious of Western artists for the intense expression of ego, or self, that I sense in their works. I am overwhelmed by their power of expression since I am unable to achieve as much" [28].

Six

Because these ten sculptors have so much in common and are associated in a traveling exhibition, they constitute a group. With reference to the exhibition name, we can call them *primalists*.

On the international contemporary scene, they are situated in the conceptual current: the object is primarily a vehicle for communicating ideas, and secondarily an object that retains our visual attention. In other words, the object is more important as symbolic than as aesthetic.

Some works made in the 1980s by Richard Serra and Ulrich Rückriem in Germany [29] and by Chris MacDonald in the United States [30] use also wood and stone in their elemental forms to communicate meanings. They have a resemblance with some of the primalist installations but they are not the supports of the same "concepts."

Primalism is yet too recent to know if it will be recognized by the international art world as an innovation. It probably will as it offers a cluster of techniques and concepts that is new. The configuration is new, not its elements. As the French impressionists, they did not invent their working procedures. Their techniques for assembling the elements of installations, for burning wood, for preparing copper and other metals were used by the *mono-ha* artists twenty to fifteen years earlier. Other procedures, like deep cutting into a trunk to make bent-wood forms seem completely new.

Their ideas and attitudes are certainly not novel either. They are rooted in Shinto (the deference to the trees as individual living beings) and Buddhism (the continuous change and impermanence of the world, and the absence of a substantial core in sentient beings). The primalists make it clear that they are not religious practitioners but they recognize their affinity with the Buddhist worldview.

Primalist ideas and procedures are derived, but their cluster is original. Their approach to things and materials, to artistic creation and nature, to self and the common condition of the living beings constitute a set of ideas and values that unifies the diversity of their objects and installations.

Even if their encounter in a common exhibition does not generate a recognized movement, the primalists exemplify a creative trend that is at the same time international and Japanese.

Seven

In the American artworld, the identity crisis of the turn of the century has been solved after World War II. Artists do not wonder any longer if they are Europe-oriented or independent: they operate on the "one-world" scene of the arts, and are American.

Conceptual art is international. American and Japanese conceptual artists working with wood, stone, and metal use similar techniques, and share the belief that artworks are primarily vehicles for the communication of ideas. Yet their objects and installations are somewhat different, and they stand for different ideas and values.

The Japanese primalists perceive a Buddhism-derived ethos as a solution to the contemporary destruction of nature. The American conceptualists express the American ethos as they experience it in the threatening last decade of the 20th century.

To conclude, when a group of artists proposes a configuration of procedures and ideas that is socially recognized as new and significant, it is an innovation. The group may have invented the configuration or adopted it from elsewhere. In either case, the new style offers only a frame for the individual creativity of each artist.

Notes
1. When this paper was delivered as a lecture on December 12, 1990, in Tucson, it was illustrated by eighty four slides. In some of the following notes, we indicate books in which photographic reproductions of works of the artists here discussed may be found.
2. Kawakita 1974: 12.
3. For works by Shiba Kôkan, see French 1974.
4. Origas 1986: 41.
5. See Befu 1971: 100.
6. Kawakita 1974: 34.
7. Ibid.: 35.
8. Takashina 1986: 24.
9. For works by Asai Chu and Kuroda Seiki, see Asai Chu, Kuroda Seiki 1973; and Kuroda Seiki 1975.
10. For works by American impressionists, see Boyle 1974, Gerdts 1984. For works by Twachtman, see Cincinnati Art Museum 1966. For works by Hassam, see Hoopes 1979.
11. For works by French impressionists, see Rewald 1946; and Herbert 1988.
12. Gombrich 1978: 395-424.
13. Benedict 1967: 58.
14. Ibid. See also the classic Nakane 1970.
15. Takamura 1910: 142.
16. Osborne 1970: 36.
17. Maquet 1986: 33, 68-70.
18. Maquet 1979: 20-23.
19. Fox 1990: 27. For works by the primalists, see Fox 1990.
20. Minemura quoted in Fox 1990: 13.
21. Lee 1969: 380.
22. Fox 1990: 42, 43.
23. Ibid.: 55.
24. Ibid.: 73.
25. Ibid.: 105.
26. Ibid.: 87.
27. Ibid.: 50.
28. Ibid.: 92.
29. For works by Rückriem, see Honnef 1988: 246, 247. For works by Serra, see Pasadena Art Museum 1970; and Honnef 1988: 248, 249.
30. For a work by MacDonald, see Freeman 1990: 128, 129.

References & Bibliography
Asai, Chu 1973. *Asai Chu, Kuoda Seiki*. Tokyo: Shueisha.
Befu, Harumi 1971. *Japan: An Anthropological Introduction*. Tokyo: Charles E. Tuttle.
Benedict, Ruth 1967. *The Chrysanthemum and the Sword*. Cleveland: World Publishing (Meridian). [1st ed. 1946]
Boyle, Richard J. 1974. *American Impressionism*. Boston: New York Graphic Society.
Centre Georges Pompidou 1986. *Japan des avant gardes*. Paris: Editions du Centre Pompidou.
Cincinnati art museum 1966. *John Henry Twachtman: A Retrospective Exhibition*. Cincinnati: Cincinnati Art Museum.
Fox, Howard N. 1990. *A Primal Spirit: Ten Contemporary Japanese Sculptors*. Los Angeles: Los Angeles County Museum of Art; New York: Harry N. Abrams.
French, Calvin L. 1974. *Shiba Kokan: Artist, Innovator, and Pioneer in the Westernalization of Japan*. New York: Weatherhill.
Freeman, Phyllis and others 1990. *New Art*. New York: Harry N. Abrams.
Gerdts, William H. 1984. *American Impressionism*. New York: Artabras.
Gombrich, E. H. 1978. *The Story of Art*. 13th ed. Oxford: Phaidon.
Herbert, Robert L. 1988. *Impressionism: Art, Leisure, and Parisian Society*. New Haven and London: Yale University Press.
Honnef, Klaus 1988. *Contemporary Art*. Hamburg: Taschen.
Hoopes, Donelson F. 1979. *Childe Hassam*. New York: Watson-Guptil.

Kawakita, Michiaki 1974. *Modern Currents in Japanese Art.* New York: Weatherhill; Tokyo: Heibonsha.
Kuroda, Seiki 1975. *Kuroda Seiki.* Tokyo: Chuo Koronsha.
Lee, U-Fan 1969. Le monde et sa structure. *In* Centre Georges Pompidou 1986: 380.
Maquet, Jacques 1979. *Introduction to Aesthetic Anthropology.* Malibu, California: Undena Publications.
Maquet, Jacques 1986. *The Aesthetic Experience: An Anthropologist Looks at the Visual Arts.* New Haven and London: Yale University Press.
Nakane, Chie 1970. *Japanese Society.* Berkeley and Los Angeles: University of California Press.
Origas, Jean-Jacques 1986. Passage de la ligne. *In Centre Georges Pompidou 1986: 41-49.*
Osborne, Harold, ed. 1970. *The Oxford Companion to Art.* Oxford: Oxford University Press (Clarendon Press).
Pasadena Art Museum 1970. *Richard Serra.* Pasadena, California: Pasadena Art Museum.
Rewald, John 1946. *The History of Impressionism.* New York: Museum of Modern Art.
Takamura, Kôtarô 1910. Le Soleil vert. *In* Centre Georges Pompidou 1986: 142.
Takashima, Shûji 1986. Introduction. *In* Centre Georges Pompidou 1986: 23-28.

MODELS OF INNOVATION AND THEIR POLICY CONSEQUENCES

Stephen J. Kline, Stanford University

Two top-level models of innovation in industrial societies are described and compared: The conventional (Linear) model and a newer model called The "Chain-Linked model". The comparison indicates the linear model has a number of major deficiencies.

The paper also suggests broader definitions for: (i) the systems in which innovations are embodied; (ii) innovation. In addition, the paper provides a numerical measure of the complexity of systems. The linear model, taken with the associated conventional definitions, is then compared with the chain-linked model taken with the suggested broader definitions and the implications of system complexity. The comparison suggests that the chain-linked model, with the broader definitions, seems to provide a significantly better basis for thinking about policy issues.

Preliminary Concepts

In order to carry out the comparison of the linear and chain-linked models as they apply to policy issues, it is necessary to re-examine two basic concepts: (i) the nature of the systems in which innovations are carried out; (ii) the definition of the word "innovation".

The Nature of Systems in which Innovations Are Carried Out

When we look at complete systems in which innovations are carried out, we find that they involve not only hardware but also people, organization, financial arrangements, legal and ecological constraints, and often other factors. Usually, there is strong coupling between the many components, particularly between the social (human) and the technical (hardware) components. For this reason, I like to call such systems "SOCIOTECHNICAL SYSTEMS". We will discuss below a few characteristics of sociotechnical systems which affect the understanding of policy issues.

Definition of an "Innovation"

If we think about sociotechnical systems and how we might improve them, it is not difficult to realize that improvements are possible in each of six major areas:
- o Product
- o Technical process of manufacture
- o Social arrangements in the system of manufacture
- o Fiscal or legal
- o Marketing (the sociotechnical systems of distribution and/or use)
- o The system as a whole

Each of these areas has, at times, been the locus for major improvements in industrial systems. Frequently, more than one area has played a role in a given advance. Examples exist throughout the literature of the history of technology and management; a few are given in Kline (1985, et seq).

---

* This paper is reprinted with minor modifications from "Science and Technology Policy: "What Can Be Done?" "What Should Be Done?" Proceedings of the NISTEP International Conference on Science and Technology Policy Shimoda City Japan, Feb 2-5, 1990 with permission from NISTEP and MITA Press.

Conventionally, economists have defined an innovation as "the introduction of a new (or improved) product into the market". The list just given indicates that this conventional definition is too narrow. A definition which includes all six of the possible areas of innovation seems needed.

Suggested Definition of "Innovation"

An innovation is any change in the sociotechnical systems of design, manufacture, distribution, and/or use which improves the performance of the entire system with regard to cost and quality of product, or of service to users and/or employees.

Models of Innovation

The Linear Model

The long used (Linear) model of innovation is shown in Figure 1. Linear in this context denotes sequential (not linear in the sense of linear equations).

The linear model has been used widely for a long time, at least in the western world. However, the use has usually been implicit; that is, the linear model has been held in the minds of many, but seldom written down. However, we must not let the fact that the use has been implicit delude us into thinking either that the linear model has not been used or that it has been unimportant. The linear model has been invoked in nearly all the arguments for the support of science to governments right up to the present time. We all know well the references to science as the "seed corn" on which technology draws. Those of us who have lived in one or another research community have all heard many after-dinner speeches appealing to the linear model as a justification for doing basic science. More important, thinking in terms of the linear model seems to have strongly influenced the institutional forms and the funding policies for R & D in many governments and corporations. Moreover, it is important to note that models which we hold in our minds implicitly can have more power than models we recognize explicitly as models; when we do not recognize our ideas are models of reality, we tend to think of them as "the truth", and do not re-examine and improve the models over time. In sum, the linear model has been the way many people have thought about innovation since World War II. Since the linear model was first challenged explicitly only recently by F. Kodama (1988), by the writer (Kline, 1985 et seq), and by Gomory and Schmidt (1988), we must expect that many people still use the linear model as a basis for thinking about innovation.

I will argue in this paper that the linear model suggests more inappropriate than appropriate actions in planning and managing innovations in the sense defined in the preceding section. In order to make this demonstration, it will be economical and effective to introduce what appears to be a more appropriate model which I call "The Chain-Linked Model"

The Chain-Linked Model

A schematic of the chain-linked model, as originally proposed, is shown in Fig. 2 with some improvements in words and emphasis suggested by several early commentators on the model, most notably Harold Hall of the Xerox Corporation. An important further addition to the chain-linked model owing to Yahagi and Morimoto (1990) is shown in Fig. 3.

With regard to Fig 2, we note first it suggests 6 paths as being important in innovations. A dozen historical examples of important innovations involving each of these six paths is provided in Kline (1985 et seq) to indicate their reality. Each of these six processes is described next.

Path 1: C-C-O-I: The links marked C-C-O-I in Fig. 2 indicate the Central Chain Of Innovation. This is the well recognized path by which designs are created and moved downstream into the market. However, there is a critical difference from the equivalent path as indicated by the linear model of Fig 1. In the chain-linked model, the initiating

FIGURE 1: Linear Model of Innovation

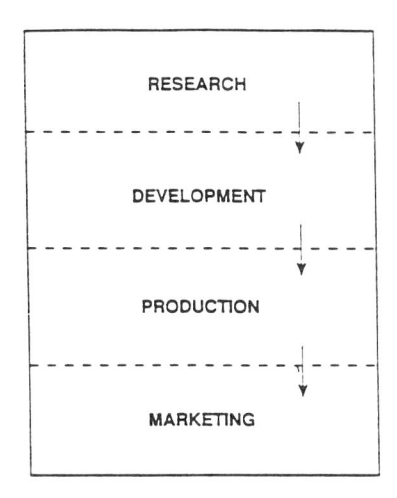

FIGURE 2: Chain-Linked Model of Innovation

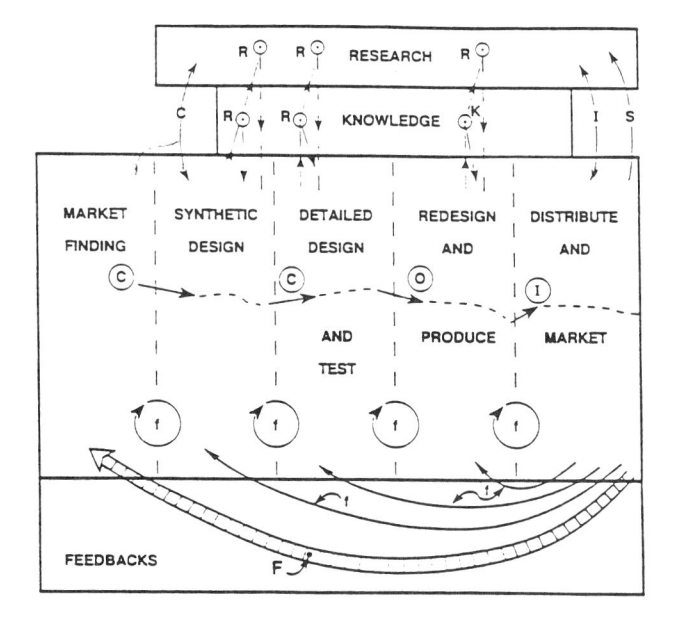

event is not research, but rather need-finding followed by synthetic design. This important difference is discussed in more detail in the section contrasting the linear and chain-linked models below.

Path 2: Links f and F. The processes indicated by the links f and F are feedback links within the system of innovation and in the case of market links to and from the external world. The link F is shown broader because of its critical importance in creating new products.

Path 3: Link I. The process represented by the Link I is the supply of instruments from the manufacturing sector for use in scientific research. A few years ago, Derek de Solla Price argued that the support of science by technology through the link I was greater, on balance, than the contribution of science to technology. In the writer's view, de Solla Price does not fully establish his case; the question of which is more important to the other remains moot. The argument is, in any case, undecidable since the contributions are qualitatively different. So perhaps we can agree that each of science and technology are important to the other and have been over at least the past century. However, we might note in passing that without Galileo we do not have modern science and without the telescope which came from the lens-grinding craft, we do not have the critical work of Galileo. The same remark applies to the roots of modern medicine. Without the work of Pasteur, we do not have the origins of the micro-organism theory of disease, and the critical work of Pasteur demands the microscope which also came from the lens-grinding craft. Hence, the historical import of the Link I is very clear; Link I is nothing new. Moreover, very few of us, I suspect, would argue that, in present time, instruments are unimportant in the current physical sciences: biology, medicine, agricultural research, etc.

Path 4: Link S. The link S represents the support of fundamental science in the research laboratories of industrial corporations. This path is about a century old. It begins only after some corporations are large enough in terms of resources and market size so that the processes become economic -- notably in General Electric and the Bell Telephone system in the U.S. and in the German Dye industry, both in the last quarter of the 19th century (see Chandler, 1977; Reich, 1985). At present, nearly all industrial corporations with more than 500 employees carry out research of this sort. We need to add that such corporate research is normally confined to areas which are seen as possibly contributing to knowledge that will improve company products or processes and rarely extends to the entire front of scientific knowledge.

Path 5: Link C. The link C represents the two way flow of ideas between scientific research and synthetic design. The link C is two way because new synthetic designs often generate basic problems in science. Some examples include Prandtl's boundary layer theory (for application to wings of aircraft), Edison's invention of the parallel circuit (in order to make his lighting system possible), and the concept of choking in compressible flows (empirically from turbine practice and 50 years later theoretically from needs in high speed aircraft design). These are only a few examples from a much longer list. The flow from science to synthetic designs is perhaps the most widely known kind of innovation since it now and then (once every five or ten years in recent decades) has created a spectacular event -- the rise of an entire new industry as in radar, lasers, atom power, antibiotics, genetic engineering and so on. Such events often provide a jump in human powers and the human condition. Nevertheless, we must immediately note two things. First, this SCIENCE ENABLED synthetic design is relatively rare, particularly the spectacular big events; this is the rare mode, not the common one. Perhaps as much as one product innovation in a thousand comes from this source. The remainder come from a different source, from what the historian of technology, Edward Constant (1980), has called "technological paradigms". As Constant notes, a technological paradigm is last year's model which embodies practice and accumulated learning, in most cases over years (sometimes centuries) and over generations of designs. There are many areas still where we cannot do science enabled designs at all, and all designs are necessarily based on the appropriate technological paradigms -- all combustions systems remain, at present, in this class, to give only one clear example. Much more common are cases where some of the design is science enabled and the rest is based on a technological paradigm: stationary power plants, automobile engines, machine tools, and many other systems remain in this class. Second, science in no case finishes an innovation; at most, science enables product innovations since production and marketing lie still downstream, and in no case are the roots of production and marketing science; this non-science base is true, not only historically, but remains necessarily true today for reasons which will be elaborated below in discussing the nature of sociotechnical systems. The closeness of the research output to the market varies greatly across industries. In drugs, the research provides the final product. In mechanical-optical-electrical devices with high piece count, there are often many intermediate steps. The differences show in the continuation rate of R & D projects (Kodama, 1986).

Path 6: Links K AND R. The links K and R represent the links to knowledge and research utilized in innovations. Several comments about these links are needed.

First, knowledge is a state function in the mathematical sense; knowledge is storable, cumulable and, in part, transferable. Research, on the other hand, is a process; it is transient and not storable. For this reason alone, we ought not lump knowledge and research since confusing state and process functions mathematically leads to errors.

Second, on a more physical level, we do not design from this year's research ALONE anymore than we live ONLY in the houses we have built this year. In the design process, the designer typically calls on knowledge in the following order: (i) on his/her own knowledge, (ii) on the knowledge of immediate colleagues, (iii) on the published literature, (iv) perhaps on an expert from a distant location. Only when ALL four of those sources of knowledge fail to solve the problem thrown up by the design in hand, do we resort to research for the obvious reason that research is nearly always far slower, more costly and more problematic than using existing confirmed knowledge.

Third, in the chain-linked model, research is seen to operate all the way down the central chain of innovation. This is obviously so since research in production methods and the solution of problems arising in the field after use of a product has begun are both frequent sources of important research problems in all mature industries.

First Level Comparison of the Linear and Chain-Linked Models

Five years experience with the chain-linked model has not indicated the need for processes not shown in the chain-linked model. A useful way to compare the linear and the chain-linked model is, therefore, to examine what each model says about the six links of the chain-linked model one-by-one.

Link C-C-O-I. The linear model suggests the initiating step for product innovation is research, that is either science or some form of basic research as contrasted with more applied development work, since development is the next step in the linear model. The chain-linked model suggests something not only quite different but also more appropriate in two ways. First, the common initiating step of product innovation is study of the technological paradigm embodied in existing models of the product or system as marketed by various companies. Second, there is a call, in any case, on existing knowledge before one resorts to the research indicated by the Link C of Figure 2, as noted in discussing the links K and R of Fig. 3. To put this differently, most product design changes originate in what we call human creativity -- a little understood process which is involved in much science but is not itself what we ordinarily call science or research.

For the Links C-C-O-I, the chain-linked model thus appears more appropriate than the linear model in the sense that it seems to represents reality with significantly more accuracy.

Links F and f. The Linear model contains no feedback links whatsoever. The chain-linked model shows feedback links everywhere. This not only conforms to the reality

FIGURE 3: Knowledge Interface of Technology and Science (KITS)

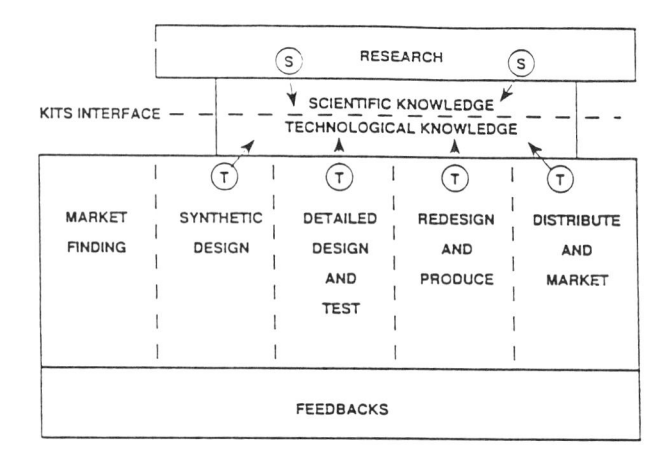

FIGURE 4:

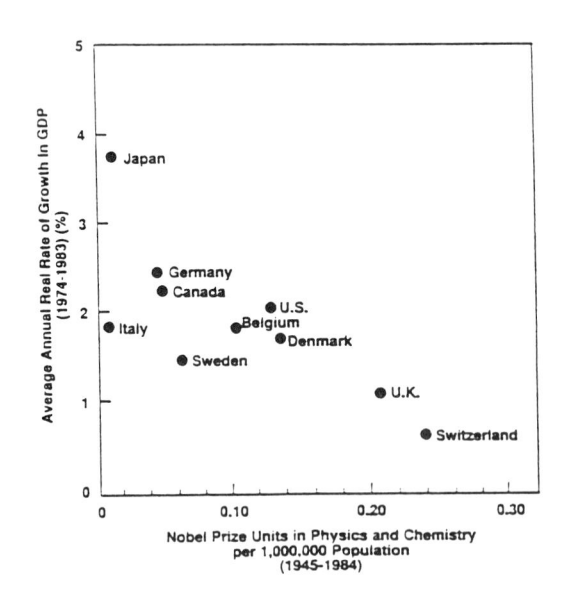

* C. Hill, Congressional Research Service Study, April 16, 1986.

more accurately, but as we will see in discussing system complexity in a later section, feedback links are vital for the large sociotechnical systems which constitute major, mature industries.

Link I. This link is also missing from the linear model. Its importance both historically and in present time is obvious and easily documented by as many examples as one chooses to seek. A few important examples have already been given above. Thus, the chain-linked model appears more appropriate in this regard.

Link S. The linear model has this link but only at the front end of the product innovation process. The chain-linked model shows the link acting along the entire central chain of innovation by distribution of research efforts and storing of the resulting knowledge in the appropriate place in the knowledge box of Fig. 2. Unless we want to conclude that research on design, production, field experiences, and consumer response are all irrelevant to product innovations and system improvement, we must conclude that the chain-linked model is more appropriate than the linear model regarding this link.

Link C. This link has already been discussed in connection with the Links C-C-O-I above where we concluded the chain-linked model provided a more appropriate picture.

Links K and R. These links are also missing from the linear model. The linear model thus confuses this year's research output with the storehouse of "technical" knowledge which has been accumulating since the emergence of the human race about 2 million years ago. The word "technical" is placed in quotes because it is used in this paper to denote not only scientific but also technological knowledge. The confusion between this year's research and the accumulated storehouse of scientific knowledge has already been discussed in remarks about the Links K and R above, and need not be repeated. We do need to add some remarks about "technological knowledge."

What is technological knowledge? Early historical examples of import might include the wheel, clothing, boats, writing and the tools for writing, printing, the processes of agriculture and herding in their early forms, etc. Since science as we know it is only about 300 years old, and human sociotechnical systems have been evolving for at least 2 million years, it is evident that innovations over all but the most recent epoch were not based on science as the historian of technology Ed Layton has noted. As we already noted, some of these technological innovations were essential to the rise of both modern science and modern medicine. So historically, the impact is clear. What then about the present? My Stanford colleague W. G. Vincenti has recently documented some important modern innovations in technological knowledge that are specifically not science. His examples include the derivation of "control volume theory" (Vincenti, 1982). Control Volume theory is the rational basis for analysis in nearly all prime movers, and any other device where mass flows in and out rather than staying in the system -- this includes perhaps 95% of all applications in many engineering fields. Another example by Vincenti (1984) documents the rise of flush rivets for aluminum construction. Vincenti shows the process arose from need for drag reduction on airframes, and that it arose in at least three companies independently in the same time frame. In all three cases it arose directly from work on the shop floor, not from any research labeled by that name. However, Vincenti's examples deal only with engineering practice and omit what can be called "neuro-muscular skills" embodied (literally) within skilled workers, for example: welding skills, tool and die making skills, crane and bulldozer operating skills, machine-tool operating, etc. There are also many forms of important industrial knowledge embodied in codes, engineering methods, experience in operating large projects (see for example, Squires, 1986), and other forms. These forms of knowledge are, in fact, so important in engineering practice that I argue in a recent paper (Kline, 1989) that over all time up to and including the present, technological knowledge has been and probably remains more important than scientific knowledge in industrial competitiveness. We will revert to this matter in the discussion of system complexity below where we will see that shop floor origins of critical technological knowledge is not a choice but a necessity.

The first-level comparison of the linear and chain linked model can now be summarized as follows. Each of the six paths of the chain-linked model embodies a significant criticism of the linear model. Of these criticisms, five appear to be of great import: (i) the lack of feedback loops in the linear model; (ii) the failure to indicate the importance of research in design and production processes in the linear model; (iii) mistaking the rare case of how product innovations are initiated for the common one; (iv) taking this year's research as the source of innovations rather than the more primary sources of stored knowledge and technological paradigms. The fifth is elaborated in Section VI titled The Complexity Of Systems.

The Role of Science in Product Innovations

Given the discussion of preceding section, the question arises, "What is the role of science in product innovations?" Certainly there is agreement that science has played and continues to play an important role at least over the last century; however, we need to ask, "What Science?, and "In what way?

We can begin to form an answer to these questions by looking at Fig. 3 which arises from a recent comment on the chain-linked model by Yahagi and Morimoto (1990). Yahagi and Morimoto rightly point out that the common source of technological knowledge in the present era is industrial corporations. This was not the case until a century ago when formal R & D began, and it is not the only source currently, but is probably the overwhelming source in the industrialized nations today. Industrial corporations routinely create technological knowledge about production processes, product designs, special analytical methods, neuro-muscular skills, etc. as part of their basic business. Much of this technological knowledge, particularly neuro-muscular knowledge, is not transferable via speech or writing, but must be slowly learned in situ, on the job, by individual workers. Even that fraction of technological knowledge which can be transferred in writing often is not, but is instead held as company proprietary knowledge in support of the competitive position of the company. This lack of publication of technological knowledge probably explains, in part, why technological knowledge has been widely considered inferior to scientific knowledge despite its critical import in design and production which are the central functions of engineering and form the core of industrial competitiveness.

For these reasons, Fig. 3 divides the knowledge box of the chain-linked model into two parts. Technological knowledge (circled T's) is shown as coming from below, that is, from the work along the Links indicated by C-C-O-I primarily by corporations. Scientific knowledge (circled S's) is shown as coming from above, that is, from research in corporations and via science. The Knowledge Interface of Technology and Science (KITS interface) is where the two forms of knowledge come to bear jointly on a problem in design, development, production, or from field operating experiences and market requests.

Thus the KITS interface provides a test for what research MAY assist in innovations of commercial import. If a piece of science cannot be brought to bear on the KITS interface for some application, it is not of commercial significance. If the research (or science), arises at the KITS interface it is of commercial importance by construction. This concept fits well within the framework of "Demand Articulation of Targeted Technology Development" developed at length recently by Kodama (1989). Such problems have usually been labeled "applied", and are therefore often viewed as easier, less important, and less fundamental than problems arising along the boundaries of science; however, these views are all over-simplifications. Problems arising at the KITS interface, on average, are harder than those arising at the boundaries of science since they typically involve more constraints and deal with more complex systems. By definition, problems at the KITS interface are more important commercially than any other class of problems in science. Problems arising at the KITS interface are, on balance, less fundamental than those arising on the boundaries of science virtually by construction, but some problems which have arisen at the KITS interface have given rise to very important fundamental advances in science; some examples have already been given. There are many more.

Two more remarks are needed to complete the discussion of the role of science. First, the methods of science and the world view of science (as distinct from the knowledge content of science) are indispensable, the sine qua non, of all technical work in the current world. The ancient "traditional" world view (see Inkeles and Smith, 1976) does not suffice for technical work. Second, the knowledge of the world, embodied in the principles and equations describing the MACROSCOPIC world and, to a lesser extent, quantum mechanics, are the foundation stones on which rational analyses in engineering designs are grounded and organized. Modern engineers, the prevailing central body of "technologists" must have these materials. However, these principles, with certain notable exceptions (materials, combustion for example), have resided since the early 20th century primarily in the knowledge box (not the research box). Hence, the current problems for industrial innovations lie more in teaching engineers mastery (not mere acquaintance, but true mastery) of these materials with suitable design examples than in researching the principles. Moreover, the knowledge which will be gained from research in very high energy physics has a much lower probability of producing knowledge which can be brought to the KITS interface of Fig. 3 than the principles of macroscopic systems which were already largely in place by the end of 19th century. The energy levels needed for access of the domain of sub-nuclear particles is so high that costs are likely to severely limit applications. The same is true of astro-physics which deals with inert systems of such size and such exotic states that it is hard to visualize terrestrial, commercial applications. Thus, the connection between the advances in physics and industrial innovations suggested by the experiences of the late 19th and early 20th centuries may have altered in recent time, and history may therefore be a poor guide.

The Drivers of Innovations

As Howard and Guile (1989) have noted, the drivers of innovation shift as one moves down the central chain of innovation and thus shift with the stage of an industry. This has important bearing on the appropriateness of models of innovation. For discussion, we can arbitrarily divide the maturation of an industry into five stages: nascent, infant, maturing, mature, and obsolescent.

As we move along this chain of maturation, the drivers of innovation move along the Central of Innovation. As many individuals have pointed out, as we move along this chain of maturation, shifts in the nature and number of firms involved and in the rate of change and magnitude of innovations also occur. In the infant stage, 100 companies may join the race for market dominance. As the industry matures, one or a few dominant designs appear. These designs take most of the market, and the number of companies is sharply reduced, but their size becomes, on balance, much larger. These changes with stage of an industry need to be reflected in an appropriate model of innovation.

For this discussion, it is important to note that, in the nascent stage, only one or a handful of workers are involved in the radical innovations created - since what is needed is a few bright dedicated individuals who can stand outside conventional paradigms, and rethink possibilities. This is true both historically and currently and in large companies as well as small. Radical innovation is therefore possible in small companies. However, as an industry matures, more people become involved. In the mature stage of major industries, very large organizations are needed to carry out all the tasks of design, production, marketing, services to customers, feedbacks for new innovations, etc.. Much more can be said about these issues, but these remarks are enough to set the stage for the next section.

TABLE 1 -- DRIVERS OF INNOVATION

STAGE OF INDUSTRY	COMMON DRIVER(S) OF PRODUCT INNOVATIONS
NASCENT	RADICAL INVENTION ENABLING SCIENCE
INFANT	PRODUCT DESIGNS CREATING PRODUCTION PROCESSES DEVELOPING A MARKET
MATURING	STABILIZING DESIGNS IMPROVEMENT IN QUALITY REDUCTION IN COSTS
MATURE	FINE TUNING MODELS TO MARKET FURTHER COST REDUCTIONS
OBSOLESCENT	FINDING NEW MARKETS RETRAINING WORKERS

The Complexity of Systems

We often speak of system complexity, but we have had no numerical measure for it. A simple numerical measure of system complexity has been proposed recently by the writer (Kline 1989); it is repeated here because it supplies important conclusions concerning innovation processes.

The measure requires the following symbols:
C = COMPLEXITY INDEX OF A SYSTEM (OR CLASS OF SYSTEMS)
V = NUMBER INDEPENDENT VARIABLES NEEDED TO DESCRIBE THE STATE OF A PARTICULAR SYSTEM (OR CLASS OF SYSTEMS)
P = NUMBER OF PARAMETERS NEEDED TO DELINEATE A PARTICULAR SYSTEM FROM OTHER SYSTEMS INCLUDING THOSE IN THE SAME CLASS
L = NUMBER OF FEEDBACK LOOPS WITHIN THE SYSTEM AND CONNECTING THE SYSTEM TO ITS ENVIRONMENT.

Using these definitions, we can take as the upper and lower bounds of C the following (where * denotes multiplication):

$$V + P + L < C < V * P * L \qquad (1)$$

For a given system, or class of systems, the approach of C to the upper or lower bound of Eqn. (1) will depend on the degree of connectivity (coupling) within the system and to the environment. C is a rough measure, but nevertheless appears to provide important information. We can see this by looking at a few examples.

Consider first the paradigmatic problems analyzed in elementary classes in physics, chemistry, or engineering. Typically, in such systems, we first examine one system, by fixing the values of P thus eliminating that source of complexity. In naturally occurring inert systems (unlike living systems), L is zero. If we examine the equations of any or all branches of physics or chemistry, we will find no terms representing gathering information and using it in feedbacks loops which connect to the levels of system control. This is as it should be because inert systems relax toward equilibrium but do not measure, process or utilize information in the sense living systems and some human-made artifactual systems do. Thus, if the equations of physics (or chemistry, geology, etc.) contained feedback loops of information, they would erroneously predict behaviors which do not occur in the systems these fields have been created to describe.

Thus, C for these foundation paradigmatic systems of physical science deals only with V. Moreover, the number for V in such systems is usually 1, 2, or 3. This is true for the equations of mechanics, electromagnetism and thermodynamics. Indeed, the justly famous "Phase rule" of J.W. Gibbs is precisely a specification of the number V for systems of various forms of chemical constitution and physical aggregation. Moreover for "a simple compressible substance", the phase rules indicates V = 2. And the "simple compressible substance" is the central paradigmatic case on which thermodynamics is built. In simple electrical circuits, we have simply current, I, (or voltage, V) as a function of time and thus, V = 1. For the building block circuit P = 3, but, as noted above, we begin by fixing the values of the parameters, and later we study the effects of variation of the parameters on system behavior. A single instance of turbulent flow has V = C = 4, and turbulence has often been called the most difficult problem in physics.

The point for this paper is that C, for paradigmatic systems of the physical sciences, is usually 3 or less; C = 6 is considered beyond solution typically. Nor do computers solve these difficulties for systems of high C. To do the easiest cases of turbulence (where C = 4) even for the most degenerate problems, requires of the order of two-man years of programming and 2-6 months CPU time in the largest available supercomputers in 1989.

Because analyzing systems with C > 3 is so difficult, in the physical sciences and engineering, we have evolved many powerful tricks for reducing the number C in order to make analyses simpler, or, in many instances, even possible. The scope and range of these "tricks" pass beyond the needs of this paper. The only point we need is that science and paradigmatic engineering analyses deal with systems where C is small. More precisely, we consider C = 4 a very difficult problem, and C = 6 is usually thought to be beyond analysis and in need of developmental methods for study (that is, design, build, and test). For these reasons, simple paradigmatic systems with C < 4 are central building blocks for system designs; they are the way physical scientists and engineers think about the world for the most part.

What is the value of C for more complex systems?

For the complex hardware designed by modern engineers, the number P can be approximated by the number of individually called out items on the drawings needed for manufacture. This can be quite large, in the hundreds, thousands or tens of thousands.

For a single individual human being the lowest number assignable for C is a billion (10^9); this is a purposeful underestimate, but will serve for our purposes.

For a single sociotechnical system of the sort which constitute large, mature industrial organizations, the lowest number that can be assigned (again a purposeful underestimate) is $C = 10^{13}$.

If C = 6 is too large a number for analysis, what then can we say about a system where $C > 10^{13}$? The obvious answer is that we have no predictive paradigms of any reliability whatsoever for such systems as wholes. We often have predictive paradigms for parts of such systems, but certainly not for such systems as wholes. Two very important implications for innovation follow immediately.

First, we can create and manage sociotechnical systems only by building and observing the systems; analysis and/or computation will not suffice.

Second, we can improve sociotechnical systems, that is, create innovations, only by observing the system, creating perturbations which we hope will improve performance, and then observing what happens. That is, we control and improve such systems by open-loop feedback processes using human intelligence in a learning-by-doing mode. Computers won't do the job because we lack predictive paradigms without which we cannot instruct the computers. Computers may aid the open loop process by integrating information; they may aid in analysis of critical parts; but analysis of the complete systems lies beyond the power of current computers. Given what we know about "computability", this is likely to remain true forever.

As a result, when sociotechnical systems grow large, we must create and keep open the feedback links which provide information on performance both for operating and for improving the system. In such systems, as soon as the feedback links begin to atrophy, the system is tending toward a rigid bureaucracy. If a system becomes a rigid bureaucracy and the environment then shifts, as nearly all environments do today, the system will tend toward death. We have observed these phenomena in many now expired corporations.

For this paper, it is critical to note that the remarks of the previous section apply to all production systems in industrial corporations, and this has been true over the entire period of the industrial age; (see Kline, 1989). This point has major bearing on what we will see in the final section on implications for policy. The same remarks apply to design criteria for complex hardware since those criteria reflect the preferences and needs of humans and humans institutions; they thus invoke the complexity level C appropriate to many humans even when the value of P is low for the product involved. Here again, we have no predictive theorems and no computer programs, only fallible human judgements.

Implications for Policy

This section is an initial foray into the relation between models and technology policy. It is intended to provide ideas more than to reach conclusions.

The first level comparison of the linear and chain-linked model in the section above seems sufficient to establish the severe shortcomings of the linear model. In this section, we will therefore consider a broader basis for comparison; We will compare two clusters of ideas. The first cluster will consist of the linear model plus three ideas often associated with it: (a) science is the basis for research, (b) the systems in which innovations are carried out are technical: (c) technical systems can be understood through science taken largely in the sense implied by physics and chemistry as a methodological paradigm. We will call this first cluster, "the linear/science cluster". The second cluster will consist of the chain-linked model plus the opposing three ideas: (a) research via observe the system in situ, perturb and feedback is an important and valid form of research; (b) the systems in which innovations are carried out are sociotechnical; (c) sociotechnical systems have complexity numbers so high that we have no valid predictive theorems for the systems as wholes. We will call this second cluster the "chain-linked/system cluster". In this way, we will be able to contrast two very different schema for formulating policy, and see what is found.

At the corporate level, the linear/science cluster suggests directly only one thing for policy, "Do science" since science is the genesis of innovations. This model misses five of the six links of the chain-linked model completely, and thus tells us nothing about many important opportunities for innovation. In addition, it mistakes the rare enabling event for common primary sources of innovation -- creative design and technological paradigms. Moreover, the linear layout of the linear/science cluster suggests adoption of "over-the-wall" organization of innovation where separate sections (or departments) do research, design, development, production and, marketing with handoffs from one group to the next not involving transfer of people. It also suggests that, in view of its importance, research should be carried out in a separate laboratory isolated from design, production, and marketing functions. Do these ideas lead to effective innovation?

In the writer's experience, consulting in a number of industries, the over-the-wall mode of organization of innovation is relatively ineffective and slow. It tends to lead to higher cost designs of lower reliability products. It tends to set up "blaming" of other sections, thus retarding solution of problems. Finally, the over-the-wall institutional form loses information in the handoffs between sections because some information, often critical information, moves only with people and not via drawings, text material or computer files.

In no instance known to the writer has a truly isolated corporate research lab been effective, and in some instances such labs have been striking failures. Research labs seem to be effective only when they are appropriately articulated with downstream functions by feedback loops. In successful labs, this often has included movement of personnel up and downstream along the central-chain- of-innovation. Admittedly, these data are personal and represent a small sample within the mechanical, aeronautical, chemical, paper, automotive, and textile industries. The drug industry may show different results. However, the experiences are uniform, and they suggest the linear/science cluster misleads us in several significant ways. At a minimum, we seem to need a broader based study and further thought on these issues.

Beyond all these matters, the linear/science cluster ignores the importance of technological knowledge, and thus assigns it a lower priority than science (if indeed any attention). Given all these difficulties, one might ask, what is the correlation between recent advances in physics and chemistry and growth of the gross domestic product within given nations. If the linear/science cluster is a good model, one ought to find a positively sloped correlation. The data compiled by Hill (1986), shown in Figure 4, indicate precisely the opposite trend for the period 1974-83.

At the governmental level, the linear/science model suggests the central agencies should fund science, but not technological innovation. That has, in fact, been the pattern in the United States, India, Britain, and some other countries since World War II. Recent studies of competitiveness in the United States as, for example, the report of the MIT Commission (Dertouzos, et al 1989) and Kline (1989) suggest this mode misses the central core of competitiveness and is therefore insufficient. Nashad Forbes (1989) suggests that a major source of failure in innovation systems in India has been the establishment of isolated government research labs unconnected to industrial enterprises. As a result, Forbes' survey of companies indicates they have gained very little from the government's researches even though India publishes as many scientific papers per worker as western nations. This use of the linear/science cluster as a basis for policy seems particularly unfortunate in the third world since the economies involved are in a far less adequate position to utilize leading edge science than those of the already industrialized nations.

The chained-linked/system cluster suggests quite different actions at both the corporate and the governmental level.

At the corporate level, the chained-linked/system cluster suggests examination of the whole system to assess priorities for the locale of promising innovations. It suggests specific attention to creation and maintenance of rapid, effective feedback links at many points in the system. The chained-linked/system model suggests utilizing research all along the central chain of innovation (not just at the front end), and it suggests doing research only after existing knowledge has been exhausted. It implies examining not only scientific paradigms but also technological paradigms as produced by the company, by competitors and by leading university engineering research groups. The chain-linked/system model suggests focusing research on those problems which are thrown back from the work to the KITS interface as first priority (in distinction with problems generated by the current frontiers of science). It implies recognizing that the systems in which innovations are embodied are so complex that we have no predictive paradigms for such systems as wholes. This, in turn, implies that we manage and improve such systems by observing, perturbing and feeding back the results in an iterative fashion, that is, through learning-by-doing. This applies particularly to design of complex hardware and to systems of production. Indeed, the history of effective innovation in production shows this has been the effective modality over the entire industrial age (see Kline, 1989). It has been true once again in the recent important advances in production methods in Japan. The chain-linked/system model recognizes the importance of the creative work of small teams (or one individual) in radical invention as a source of major product innovations. It also recognizes that science at most enables product innovations, and that this enabling occurs only relatively rarely. The chain-linked/system model also points to the fact that innovations in production and marketing require many people, and hence need cooperative effort and clear, rapid feedback of communications.

None of the ideas listed in the previous paragraph are new. Each of them has been used by many corporations, but all of them seem to have been used by only a few corporations. Corporations that have used these ideas at least appear on average to be more successful innovators, for example, in the U.S., General Electric and the Bell Telephone system before its breakup, and a number of Japanese companies as well.

We can now reach two highly tentative conclusions about the chain-linked/system cluster. First, it appears to be a useful source for thinking about effective management of corporate innovations. Second, while no model of processes as complex as innovation should be considered perfect or complete, the chain-linked/system cluster does seem a significant advance over the linear/science cluster as a basis for thinking about policy and institutional forms for R & D in corporations.

At the government level, the chain-linked/system cluster suggests governments need not only a science policy but also a technology policy which aids industry in various ways without directing industrial efforts. Some of these ways for aid include: partial support for very expensive, commercially important generic R & D projects which will be shared by companies; coordination of study of the needs of industry; support for critical industries such as machine tools when that is needed; supply of extension services in manufacturing similar to those long-familiar in agriculture to help small manufacturers who can not afford global searches of the rapidly advancing technological paradigms. In this regard, Japan in recent decades seems to have followed these procedures; few other nations have. This may, in part, account for the many successes in innovations and gains in market shares by Japan in recent decades.

In the United States, the formal policy of the federal government has been to fund science and leave technological innovations to companies. But this formal policy has, in fact, been only part of the story because mission oriented agencies of the federal government (for example, NASA, DOD, DOE, Agriculture, and NIH) have followed policies far closer to those suggested by the chain-linked/system cluster than has the National Science Foundation.

In conclusion, it needs to be repeated that this final section is an initial investigative foray intended to suggest ideas. It is not complete, definitive, or backed by a sufficient empirical data base. Despite these current shortcomings, the discussion seems sufficient to warrant further study of the chain-linked/system cluster as a way of thinking about innovation policy at both the corporate and the governmental levels which is an improvement over earlier ideas and models. Applications of the chain-linked/system cluster in the third world economies may well be important for effective world development.

References

Chandler, Alfred D., *The Visible Hand: The Managerial Revolution in American Business*, Belknap, Harvard University Press, Cambridge, MA 1988.
Constant, Edward W., *The Origins of the Jet Revolution*. The Johns Hopkins University Press, 1980.
Dertouzos, Michael L., Lester, Richard K., and Solow, Robert M., "Made in America: Regaining the Productive Edge", *MIT Commission on Industrial Productivity,* MIT Press, 1989.

Forbes, Nashad, "Technology and the Industrial Environment in India - A Review", unpublished summary of Ph.D. dissertation, Department of Industrial Engineering, Stanford University, 1989.

Gomory, R.E., and Schmidt, R.W., "Science and Product", *POLICY FORUM*, May 27, 1988, pp. 1131-32 and 1203.

Hill, Christopher T., "The Nobel-Prize Award in Science as a Measure of National Strength in Science", *Science Policy Study Background Report No. 3*, Congressional Research Service, U.S. Library of Congress, 1986, p. 30.

Howard, William G., and Guile, Bruce R. (NAE), "Profiting from Innovation", presented at *Economic Growth and the Commercialization of New Technologies*, Center for Economic Policy Research, Stanford University, September 11-12, 1989.

Inkeles, Alex, and Smith, D.C., *Becoming Modern*, Harvard Press, 1976.

Kline, S.J., "Research, Invention, Innovation and Production: Models and Reality". *Report INN-1*, Mar. 1985. (Revisions: INN-1B, Nov. 1985; INN-1C, 1987; INN-1D, Sept. 1989), Mechanical Engineering Department, Stanford University, (Shorter versions appear in "Research is Not a Linear Process", *RESEARCH MANAGEMENT*, Vol. XXVIII, July-Aug. 1985, and in "An Appropriate Model for Industrial Innovation", *SCIENCE, TECHNOLOGY AND SOCIETY NEWSLETTER*, Sept. 1985, Published by the Science, Technology and Society Program, Lehigh University, Bethlehem, PA). Report INN-1D available upon request from Department of Mechanical Engineering, Stanford University, Stanford, CA 94305-3030.

Kline, Stephen J., "Innovation Styles in Japan and the United States: Cultural Bases; Implications for Competitiveness", American Society of Mechanical Engineers, *THURSTON LECTURE*, December, 1989, Available on request as Report INN-3, Department of Mechanical Engineering, Stanford University, Stanford, CA 94305-3030.

Kodama, Fumio, and Honda, Yukicki, "Research and Development Dynamics of High-Tech Industry - Towards a Definition of High Technology", *JOURNAL OF SCIENCE POLICY AND RESEARCH MANAGEMENT*, Vol. 1, 1, 1986, pp. 65-74.

Kodama, Fumio, "U.S. Call to Share Technology Based on Faulty Premise", *THE JAPAN ECONOMIC JOURNAL*, 26, 1307, April, 1988.

Kodama, Fumio, "Demand Articulation: Targeted Technology Development", presented to seminar on options and priorities of future research and technology policies, sponsored by Wilhelm Heinrich Heraeus and Else Heraeus-Stiftung, December 11-13, 1989, Bad Honnet, Federal Republic of Germany.

Reich, Leonard S., *The Making of American Industrial Research: Science and Business at GE and Bell 1876-1926*, Cambridge University Press, 1985, p. 27.

Squires, Arthur M., *The Tender Ship: Governmental Management of Change*, Birkhauser Publishing, Boston, 1986.

Vincenti, Walter G., "Control-Volume Analysis: A Difference in Thinking Between Engineering Technology and Culture", Vol. 23, 2, April 1982, pp. 145-174.

Vincenti, Walter G., "Technological Knowledge Without Science: The Innovation of Flush Riveting in American Airplanes 1930-1950", *TECHNOLOGY & CULTURE*, vol. 25, July 3, 1984, pp. 540-576.

Yahagi, Yoshiaki, and Morimoto, Hidetobe, "Non-Linear Model of Japanese R&D Management -- Cases of Toyota Central R&D Labs, Etc.", to be presented at *2nd International Conference on Management of Technology*, February 28-March 2, 1990, University of Miami, Dept. of Industrial Engineering and School of Continuing Studies.

SOME CONJECTURES ABOUT INNOVATION IN NASCENT AND INFANT ADVANCED MATERIAL TECHNOLOGIES

W. DAVID KINGERY, University of Arizona

The growth, maturing and senescence of a technology is very much like that of anything else. There is an initial nucleating or germinal event followed by an increasing growth rate that reaches a steady state and gradually slows to give an S-curve such as represents the growth pattern of snow flakes from the vapor, of crystals from solution, of plants, animals and technologies. A variety of different social structures, perceptions and environments are important nutrients and contexts at different stages of the process and require diverse sorts of analysis. Stephen Kline has focused most of his analysis of innovation on mature industries and technologies. I want to focus on the early stages of the growth process, on the nucleating, nascent and infant technologies which are the central focus of this conference.

Economists have traditionally defined *invention* as the conception and reduction to practice of a new idea sufficiently different that it would not have been obvious to a practitioner skilled in the art. Then *innovation* is the successful introduction of a new or improved process or product into the market. As shown in Fig. 1, we have pointed out that

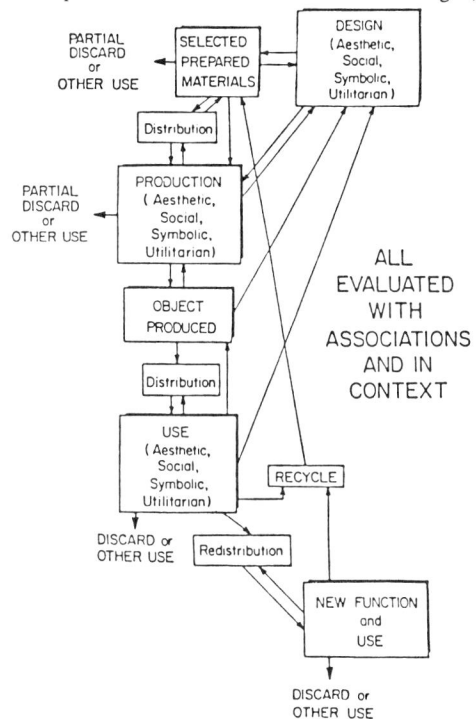

Fig. 1. A number of interrelated technologies are associated with any manufactured artifact. These technologies are interrelated and interact to form a socio-technological system.

an artifact or product is embedded in a context of technologies in which materials selection, design, methods of manufacture, distribution and use are interconnected with a variety of feedback mechanisms (Kingery, 1987) in use as well as design or manufacturing. Based on a variety of examples from the history of ceramics, it seems that the most prominent feedback loop affecting design and manufacturing technologies arises from changes in or demands from user technology, that is perceived modifications of performance or performance requirements as shown in Fig. 2. (Kingery, 1988).

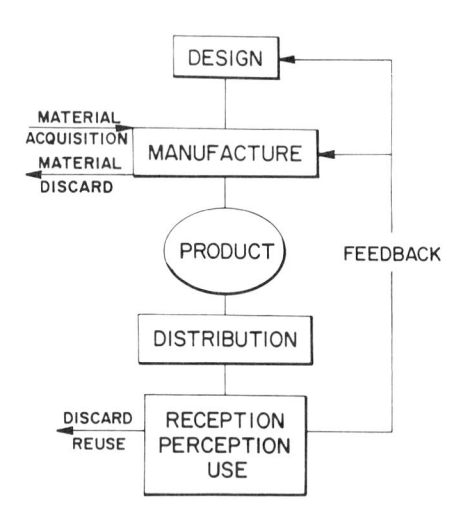

Fig. 2. The most important feedback loop affecting design and manufacturing technologies in advanced materials seems to arise in performance requirements.

In thinking about technological innovation in advanced materials, it is helpful to keep in mind the widely used Materials Science and Engineering paradigm relating processing to structure, properties and performance (Fig. 3). Structures and properties that characterize a product are inanimate attributes that can be precisely measured and compared. In contrast, materials synthesis, preparation, processing and manufacturing are socio-technological *activities* that involve not only artifact but also human behavior, human perceptions and social organization. Likewise, product or process performance involves not only the process or product itself, but human behavior, human perceptions and social organization as well as fiscal, legal and political considerations. These in turn are embedded in a larger cultural and social context. Kline lists six broad areas in which an innovation may improve performance of a manufacturing system: product, process of manufacture, social rearrangements in the system of manufacture, fiscal or legal, marketing (the socio-technical systems of distribution and/or use), and the system as a whole. Kline suggests a broader definition of innovation: "An innovation is any change in the sociotechnical system of manufacture, distribution, and/or use which provides an improvement in cost, quality and/or match to customer requirements."

In discussing advanced materials innovations we often discuss technological changes in which new discoveries and inventions are the starting point. As we investigate deterministic origin stories, we find it increasingly difficult to have much confidence in separating out any particular discovery or invention as an intrinsic starting point. In the new technologies discussed here — high temperature superconductivity, low pressure diamond synthesis,

                      ARTIFACT
                      STRUCTURE

PRODUCTION   PERFORMANCE
ACTIVITIES: ACTIVITIES:
                                              DISTRIBUTION
MATERIALS SELECTION   USE
MATERIALS SYNTHESIS   TECHNO-FUNCTION
MATERIALS PROCESSING   SOCIO-FUNCTION
DESIGN   IDEO-FUNCTION
MANUFACTURING

                      ARTIFACT
                      PROPERTIES

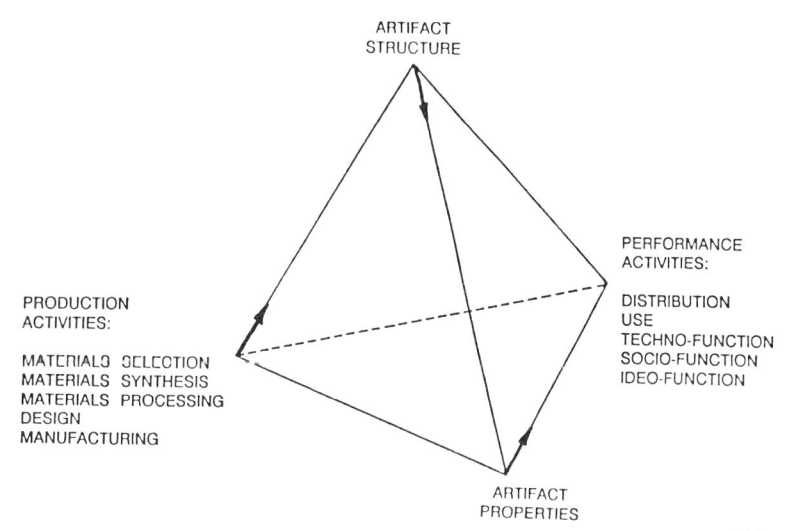

Fig. 3. Artifact attributes - structure and properties - are different in kind from the production activities which give rise to structure and properties and the performance activities which make use of structure and properties.

silicon nitride structural ceramics — that will be seen to be the case. For example, in the development of high T_c oxide superconductors, Bednorz and Müller received the 1989 Nobel prize for their 1986 discovery of an oxide superconductor with a critical temperature of 36°K. However, there were a number of known similar related oxide superconductors with lower values of T_c and the increase from 23°K to 36°K did not have any particular technological advantage. It was only when new compositions were discovered with T_c greater than 77°K, the boiling point of nitrogen, that technological innovation and a fever for new discovery developed. Basalla (1988) has pointed out that in the evolution of technology we can always find antecedents for new inventions and new techniques. There is a wide range of novelty and it is the processes of selection that are critical to the unfolding of technology's path.

Commercial innovations require acceptance in the market place of customers or design engineers or factory workers to become a *fait accompli*. In much the same way, research or technological accomplishments only become a *research innovation* or a *technological innovation* when they are accepted as such in the market place of a research community. This can only occur when a novel accomplishment is put into the public domain by means of a published patent disclosure, conference presentation, distributed preprint, news conference or article describing the accomplishment. A very small fraction of such achievements come to be accepted by a community of practitioners, diffused throughout the community and serve as *paradigms*, model achievements (in the sense of Thomas Kuhn, 1970) which are widely recognized, adopted and used within a technical community. A successful research innovation is recognized by its success in the market place.

As discussed by M. Polanyi (1958) and T. Kuhn (1970), the essential element for market acceptance of a novel research or technological innovation is the anticipation of future promise as is exemplified by high temperature oxide superconductors. Recognition of future promise tends to be muted in most scientific publications; researchers have found that speedy recognition is more often achieved with a story in the Wall Street Journal or the New York Times than in Physical Review. Name recognition of the researcher within the technical community can also accelerate or impede the transition from research accomplishment to a recognized research innovation. A part of the skepticism about diamond synthesis in the Deryaguin laboratory can be attributed to that group's affiliation with the earlier thoroughly discredited claim for a new form of water, polywater, which was found to be nonexistent. A third factor is related to the level of frustration that lack of success has engendered in a research community. Metallic superconductors seemed to have reached an asymptotic critical temperature limit of 23.2°K. Finally, there are cultural factors affecting the various communities concerned. Some communities have a strong "only if invented here" approach to novelty; others are eager to embrace and expand on the work of others. Research innovations and technological innovations occur within a cultural and social context in which human behavior, human perceptions and social organization are as important as physico-chemical processes and product attributes. Recognition and acceptance of an innovation in the technological marketplace by the involved community is not absolute, but involves perceptions and judgments about which informed observers may differ.

In the history of invention and innovation Thomas Hughes (1989) has pointed out that successful inventors and innovators have identified and focused their attention on critical problems which he has called, in analogy with a military front, reverse salients of technological systems. Sperry's invention of the gyro-compass as a basic component of navigational systems for use on steel ships resulted from Sperry's perception that compass technology was a reverse salient in the change from wooden sailing ships to steel steamships. It is widely agreed that Edison's development of a lighting system required an effective integration of all system components: generating plant and distribution lines as well as an effective incandescent light bulb. Basalla (1988) has pointed out that innovation always consists of a replacement or substitution of a new material, device or process having some analogical relationship to a predecessor. This is true even of those inventions that we think of as revolutionary new ways of doing things. It explains why revolutionary inventions have occurred so often as multiple events and have so frequently been predicted in science fiction. It's not so much imagining what to do but rather how to do it within an effective integrated system of technology.

For advanced materials we may wonder if these historical insights are good analogies because material innovations are driven not so much by reverse salients in an existing materials technology, but rather by reverse salients in the development of systems *incorporating* new materials. However, in large measure the differences between opportunity and need lie in the eye of the beholder. Silicon nitride was perceived by the British Admiralty in the 1950s as a reverse salient, a critical necessity to achieve the vision of a future high temperature light weight gas turbine. The need for this existed in what we may see as a cultural imperative for improved gas turbines for advanced weapon systems. This was clearly perceived in Britain but generally overlooked in other military cultures. Two decades later in the U.S., DARPA, an agency created for the special purpose of identifying and developing opportunities related to weapons systems, perceived an opportunity for a ceramic automotive gas turbine as worth pursuing. A few years later with the oil shock of the 1970's there came into being a widely perceived *need* for more efficient engines and gas turbines which was combined in Japan with the perceived opportunity (and need in their island economy) for the economic advantage of being on the forefront of cutting new technologies; this led to MITI sponsorship of silicon nitride research. Sometime later, Isuzu and Kyocera as well as Nissan and NGK Spark Plug Co. saw an opportunity to develop marginally improved automobile engine performance as a way of improving processing capabilities. They perceived this opportunity as providing long term advantages, starting with small and almost insignificant markets; a necessary way of learning by doing. The anticipation of a significant profitable market for silicon nitride structural ceramics remains an anticipation after fifty years and several hundred million dollars of investment in research and development programs. (But automotive parts including supercharger turbines are now a break-even business of more than one hundred million dollars per year).

It was the discovery of low pressure diamond synthesis and of oxide superconductors by the scientific community that created an opportunity for innovation. It is the anticipation of improved performance in a number of possible devices and systems rather than any existing system's critical need that has led to the scramble to achieve research innovations that in time -- my guess is quite a long time -- may result in a significant commercial market.

The measure of strength of an opportunity for nascent innovation in advanced materials lies not in the materials themselves but rather from the fact that these materials may be the critical component, the reverse salient, in an existing or imagined device or system having a much larger value than the potential cost of the advanced material. While it seems extremely unlikely that the discounted future value of silicon nitride as a commercial innovation will ever approach the hundreds of millions of dollars and forty years of research and development already invested, that is not necessarily true of higher temperature low weight gas turbine engines. Even so, engine manufacturers have not been betting their own money on this proposition. It is rather the potential users of this technology, military establishments or power generation systems focussing on an even more expensive system than the engine itself, for whom the potential benefits may possibly match the cost.

In contrast to silicon nitride, there has been extensive industrial investment aimed at developing manufacturing processes for diamonds and oxide superconductors. This represents the judgment, perhaps the fear, that commercial innovation of these materials will have a significant impact on computer systems (for the likes of AT&T, IBM and Hitachi), military instrumentation (for DARPA) and now in Japan for long term programs for power generation and perhaps even magnetic levitation systems. In a sense we have come a full circle in that opportunities are also seen as perceived needs of system designers and system users who have a sufficient stake in the outcome to justify the discounted costs of present research in advanced materials. *Push-pull models of the innovation process are inextricably intertwined.*

If we accept that advanced materials are of value because they are incorporated into larger, more valuable devices or systems, we can conjecture that the current existence of such a device or system, or the precision with which it can be designed and the necessary performance factors predicted, or the extent to which it is merely a blurred vision of the future should affect the rate of advanced material technological innovation. In 1896, Walther Nernst discovered electrolytic conduction in solids and invented the concept of a light bulb operating in air, without the necessity of a vacuum enclosure, using a refractory zirconia-yttria glower as the electrically conducting incandescent element. The device was clearly envisioned, a satisfactory glower was the critical achievement necessary and the glower fit into an existing system of power generation and distribution with no system changes required. The technology to achieve this — forming the glowers, providing circuitry for preheating the incandescent element and adding a necessary ballast resistance — were rapidly developed along with the processing of the advanced ceramic material suitable for manufacturing the Nernst glower. This new light bulb was a successful commercial innovation achieved in less then two years. (But it also had a short life, soon being replaced by superior tungsten filament bulbs). High dielectric constant barium titanate was able to substitute directly for other materials as a capacitor dielectric; soon after its discovery it was introduced as a successful advanced material innovation. More recently, the rediscovery of solid ionic conductors such as stabilized zirconia and beta alumina has created new opportunities for developing energy storage systems, solid electrolyte batteries and fuel cells. None of these devices would directly substitute for part of an existing system. Not only advanced materials and new devices are required, but also substantially modified systems would have to replace or substitute for complex existing systems. This is obviously a task requiring a much greater activation energy and longer time constant than merely replacing a component.

The conjecture that evaluation of the potential rate of commercial innovation for an advanced material must begin by considering required modifications to the system in which it is implanted or the creation of a new system, the complications of new device development, and only then working back to material attributes such as structure, properties and processes of material synthesis seems quite likely. This conjecture requires that the use technologies of the system, device, and material plus legal, fiscal, political and cultural perceptions as well as social organizations associated with all these components of a system are essential constituents for any analysis of advanced materials technology innovation. Advanced materials technologies are essentially *enabling* technologies.

As we have seen with silicon nitride, a consequence of the requirement for transforming a discovered opportunity into a novel component giving rise to a changed device which is part of a new product in a modified system is that the time required from discovery to significant commercial innovation may be very long. From the 1911 discovery of low temperature superconductivity in metals some five decades passed before commercial innovation was achieved. It has been suggested that the half life for materials innovation is becoming shorter (Hench, 1990), but we have doubts about that as a general proposition. It depends on the nature of the system and the advanced material. For a direct replacement of one material by another without changing the product or device very much, change can be rapid. When whole new systems need to be developed, we see no short rapid path to commercial innovation.

A consequence of this primary conjecture about the importance of user systems would seem to be that the rate of progress toward innovation is proportional to the strength of the feedback loops between system users and designers, device users, designers and manufacturers, and materials users, designers and manufactures. Materials developed for internal use achieve commercial success much more rapidly than those searching for markets (Economy, 1988). In electronic ceramics the maintenance of close interactions with users and rapid feedback was a principle characteristic of the growth of Kyocera as a leading electronic ceramics manufacturer (Clark and Rothman, 1986). More recently the close relationship of Nissan with NGK Spark Plug Co. and Isuzu with Kyocera in developing ceramic components for automobiles seems to have been an essential constituent of their successful innovations.

The degree of change required in downstream components, devices, products and systems for an advanced material innovation to occur is a function of the advanced material and the system, not related to culture *per se*. The long time constants imposed by system changes resulting from advanced materials innovation place the value of such innovations in the realm of system developers and system users, i.e. defense departments, MITI and large corporations with a long time frame. The requirement for effective user technology, design technology and manufacturing technology feedback would seem to favor corporate cultures with close relationships between users and manufacturers. In the U.S. these relationships exist in the field of military procurement in spite of nominal arms length negotiation. As a result, the rate of innovation in military systems has been very high. Otherwise, the American culture of purchasing agents playing off one supplier against another, low cost bid procurement procedures and price-determined procurement would seem to mitigate against the close feedback loops required for effective innovation. The vertical structure and closer relationships of large manufacturers with customers and with client suppliers in the Japanese corporate structure would seem to be a much more fertile environment for advanced materials innovation in the commercial market.

In his discussions of the progress of science, Michael Polanyi (1958) has pointed to the importance of tacit knowledge. We conjecture that there is a large element of tacit knowledge involved in the development of new or modified methods of synthesizing, processing, manufacturing and using advanced materials. This means that there must be a large amount of learning by doing and implies the need for close interaction and strong feedback between users of products with material-enhanced performance, workers and engineers actually making things and the scientist-engineers designing them. If this conjecture about tacit knowledge is correct, then the transfer of tacit knowledge must go in both directions along a chain of interactions — in materials manufacturing from the process designers to the production engineers and also from production engineers to process designers. This was certainly the case in 1900-1902 at the Nernst Lamp factory in Pittsburgh where chemists and engineers were active participants at the factory engaged in the invention and production of the Nernst glower as a unified activity. Increasingly, as a result of scientific management, of Taylorism, and the development of mass production, there has grown to be a chasm between management, engineers and workers in the U.S. A corporate culture has developed in which management directs the team effort and often considers workers as cogs in the manufacturing process. This culture of strong specific direction makes the acceptance that tacit knowledge flows in both directions difficult and hampers successful innovation.

Effective internal communication within a manufacturing corporate culture between designers, engineers and manufacturing workers seems essential to the commercial innovation of processes involving tacit knowledge and requiring learning-by-doing. This also implies that there be a reasonably long time frame and steadiness of purpose in which learning by doing can be accomplished. In the U.S. corporate culture, the communication requirement seems to be best achieved in the environment of small capital venture organizations in which bureaucratic rigidities and chasms between management and hand-on workers have not had a chance to develop. The cultural environment of vertically integrated groups within Japanese corporate culture, and perhaps the absence of an historical imperative toward Taylor's "scientific management" seems to be more conducive to advanced material processing development.

Nascent and infant technologies are properly seen as being nucleated or germinated by discovery or invention. Increasingly, with extensive government support of science, we find the number of discoveries growing at an exponential rate independent of any conscious aim at innovation. Nonetheless, many potential opportunities for nascent technological innovation are created. We conjecture that these discoveries lie fallow until there is a perception or recognition of their being needed for the improvement or development of a technological system. Silicon nitride was first patented in 1895. It was first proposed as a refractory bond in 1905. It was not until the 1950's that the first tentative commercial innovation occurred. In contrast, the discovery of oxide high T_c superconductors in 1987 was immediately perceived as being a source of potentially critical components for systems seen as cultural imperatives.

SUMMARY

A key conjecture necessary to understand the rate of innovation of advanced materials technology is that these materials are valued as they are incorporated in more valuable devices and systems. Advanced materials innovations depend on the nature and extent of innovations required for modifying these devices and systems. The rate of innovation depends on the availability of inventions and discoveries, the effective use of feedback loops between systems users and designers, product users and designers, component users and designers and materials users and designers. In order for rates of process innovation to be high, the transfer of tacit knowledge by effective personal interactions in these feedback loops is essential.

The key elements seem to be (1) the number of inventions and discoveries available for exploitation, (2) the degree of change required in components, devices, products and systems to take advantage of a nascent advanced material technology, (3) the effectiveness of the feedback loops and information exchange between user technology, design technology and manufacturing technology and (4) the effective transfer of tacit technological knowledge between design engineers and production workers in the required process of learning by doing.

REFERENCES

G. Basalla, 1988, *The Evolution of Technology,* Cambridge Univ. Press.

K.B. Clark and E. Rothman, 1986, "Management and Innovation: The Evolution of Ceramic Packaging for Integrated Circuits" in W.D. Kingery, ed. *High Technology Ceramics: Past, Present and Future,* Amer. Ceramic Soc., Westerville, Ohio, pp. 335-350.

J. Economy, 1988, "A Spectrum of Advanced Materials Innovations", *The Chemist, 9,* pp. 15-20.

L.L. Hench, 1990, "From Concept to Commerce: The Challenge of Technology Transfer in Materials", *MRS Bulletin,* August, pp. 49-53.

T.P. Hughes, 1989, *American Genesis,* Viking Penguin Press, New York.

W.D. Kingery, 1989, "Science des Matériaux et Société", *Polyrama,* Ecole Polytechnic Federale de Lausanne, pp. 65-67.

W.D. Kingery, 1989, "Ceramic Materials Science in Society", *Annu. Rev. Mater. Sci, 19,* pp. 1-20.

T.S. Kuhn, 1970, *The Structure of Scientific Revolutions,* Univ. of Chicago Press.

M. Polani, 1958, *Personal Knowledge,* Univ. of Chicago Press.

A COMPARISON OF JAPANESE AND U.S.* HIGH-TECHNOLOGY TRANSFER PRACTICES

Robert S. Cutler

Introduction

Much has been written about recent Japanese commercial success and its economic impact in international markets. A number of reasons are offered in explanation. One frequently cited is the Japanese ability to assimilate and apply new technologies derived from basic research done in the United States. Another reason is the policy of Japanese companies to develop and produce quality products based on new technology.

In this paper, I present some empirical results and observations which describe the principal ways in which a sample of industrial researchers in Japan and in the U.S. utilize certain new technologies resulting from university research. The findings are from a survey conducted in Japan and the U.S. between October 1986 and December 1987.

I conclude that personal communications and technical collaboration are the key factors in the rapid diffusion of research results in both countries, and that in Japan, government agencies and professional societies take a much more active role in organizing and energizing the civilian technology transfer process than do counterpart organizations in the United States.

The Technology Transfer Survey

The investigation involved a comparative study of Japanese and U.S. high-technology transfer practices, particularly regarding the utilization of university research in three fields: robotics, biotechnology, and ceramic materials.

The focus was on three fundamental engineering fields where Japanese and U.S. firms appear to be comparable in terms of technological capability. I had read in the press [1] that former attitudes about technology transfer were beginning to shift and, in some new fields, the Japanese were beginning to innovate, rather than import patented technology, and to export and license their latest technology to international markets.

During the early part of my nine-month stay in Japan, I recognized the fact of the so-called "Japanese miracle," the rapid economic development over the past two decades based on technology. I then set out to investigate and compare the ways in which new technologies are acquired and commercialized in Japan, and hopefully to learn how it is done so well and so fast.

Technology transfer involves many functional as well as cultural factors. When interpreting the differences observed between Japanese and American technology practices, I believe it important to view the Japanese—their institutions and their behaviors—from a cultural perspective.

Simply stated, the Japanese have a different language, a different thought process, and different social and business process than Americans. To attempt to observe technology separately from its environment is to lose sight of this larger picture. I was soon to discover that there are strong cultural elements in the ways the Japanese acquire, evaluate, and transfer new technology. I elaborate on those elements later.

---

*Reprinted from IEEE Trans. Eng'g. Management, 36, 17-24, 1989.

Scope of Survey

My research in Japan primarily involved a survey of Japanese university and industrial researchers who are working in three high-technology fields. Fifty-five interviews were conducted at twelve universities, nine companies, and at six government R&D organizations. The parallel survey in the U.S. included 51 researchers at eleven universities, eight companies, and three government organizations. In total, 106 researchers were polled in the two surveys (Fig. 1).

The questions focused on the professional behavior of the researcher himself, rather than on the research *per se*. The objective was to identify the principal transfer mechanisms used by the particular researcher in Japan or the U.S. and his professional colleagues at universities and other R&D organizations.

In addition, information was sought about career objectives and hiring practices, awareness of significant research advances, and attitudes toward collaborative arrangements with foreign counterparts.

Approach

First, let me offer the succinct definition of "technology transfer" which Jacques Bagur of Gulf South Research Institute presented on June 21, 1987 to members of the Federal Laboratory Consortium: "Technology transfer is the process by which knowledge concerning the making or doing of useful things contained within one organized setting is brought into use within another organization context."

The concept of technology transfer which I use in this study consists of several functional mechanisms which are classified into three domains:
- Publications—journals, technical reports, trade press;
- Patents—invention disclosures, patents, and licenses; and
- People links—meetings, collaborations, joint projects.

These domains are operationally defined by the principal mechanisms used for communicating, facilitating, or otherwise moving the results of university research into industrial application (Fig. 2). The approach supports the multiple factor philosophy, wherein technology transfer is seen as a process involving many functional and environmental factors working in concert.

An Appropriate Research Model

As a researcher myself, I was compelled to devise an appropriate model and to collect relevant information and data. From the kinds of program evaluation studies we do at the National Science Foundation (NSF), I have learned that a proper evaluation design involves a simple model which describes the principal factors and the relationship of the data to the results. My reasoning for choosing the three domains of the model are as follows.

1) When one attempts to compare research activities, although there is some professional controversy as to what is significant, it is generally accepted among science policy researchers that the publication of journal articles and citations to those articles in other publications are reasonable measures of scientific advancement and research productivity.

2) Patent counts are now becoming useful to econometricians who study the process of technological innovation. The use of such numbers is less exact than citations to the literature, nonetheless some carefully selected patent statistics reflecting large quantities can be a useful indicator. There is also new interest in university patents because such patents can attract industrial support [2].

3) The third domain of the model is what I call "people links." From talking with several policy analysts before going to Japan, and from my experience as an R&D engineer in industry and a research administrator at NSF, I have learned that technology transfer also occurs in activities such as professional societies, workshop seminars and employee mobility.

Japan   United States

Universities:

No.*   No.
(1) Hokkaido University   (4) Carnegie Mellon University
(1) Kyoto University   (3) Mass Inst. of Technology
(2) Nagoya Technology Inst.   (3) Penn State University
(1) Nagoya University   (3) Stanford University
(1) Osaka University   (2) Rutgers University
(2) Saitama University   (2) Univ. of Arizona
(1) Sophia University   (2) Univ. of CA/Santa Barbara
(2) Tokyo Inst. of Technology   (2) Univ. of Delaware
(3) Tohoku University   (3) Univ. of Massachusetts
(3) Tsukuba University   (3) Univ. of Nebraska
(3) University of Tokyo   (3) Univ. of Utah
(1) Maseda University   ___
21   29

Industrial Laboratories:

(3) Hitachi Central R&D Laboratory   (2) Ceramatec, Inc.
(3) Hitachi Production Automation   (3) Eaton Corporation
(5) IBM Tokyo Research Laboratory   (3) IBM Corporation
(1) Kyocera Corporation, Inc.   (1) ICR Associates, Inc.
(3) Mitsubishi Metal Corporation   (2) Monsanto Company
(2) NEC Central Research Labs   (1) J.D. Searle, Inc.
(2) Nippon Steel Company   (1) Repligen, Inc.
(4) Nissan Motors Co., Ltd.   (1) United Technologies Corp.
(1) Smith Klein Beckman Japan, Ltd.

___   ___
24   14

Government R&D Organizations:

(4) Mechanical Engineering   (4) National Bureau of Standards
    Laboratory (MEL)   (2) National Institutes of Health
(2) Electrotechnical Laboratory (EFL)   (2) National Science Foundation
(1) Institute for Agricultural
    Research (NIAR)   ___
(1) Ministry of International   8
    Trade and Industry (MITI)   **********************************
(1) Ministry of Education and
    Culture (Monbusho)   Japan   U.S.
(1) Japan Research Development   Univ.   21   29
    Corporation (JRDC)   Ind.   24   14
                                              Govt.   10   8
___   55   51
10   *(n) - number of interviews

Fig. 1. Survey sample

1. Meetings, Seminars, Intensive Conferences
2. Professional Society Meetings
3. Journal Publications, Newsletters
4. International Conferences
5. Advisory Boards, Councils, Committees
6. Study Missions, Site Visits, Trade Shows
7. Patent and Licensing Agreements
8. Consulting Arrangements
9. Joint University/Industry Research Projects
10. Visiting Scientists and Resident Researchers

Fig. 2. Principal technology transfer mechanisms.

	Japan	U.S.
Robotics:	• Journal of the Robotics Society of Japan • International Journal of Robotics Research • IEEE Journal of Robotics and Automation	• IEEE Journal of Robotics and Automation • ASME Journal of Dynamic Systems Measures and Control • International Journal of Robotics Research
Biotechnology:	• Journal of Biotechnology • Science • Journal of the American Chemical Society	• Journal of Biological Chemistry • Biochemistry • Journal of Plant Physiology
Ceramic Materials:	• Journal of the Ceramics Society of Japan • Journal of the Physical Society of Japan • Journal of the American Ceramics Society	• Journal of the American Ceramics Society • Journal of Materials Science • Journal of Applied Physics

Fig. 3. Principal journals mentioned in survey.

The three-domain model expresses the notion that technology transfer is more than simply the exchange of technical publications, or the licensing of patents. Rather, the model of the transfer process includes various contact mechanisms and communications activities which essentially are person-to-person linkages. Such mechanisms actually serve to bring the desired technology know-how into actual use.

An interview questionnaire was designed to obtain information from each researcher on the following subjects.
- a) Publication activities—which journals are most frequently read and where authored articles were most recently published.
- b) Patent activity—whether listed as an inventory on patents issued within the last five years, and whether the patents are licensed or used.
- c) People links—whether active in professional society activities, consulting, collaborative work, conferences, and career mobility.

Findings

Based on an analysis of the surveys[1] conducted in Japan and in the United States, the following comparative results were reported.

Publications
- In Japan nearly all (94 percent) of the researchers surveyed were able to read and write in English, while in the U.S. very few (4 percent) of the Americans interviewed admitted any technical competence in the Japanese language.
- In Japan a majority of those researchers surveyed (85 percent) published and read English language journals articles as well as those in Japanese, while in the U.S. few (9 percent) said they read any translated Japanese journal articles in their field (Fig. 3).
- In Japan journal publications do not necessarily contain new work, while most U.S. journal editorial policies insist upon new and original work only.

Patents
- In Japan few university professors (14 percent) hold patents, while in the U.S. nearly 46 percent of the university researchers surveyed do.
- In Japan, between 1981 and 1985, the number of university patents reported to the Japanese Society for the Promotion of Science (JSPS) by the top five universities increased from 24 to 32 patents (66 percent), while licensing agreements increased from 2 to 6 (Fig. 4).
- In the U.S., between 1981 and 1985, the number of patents reported by the top five research universities increased from 122 to 177 (45 percent). Licensing for the same five years increased from 53 to 96 (81 percent) (Fig. 5).

People Links
- In Japan 93 percent of the university researchers and 80 percent of the industrial researchers surveyed said they attend technical meetings outside their work location at least twice per month, while in the U.S. 43 percent of the university researchers

---

[1]Survey sample (n = 106) consists of 55 Japanese and 51 U.S. researchers. The population it represents would be hard to describe fully, but I hope is an important part of the university, industry, and government R&D organizations performing advanced research in the fields of robotics, biotechnology, or ceramic materials, between October 1986 and December 1987 (see Fig. 1).

Institution	1981 App	1981 Pat	1981 Lic	1982 App	1982 Pat	1982 Lic	1983 App	1983 Pat	1983 Lic	1984 App	1984 Pat	1984 Lic	1985 App	1985 Pat	1985 Lic
Osaka University	2	0	0	5	2	0	10	6	0	11	2	0	9	4	0
Tokyo Institute of Technology	6	2	1	8	3	2	15	4	4	19	6	0	12	1	0
Kyoto University	3	2	0	2	1	0	5	3	0	9	4	1	3	0	1
University of Tokyo	7	2	1	3	2	1	6	2	2	7	1	2	6	3	4
Tohoku University	6	0	0	7	0	0	3	3	0	2	5	1	2	2	1
Total:	24	6	2	25	8	3	39	18	6	48	18	4	32	10	6

Source: JRDC (5/87)

Fig. 4. Top five Japanese universities' patent activity (applications, patents, licenses) (1981-1985).

Institution	1981 Dcl	1981 Pat	1981 Lic	1982 Dcl	1982 Pat	1982 Lic	1983 Dcl	1983 Pat	1983 Lic	1984 Dcl	1984 Pat	1984 Lic	1985 Dcl	1985 Pat	1985 Lic
Univ. of California	234	41	6	308	49	11	214	48	9	285	51	12	320	51	18
Mass. Institute of Technology	155	51	10	161	61	12	147	51	18	122	56	20	121	52	13
Stanford	112	12	19	147	7	22	196	14	25	124	35	30	134	38	35
Univ. of Wisconsin	41	18	13	69	18	6	83	27	11	82	23	17	77	30	12
Univ. of Washington	28	0	5	21	4	2	38	2	2	47	4	11	62	6	16
Total:	570	122	53	706	139	53	678	142	65	660	169	90	714	177	94

Fig. 5. Top five U.S. universities' patent activity (disclosures, applications, patents, licenses) (1981-1985).

and 17 percent in industry said they did so.
- In Japan the average proportion of the Ph.D.'s reported in the work unit (Ph.D. ratio) was 18 percent for the universities, 17 percent for industry, and 33 percent for government labs. In the U.S. the ratios were 38 percent for universities, 57 percent for industry, and 61 percent for government labs (Fig. 6).
- In Japan 62 percent of the high-technology university researchers surveyed and 46 percent of those in industry said they were involved in at least one joint university/industry project. In the U.S. the level was 84 percent for universities and 93 percent for those surveyed in industry.
- In Japan 78 percent said they have worked for their current employer since graduating from college, while 23 percent of those surveyed in the U.S. said they did.
- In Japan 59 percent reported having attended at least one international meeting during the past two years. In the U.S. the proportion was 28 percent.
- In Japan 65 percent of those surveyed said they spent a year or more in the United States or in Europe. In the U.S. 34 percent said that they had spent more than one year abroad; 4 percent had worked in Japan; and 17 percent had visited Japan for brief periods ranging from one to three weeks.

	Japan	U.S.
	(n=27)	(n=22)
Universities:	18%	38%
Industry:	17%	57%
Gov't Labs:	33%	61%

Fig. 6. Ph.D. ratio in laboratories surveyed.

Additional Observations
- In Japan 83 percent of the researchers surveyed said they were aware of current research advances made by foreigners in their field. In the U.S. only 30 percent said they knew of any.
- In the U.S. the following attitudes and interests were expressed by researchers interviewed regarding the work of Japanese colleagues:
    — 68 percent of the university researchers, 35 percent of the industrial researchers, and 60 percent of the government researchers acknowledged having had at least one Japanese research colleague or visiting researcher in his laboratory.
    — 78 percent said that they would welcome some type of research collaboration with an appropriate counterpart in Japan.
    — 63 percent said they would be willing to work in a laboratory in Japan for an extended period of time. (Most favored four to six months.)
- Among those surveyed in both countries, the mechanisms preferred most for affecting high-technology transfer are:

In Japan	In the U.S.
Meetings, seminars (90 percent)	Meetings, talks (84 percent)
Professional conferences (75 percent)	Gordon-type conferences (62 percent)
Study missions, site visits (58 percent)	Publications (55 percent)

In addition, some two-thirds of the robotics researchers surveyed in Japan said they currently exchange VCR video tape recordings with colleagues in their own country. However, it is not clear how widespread the use of video tape recordings is among U.S. researchers.

Caveat on the Analysis

Due to limitations in the data and the sampling method used, one should not draw definitive conclusions from this study. However, there are some interesting findings which are more suggestive than indicative. Moreover, the concept of technology transfer itself is complex and difficult to define precisely. This is an emerging area requiring more study and analysis.

Discussion

From the findings outlined above, it is clear there are similarities as well as some important differences in the way technology is transferred between university and industrial researchers in Japan and in the U.S.

In contrast to the kinds of scientific research performed in the U.S., most of the research I observed in Japanese universities can be described more accurately as "fundamental engineering science," rather than basic scientific research. It usually is done in groups rather than by individual investigators, and it consists largely of experimental verification work. However, there are a few senior professors doing some theoretical work at the more basic end of the research process.

Publications

To describe what the Japanese do differently, first I will discuss journal publications.

The principal sources of basic research information for the Japanese researchers I interviewed are the journal articles published by leading university researchers in the U.S. and in Europe, rather than by other Japanese researchers.

Journal editors in Japan apparently do not insist on publishing only original work. Their journals often consist of progress reports as well as reports on setting up and testing methods of experimentation which may have been published elsewhere. I am told, however, that academic societies in Japan also publish some paper journals (called "Ronbun-shi") which are used to report original research. This practice is related partly to Japanese feelings about originality,[2] which are quite different from those in the West, and in part, to Japanese research funding practices, particularly in universities which require progress reports to be published.

Japanese engineering researchers work in teams to carry through a particular project, from the initial research stage, through development, to prototyping, and even on to production and

---

[2]The traditional Japanese attitude about originality is one which prefers to follow a pattern rather than to break new ground. In Japanese, the term "learn" (manabu) is derived from "imitate" (maneru) [5].

marketing. It is difficult to track research advancement via publications, in Japan, because there are no intermediate publication points.

In the U.S., by contrast, a university researcher typically does the fundamental work and then publishes his or her findings in the journal literature. From those publications in the open literature, another researcher picks up the new knowledge and basic ideas which he/she considers to be feasible, carries it through the applied research phase, and again publishes the results either in the journal literature, as a company report, or as a patent disclosure. The industrial R&D community picks promising projects out of this pool of new technology. In this process, however, users' requirements are rarely cited or integrated into the research design, as often is the case in Japan.

Patents

Although the proportion of surveyed Japanese professors holding patents is smaller (14 percent) than that of the Americans (46 percent), the top five universities in Japan reported an increase of 66 percent between 1981 and 1985, and the American top five universities reported a 45 percent increase.

The difference between the two groups in the number of patents acquired stems largely from the traditional belief in Japan that universities are primarily for the teaching of students, rather than for commercializing research results which is the domain of industry. However, this picture is now changing.

Although the number of Japanese university inventions since 1981 is smaller than that for U.S. universities, the JSPS data show a remarkable increase in the licensing of those patents during the past five years.

This increase appears to reflect the recent shift in the patent policy of both countries (since 1978 and 1980) which authorized universities and research laboratories to promote inventions resulting from government funded projects. Both in Japan and the U.S. there are programs now in place to assist university professors to transfer their inventions to commercial use. The Japan Research Development Corporation (JRDC) is the agency responsible for promoting the transfer of university patents to industry. In the U.S. there is no central government responsibility for this activity, rather each research university has its own patent licensing office.

People Links

I observed throughout this study that the most preferred and also the most effective technology transfer mechanisms are "people intensive," rather than "paper intensive."

This conclusion became clear to me from the amount of time (two-thirds) the researchers said they devote to exchanging new ideas by participating in talks, meetings, and working with leading colleagues, as compared to the remaining one-third of their time spent reading, extracting or preparing new information for publication or for patents. This allocation of time appears to be as true in the United States as it is in Japan, at least for the three high-technology fields surveyed.

Apparently there are strong personal needs for face-to-face discussions leading to bench-to-bench collaboration in order to better communicate new complex ideas from one person to another, and then to utilize them elsewhere in the research lab or in another organization or institutional setting. I conclude that high-technology transfer is largely a "contact sport": meeting with people, carrying new ideas forward, and joining individual efforts toward a common goal.

The rapid transfer of university research to industrial technology also requires the necessary know-how which is a skill attribute of a researcher [3]. In tracing the transfer paths within the fields of robotics, biotechnology, and ceramic materials, in both countries, I find a similarity in the preferential use of person-to-person contacts for obtaining substantive information. Many of these links involve long-term collaborative work between university and industrial researchers. Examples of successful transfers of university research to industrial applications can be found in computer-vision robotics, genetic engineering, and functional ceramics.

If one is attempting to compare the principal technology transfer practices observed in the two countries, three significant factors which underlie the Japanese R&D system are worth mentioning. They are attitudes about cooperative research, the "old-boy" network, and R&D management styles.

Japanese Technology Transfer and Cooperative Research

Japanese companies achieve effective utilization of high-technology research and its transfer between laboratory and production by holding many more technical meetings on an industry-wide basis than American companies do. Professor Thomas Eagar of M.I.T. observed that "there is not just technology transfer within a company in Japan, but also between companies, and companies and universities, through the many meetings of the various professional societies" [4].

I do not believe there is such a system in the U.S. which pools, analyzes, and disseminates current information on international research activities as effectively as the Japanese system does.

The topics discussed at many of these meetings include more technical content and detail than is common in the U.S. In addition, major research laboratories become familiar with the work at other labs, resulting in rapid dissemination of new results and less duplication of effort. The meetings also permit researchers to communicate very effectively their knowledge of work outside of Japan.

There are a number of reasons why the Japanese system works. One is the strong leadership of the university professors who serve as committee chairmen. There are strong ties between these professors and their former students that do not seem to exist in the United States.

Several of the robotics engineers interviewed in Japan showed me video tapes documenting their current experiments and the work of their colleagues abroad. This low-cost highly effective audiovisual reporting mechanisms is yet another example of the way Japanese researchers rapidly exchange current research results.

The Japanese Old-Boy Network

The process was described by several speakers at a seminar on high-technology competitiveness held by the Japan Technology Transfer Association in Tokyo on March 13, 1987. That discussion helped to crystallize what I discovered during my four dozen interviews in Japan.

Japanese industry has two powerful assets: a cohesive national policy on technology development and a scientific "old-boy" network, with links to practically every board room and laboratory in the country. The government spends nearly one-third of its R&D budget (20 percent of total R&D spending) at universities and at government research institutes, and nearly all of this activity is centrally coordinated through government committees and the scientific "old-boy" network.

Here is how the two circles of power work. Perhaps you have noticed that Japanese companies seem to sell similar products, so much so that it looks like they must be collaborating on the designs and specs. That is because high-tech Japan is a small country and the top engineers in the companies know each other. For that matter, so do the company presidents, who most likely went to the same university at the same time. When one company starts something new, the president calls his friends to discuss it.

Japanese companies do not suffer from the not-intended-here syndrome, that attitude which stifles ideas from external sources. Instead, they are eager to please their customers and would rather have their people involved in making something better for the marketplace, than in trying to capture all of the profits from a new technology product. In fact, the licensing of patents from other companies and from foreign sources, including many U.S. universities, is widely practiced.

Many foreigners imagine that government officials at the Ministry of International Trade and Industry (MITI) stand over the R&D stage like grand puppeteers, manipulating private industry at will. This is not the case, particularly because the average MITI officer changes jobs every two years.

MITI's method of influence is through its committees. A mixture of industry leaders, academics, and consumers (users) are selected for dozens of committees on new technology and industry matters, ranging from restructuring a weak industrial sector to organizing a national program for advanced robotics or for manned spaceflight.

Through committee debate, MITI helps industry form a consensus on which areas of new technology it should concentrate on. By this committee method, policy is actually negotiated by industry leaders, so it is accepted naturally by all the companies. That is what I found to be the secret of Japan's cohesive industrial policy: the government acts as the organizer and coordinator of private industry action. Eighty percent of the R&D funding in Japan comes from private industry, rather than from the government.

R&D Management

A final remark about Japanese methods for running research organizations and their methods for decision making.

What I observed closely resembles what Ouchi of UCLA calls "theory Z" [5]. One main feature of Japanese society which Ouchi describes as being essential for the success of each work unit is the great trust that exists between superiors and those who work for them.

One of the best technology transfer practices of Japanese industry is the quality circle, where five to ten workers meet almost daily to discuss possible improvements in their work. This method works in Japan where it serves to give group sanction to innovative departures from the old ways of doing things.

There is a general sense of family solidarity which seems to characterize Japanese endeavor, whether at home, at work, or in professional pursuits. The personal commitment, trust, and desire for cooperation among researchers serves as a glue which keeps the Japanese R&D organization together. "In Japan it is difficult to move people, but it's easy to move ideas," one Hitachi laboratory director told me.

From an organizational viewpoint, Dimancescu, of the Technology & Strategy Group, observed that "U.S. companies still live in the world of highly compartmentalized functions and responsibilities. Many of these are staffed by people whose labors are rewarded for maintaining a very narrow definition of the task required of them. This behavior generally goes under the rubric of 'division of labor' or 'specialization,' and is valued as desirable ends. In such a cultural environment, information neither travels fast nor necessarily to the right people at the right time. Hence we find an inferior process of tech transfer (in the U.S.) relative to what is observed in Japan" [6].

Japanese companies have been highly effective in applying new concepts of project management which look nothing like what is practiced in the U.S. The Japanese concept of project management starts with the fundamental belief in the coequal importance of *all* players needed to fulfill a task and continues on with the constant and continuous process of linking these players together horizontally. This procedure goes a long way toward explaining how the Japanese have advanced so rapidly in high-tech fields during the last three decades.

In the U.S. the antitrust laws have required each competing firm to carry on its own industrial research. Technical cooperation not only is limited but is often perceived as unlawful by corporate management. Recently, however, the law has been liberalized to allow certain consortia like MCC, SEMATECH, and the Semiconductor Research Corporation (SRC) to be organized.

Things are different for industrial research in Japan. The government there actively promotes the formation of research associations among leading companies in particular fields for the purposes of developing and transferring new technologies. Patents resulting from these arrangements are pooled for participating companies to use. And there is a remarkably high degree of communication and collaboration between professors at leading Japanese universities and their colleagues who work in competitive companies [7].

How Technology Policies Differ

One way to classify technology policies is by whether they are diffusion (technology-push) or mission oriented (user-pull). For example, technology policy in Germany and Sweden is diffusion oriented, whereas the technology policy in France, England, and the U.S. is mission oriented.

Japanese technology policy, on the other hand, is both mission oriented and diffusion oriented. Like countries in the first group, Japan emphasizes a broadly based capacity for diffusing innovation-related public goods. Like countries in the second group, it also employs coordinated efforts to advance national technological goals. However, Japanese policy differs from the policies of the other nations in two respects. First, in the recent past, Japan was at a far lower level of development than other industrialized nations. Second, the consensus-based government-industry relationship in Japan involves centralized decision making and decentralized implementation. These two factors have led to technology policies that emphasize rapid upgrading of the nation's technological skills, but in a more decentralized and broadly based manner than in the mission oriented countries [8].

There are three basic elements to Japanese technology policy: 1) promoting leading edge industries through tax policy more than direct financial assistance; 2) facilitating technology transfer; and 3) upgrading of the human capital base on a more general, less industry-specific basis.

Conclusion

This study attempts to plow new ground in an uncharted and complex area: the cross-cultural comparison of technology transfer mechanisms used in Japan and in the U.S. The findings are derived from information obtained during an exploratory survey of active researchers in both countries, who were not randomly selected. However, care was taken to avoid undue geographic concentration and institutional bias. The results presented are more indicative than definitive. Nonetheless, I believe they represent technology transfer in the three high-technology fields surveyed.

I conclude that personal communication and technical collaboration are the key factors in the rapid diffusion of high-technology research results in both countries, rather than the widespread availability of scientific journal literature and recent efforts to promote university patents. The differences observed in practice stem largely from some of the cultural and institutional factors described.

The empirical findings confirm the conventional review that the flow of high-technology information is largely from U.S. university researchers to industrial researchers in Japan. However, the data also show that some of the most advanced ceramics and robotics technology used in the U.S. increasingly is derived from research initiated in Japan.

Journal publication and university patenting are more widely used in the U.S., where university professors both teach and do basic research. Meetings and intensive conferences, however, are by far the most popular mechanisms used for technology transfer among those U.S. researchers surveyed.

In Japan the results of university research are utilized primarily in industrial settings. Typically industry uses outside professional meetings and close collaboration as the means for translating the scientific knowledge and new engineering know-how into commercial use.

In the U.S., government agencies support most of the basic and applied research performed at universities primarily for public purposes such as military defense, public health, and space exploration. By contrast, most of the high-technology research in Japan is funded and performed by industrial companies for commercial purposes. Furthermore, Japanese government agencies and professional societies take a more active role in organizing and energizing the civilian technology transfer profess than do the counterpart organizations in the U.S.

The present study confirms an earlier conclusion by Herman Bieber [9] that "technology is primarily transferred by people, not via organizational charts or formal reports." This observation, made in 1969 and primarily related to the communication of new technical information within a single organization, also appears to be valid for effecting technology transfer between different institutional and cultural settings, such as for high-technology collaboration between university and industrial researchers in Japan and in the U.S.

This study should be of interest to engineering managers and researchers concerned about the nature of technology transfer and how it occurs in Japan and in the U.S.

Acknowledgment

The author wishes to thank the National Science Foundation for permission to be away for the period of the project. He also wishes to thank S. T. Cutler who contributed in significant ways to the preparation and editing of this paper.

References & Bibliography

[1] J. Pollack, "The patent as trade barrier," *The New York Times*, July 5, 1984.
[2] R. Cutler, "Patents resulting from NSF engineering program," *World Patent Inform.*, vol. 9, no. 1, pp. 38-42, 1987.
[3] F. Rossini *et al.*, "Interdisciplinary research: Current experience in policy and performance," *Interdisciplinary Sci. Rev.*, vol. 8, no. 2, pp. 127-139, 1983.
[4] W. Eager, "Technology transfer and cooperative research in Japan," *ONR Far East Scientific Bull.*, vol. 10, no. 3, 1985.
[5] G. Ouchi, *Theory Z*. Reading, MA: Addison-Wesley, 1981.
[6] D. Dimancescu, Technology & Strategy Group, Inc., Boston, MA personal communication, April 20, 1988.
[7] R. Cutler, "Impressions and observations of science and technology in Japan," Univ. Tokyo, Japan, Rep. to Japan-U.S. Educational Commission (Fulbright Foundation), May 1987.
[8] H. Ergas, *Technology and Global Industry: Companies and Nations in the World Economy* (Nat. Acad. Eng. Series on Technology, Social Priorities). Paris, France: Organisation for Economic Cooperation and Development (OECD), 1987.
[9] H. Bieber, "Technology transfer in practice," *IEEE Trans. Eng. Manag.*, vol. EM-16, no. 4, pp. 144-155, 1969.
[10] J.L. Bloom, "Japan enters the world series of technology," *The World & I*, pp. 214-223, Sept. 1987.
[11] R.G. Havelock and V. Elder, "Technology transfer in Japan: An exploratory review," Center Productive Use Tech., George Mason Univ., Fairfax, VA, Rep. June 1987.
[12] H. Kobayashi, "Scientific creativity and engineering innovation in Japan," presented at the Symp. Science in Japan, 153rd Annu. Meeting, Amer. Assoc. Advancement Sci., Chicago, IL, Feb. 16, 1987.
[13] R.J. Marcus, "Observations on science in Japan," *ONR Far East Scientific Bull.*, Mar. 1983.
[14] L.S. Peters, "Technical network between U.S. and Japanese industry," Center for Sci. and Tech. Policy, Rensselaer Polytechnic Institute, Troy, NY, Rep., Mar. 1987.
[15] E.A. Vogel, *Japan as Number One*. Tokyo, Japan: Tuttle, 1980.
[16] R. Cutler, Ed., "Japanese science and technology: The changing institutional framework," in *Proc. AAAS Symp., Science in Japan: An update, 153rd Annu. Meet. Amer. Assoc. Advancement Sci.*, Chicago, IL, Feb. 16, 1987, to be published.

THE JAPANESE RESEARCH ENVIRONEMENT - A STUDENT'S VIEW FROM WITHIN

Mary I. Buckett, Northwestern University

First Impressions
"It comes down to this. Newcomers at welcoming cocktail parties are the only people unified in their understanding of Japan. After the first month, diversity of opinion reigns."
- P. Robert Collins from "Max Danger, The Adventures of an Expat in Tokyo"

Japan - the land where the cars are white, the roofs are blue, and the women wear black. Almost certainly included in the gaijin's (outsider's) first impression of the 'Land of the Rising Sun' is the aura of uniformity - cultural as well as physical attributes. After that, it truly goes 'up for grabs'. One's impressions become influenced by one's experiences and - unfortunately for some - those experiences are sometimes less than pleasant. This can be especially true for the visiting scientist. Overcoming the cultural barrier of the homogeneous, inward-looking Japanese society can be wonderfully challenging, occasionally frustrating, sometimes shocking - but always an unavoidable problem the foreigner must face. Although the language hurdle is one aspect of this barrier, there are other more subtle influences which must be recognized before a favorable working relationship can be achieved. I am a graduate student in Materials Science and Engineering at Northwestern University. During my graduate studies, I've interned at Lawrence Livermore National Laboratory and now work part time at Argonne National Laboratory. This summer I had the opportunity to participate in the Japan Summer Institute Program, offered for the first time through the National Science Foundation. For two months, myself and 24 other graduate students in various scientific disciplines worked at national laboratories and government institutes in Tsukuba Science City. This paper presents some of my impressions of the Japanese research environment - gained partly as a result of my stay there this summer, but also in part from my experiences being involved in an electron microscope development project with a large Japanese corporation, and from my experiences of being half Japanese (mother) and having to deal with family on both sides of this cultural barrier.

In the complex interaction between Japan and the United States over the past 35 years, it is clearly the Japanese who have profited the most. And they have done so essentially because many of them have observed the United States in such minute detail. At times, indeed, the dogged manner in which Japanese scholars, journalists and businessmen pursue even the most trivial piece of information about American life borders on the absurd. [And yet it is] this tireless investigation of the tastes, habits and needs of American consumers that has so often enabled Japanese industry to outcompete American companies in their own mark.
- Robert C. Christopher from "The Japanese Mind"

No one will contest that the 'creative adaptation' of the Japanese has made them leaders in advanced technology and commercialization. They remain avid observers of their neighbors across the Pacific. So much so that shades of American influence are evident in almost every aspect of Japanese life, especially the young who have essentially devoured anything American from rock n' roll music to sports heroes and movie stars.

But in recent years the Japanese have not only been preoccupied with their scrutiny of

America but how they are perceived by America and the international community. Lately, Japanese science and technology agencies have been criticized both for their lack of support for basic research and their lack of promotion of technology and personnel exchange with foreign countries. To this they have responded swiftly. Three major science and technology future directions have been identified by the Japanese government: 1) a shift in research focus to more fundamental scientific research, 2) a reform of some of the more constraining aspects of the Japanese science and technology system, and 3) greater participation within the international scientific community. Politically, an enormous effort is being made to bolster U.S. - Japan technology transfer. The Japan Summer Institute Program is but one small part of this full-out effort by the Japanese government to remedy the imbalance. The program objective is (as stated in the application) "to provide 25 select U.S. science and engineering graduate students first-hand experience in a Japanese research environment, an introduction to the science and science-policy infrastructure of Japan, and intensive Japanese language training" - with the hope of getting American scientists interested in future collaborations with the Japanese. But political dreams don't always become scientific realities.

Well, as Max Danger predicted, diversity of opinion reigned by the end of the first month for the Summer Institute participants. My own first impression of Japan was a feeling that I had stepped back into the 1950's - the fads, the colorfulness, the avid consumerism, the importance of family, the general contentment. But these initial images soon gave way to confusion and wonder. In the broad scheme of things, Japan did not seem at all content. How were the Japanese going to accommodate this new feeling of internationalism sweeping the nation while still harboring their self-centered, inward-looking cultural values? How would Japan deal with the ever-widening generation gap between old tradition and youthful ambition? Why all the fascination with things non- Japanese (especially American), yet such a strong unwillingness to accept foreigners into their social structure? What kind of culture is it that has a word - karoshi - for 'death due to overworking'? With sky-high inflation, crowded conditions, harsh work schedules - how could the Japanese claim to be content? I found myself weighing their merits on my own cultural value scale - a dangerous mistake. Only one thing was clear. The Japanese themselves were looking for the answers to these very questions. Japan is a land undergoing immense change, knee-deep in contradictions, and difficult to comprehend. The most any outsider can hope for is to learn to accept and adjust to the inherent differences we can't understand. I look back at my time spent there as invaluable - so many things learned, so many friendships gained, so many barriers crushed. In the following sections, I hope to present an accurate picture of what a foreign researcher can expect to encounter in the Japanese research environment, with emphasis more on the things that are different and in some cases how I dealt with them. Let's begin with Tsukuba Science City.

Tsukuba Science City

In the scientific community around Tsukuba, the cultural barriers are considerably less severe than in mainstream Japan - most likely due the international exposure and awareness of this branch of society. The myths that plagued me beforehand turned out to be just that - myths. For example, never did I feel that I was being treated differently or poorly because I was a woman. (However, I was one of only three female scientists at my institute.) Nor did I notice any measurable differences in the ability, creativity, motivation, sense of humor, or integrity of the Japanese scientist compared to his American counterpart. In my field of research, radiation damage of materials, the scientific approach was strikingly similar. But while this was true of the research itself, significant cultural differences in the research environment did exist which are described in the following sections.

The Japanese have an enormous tendency to organize things in nice, neat little packages. This includes companies - which organize cities around their factories, and the government - which organized a 'comprehensive research and housing complex' called Tsukuba Science City. It is located approximately 60 km northeast of Tokyo. Planned and built by the government to be a center for high-level research and education, it now houses more than 55 national and private research institutes as well as Tsukuba University.

My first impression of the place - in a word - was "Brasilia". Indeed, very little of the traditional Japanese flavor has been preserved in the building of this city. Although, to be fair, it is a quaint and very pleasant place to live. There are many parks and bikeways. Western-style toilets abound. International foods - such as peanut butter, ketchup, potato chips - are easily obtained here. Should you choose to be the accidental tourist, you have a choice of McDonald's, Kentucky Fried Chicken, Denny's, Shakey's, Baskin-Robbins, as well as a number of ethnic restaurants - Mexican, Italian, Chinese, etc. Because of the nature of the community - scientists, their families, and students - most of the residents speak English. All in all, it's a nice 'buffer zone' between the harsh cultural and taste differences of mainstream Japan and the Western world. The foreign visitor doesn't really get a genuine feeling for Japanese living, but is comfortable.

Foreign visitors are treated well in Tsukuba. The Japanese go through great pains to please. Department store clerks tirelessly chant their welcome greeting "irrashaimase" to incoming customers. A number of interactive groups for foreigners have been organized. Nearly all of the items required for a long term stay are provided in one way or another. Service is generally prompt, courteous, and on-schedule. Convenient bus and train service from Tsukuba to Tokyo does exist, but is expensive.

Actually, it doesn't take long for the foreigner to realize that everything is generally expensive in Japan - food, clothing, shelter, gas, entertainment. Telephone installation is surprisingly outrageous in cost. The average salary is higher, as well, but perhaps not quite high enough to compensate for the inflated prices.

Some of the best (i.e. most spacious) lodging in Tsukuba is reserved for the long-term foreign visitor, while the Japanese scientists and their families are housed in smaller government- subsidized apartment complexes. But even so, the housing is more on par with student housing in the U.S.: multiple-units packed into the minimum required land space. Countertops and doorways are scaled down to 3/4 size. In-house laundry facilities, especially dryers, are rare. You will not find a hot water cycle in a wash machine anywhere in Japan.

A more difficult time is probably had by the foreign rmsearcher's family. At most of the institutes, working hours go well into the evening. Every other Saturday is a working day as well. After-hours socialization with coworkers is common practice and it does not include the spouse. Therefore, it can turn out to be a long, lonely stay for the American spouse and family. Japanese women, who generally do not work outside the home but devote their time to taking care of the children, are unfortunately not easy to get to know. They have their own network but it's difficult for a foreigner to break into, especially if she does not speak Japanese. House-husbands, such as would be the case for my family, are unheard of and would certainly not be welcomed into the network.

Aside from the International Science and Technology Expo held in Tsukuba in 1985, little evidence could be seen of the spill over benefits this localized base for science should be reaping. Although strong collaborations were seen to exist within individual institutes, interaction between scientists of different institutes as well as between the institutes and Tsukuba University was very limited in comparison to the U.S. For example, colloquia were not shared between institutes doing complementary research - very unlike similar situations in the U.S. (such as the collaborations shared between Argonne National Laboratory and a large network of local universities). Networking appeared to be extremely poor between institutes run under different science and technology arms of the government. Many of the stronger collaborations between scientists originated before they ever reached Tsukuba. The bottom line is that there is no open forum for - or means to communicate - the sharing of ideas. Perhaps a forum exists, but the scientists just don't know how to take advantage of it. In this respect, the Japanese research environment at Tsukuba is very different from the U.S. The Japanese appear to have an inherent distrust of strangers and unfamiliar surroundings, which is but one of the many cultural hurdles facing the foreign visitor. My feeling is that it has a lot to do with their rigid protocol and inefficient methods of communication, both of which are discussed further in the next sections.

A similar trend was noticed with the basic research initiatives. For as much as companies are willing to participate in these joint ventures, there appears to be an equal amount of distrust and unwillingness to openly share technology and ideas. Details of experimental work are often kept confidential, which makes it difficult for peer review. It will be interesting to watch how these basic research initiatives evolve in the next few years.

The Japanese Government Research Institute

Although the Tsukuba Science City concept has perhaps not fully matured, research within the various public and private laboratories is very healthy. My assignment was the National Research Institute for Metals (NRIM or Kin zo ku zai ryo gi ju tsu ken kyu jo to the Japanese), in the Nuclear Materials/Materials Characterization group. NRIM is sponsored by the Science and Technology Agency (STA) branch of the government. The mutual interest of this group and my own PhD work is studying the effect of ionizing radiation on materials; but where they concentrate on bulk phenomena, I'm more interested in surface phenomena. In addition, we were both interested in developing special purpose electron microscopes to carry out our research. During my stay in Japan, I was able to visit a number of the other research institutes in and around Tsukuba Science City - including the National Institute for Research in Inorganic Materials Science and Technology (NIRIM), the National Space Development Agency of Japan (NASDA), the Electro-technical Laboratory (ETL), the National Laboratory for High Energy Physics (KEK), NEC Research Laboratories, the Japan Atomic Energy Research Institute (JAERI), Tokyo Institute of Technology (TIT), both JEOL and Hitachi Corporations - leaders in the production of electron microscopes, as well as a number of other universities and companies throughout Japan.

Overall, NRIM and most of the other research institutes I visited are set up similarly to U.S. national laboratories. Facilities and equipment were of the same or better quality. They were generally smaller in size, with fewer personnel. It's a toss up trying to decide who has more bureaucracy - but with their multiple (>5) signature stamp purchase order approval form, I'd guess that the Japanese are slightly ahead.

There are four main divisions at the Tsukuba branch of NRIM - High Strength Materials, Nuclear Materials, Superconducting and Cryogenic Materials and Surface and Interfaces of Materials. Divisions/groups are set up according to a particular research emphasis - such as Nuclear Materials/ Materials Characterization in my case. NRIM has approximately 437 personnel: 333 scientists and 104 administrators split between the Tokyo-Meguro site and the Tsukuba site. However, the entire Tokyo branch will be moving to Tsukuba within the next few years.

One of the more striking initial observations to me was how equipment-intensive the laboratories were, yet how extremely short they were on labor. Apparently, it is easier to obtain funds for capital equipment than for skilled personnel. In some areas, such as surface science, the ratio of instrument/scientist was significantly greater than one - in stark contrast to the U. S. where, for example, a given surface analysis ESCA chamber often handles multiple users (of the order of 5-10 for the instrument at Northwestern University). The electron microscopy facilities were busier, but microscopy time was still more available than what I've been used to. The Materials Characterization group alone was in charge of two transmission electron microscopes, with a high voltage (MeV) instrument on order from JEOL Corporation.

[As an aside, it's also interesting to note how many high voltage electron microscope (HVEM) facilities, despite their enormous cost, exist in Japan. It can be said that at any given Japanese university, the quality of a materials science department rests on whether they have a HVEM or not. NRIM has ordered one, NIRIM, a neighboring STA institute, already has one. Considering that you can still count the number of HVEM facilities in the U.S. with one hand, the amount of Japanese research yen put into high voltage electron microscopy alone is incredible.]

Groups vary in size from less than five to not more than fifteen members. Any given group may be made up of a number of scientists from Japanese industry on loan to the laboratory through a cooperative program. This aspect of technology transfer appears to be working well to provide an effective indirect R & D link between companies through the national laboratory. Unlike U.S. national laboratories, very few students or foreigners are present in the work force. This year, NRIM has been host to two STA fellows - an American and an Italian, as well as a few scientists from a program Japan has initiated to aid in the technological advancement of underdeveloped countries. However, only one foreigner (an American) was employed permanently and he has since returned to the U.S.

The background education of the Japanese scientist and his American counterpart are somewhat different. In the U.S., most of the scientists employed by national laboratories have PhD degrees, earned prior to employment at the laboratory. In Japan there are fewer PhD degree scientists. In addition, they have a type of thesis program where scientists are hired at the B.S. level. They are trained at the laboratory and eventually write up a thesis which is submitted to a graduate program at some university. Thus, the emphasis is on the applied rather than the theoretical knowledge base. I am not sure of the percentages of employees who receive their degrees in this way. However, it's a sensible route since Japanese students in all fields - including science and engineering - generally have to pay for their graduate studies in full. Some scholarships even require reimbursement after graduation.

One cannot avoid noticing that Japanese national research facilities are short on people at all levels, but especially at the skilled technician level. There are a number of possible reasons for this shortage. One major reason seems to be the pull from areas such as banking and finance. A number of universities have already noted the desertion of a significant number of new science and engineering PhD graduates to jobs in the banking and finance areas. National laboratories may soon have a difficult time finding qualified personnel.

In the Japanese national laboratory system, people are hired with the intent of being kept for the long term. Rarely is anyone fired. One's salary is determined primarily by the number of years employed, with personal achievement indicators such as paper publications or goals achieved being of secondary importance. The end result is that a majority of Japanese scientists do not change jobs once hired.

The issue of job security is, however, a 'two-edged sword'. On the one hand, this scheme encourages - and allows time for - scientists and engineers to have mobility within the institute, thus giving them a better baseline understanding of the breadth of the research. Scientists are more inclined to broaden their expertise within the framework of the project rather than increase their expertise in a specific area, which is often done in the U.S. to increase the 'experience quotient' for future employment opportunities. (The same scheme is used in Japanese industry, where employees are moved between the various branches to give them a better overall understanding of how the different divisions within the company are related.)

On the other hand, the hiring quota is low due to the extra long-term expense of each new individual, and rarely is anyone hired at the technician level - which is not considered a permanent, long-term position. Scientists are extremely overworked. In addition to research, they must either do all their own routine maintenance or call on outside help, which is both time consuming and costly. Difficult situations can arise when disciplinary action is needed to motivate or stimulate inefficient workers. The incentives for personal achievement are less obvious in the Japanese national laboratory.

There are a number of aspects of the Japanese system which enhance the individual scientist's perspective of the research goals. It has already been mentioned that scientists usually do not change jobs frequently and that they are encouraged to spend time in various groups within their own facility. In addition, they are encouraged to spend some time (usually two years) abroad at some laboratory doing similar research. The Japanese government pays for this so it is fairly easy to have the pick of places. The Japanese scientist is also required at some point in time to do an administrative internship. For example, NRIM scientists usually do a one year internship at STA headquarters in Tokyo. This provides administrators with first-hand information of the 'goings-on' in the actual research environment as well as providing the scientist with an understanding of the management framework.

The first cultural hurdle I encountered was the rigidity of the Japanese system. Protocol and place mean everything there. You must know what your place is and where it fits in. (That way, you know how far down you have to bow.) One asks for permission to use the 20,000 yen hot plate in the same way one asks for permission to use the 20,000,000 yen electron microscope. Violation of this protocol can result in anything from long lasting grudges to permanent banishment. So, when in Japan, it is essential to play by the rules from Day 1. Daily protocol covers such areas as how the group makes decisions, how the group interacts with other groups, and how the individual interacts within the group.

Within each group, the decision-making process is slow and deliberate - never without consideration of every option in detail - which is (arguably) the most difficult aspect of the Japanese research environment to get used to. It is performed in a highly systematic fashion, with everyone involved at all stages - including myself for the short time I was there this summer. Meticulous attention is paid to the details of the research project. Perhaps somewhat unlike the U.S., a supervisor will never go ahead with something until everyone is fully informed. The scientists thus have a much better view of the scope of the research project and where it fits into the 'big picture'. And because of this long drawn out process, once a decision on a particular course of action or a particular design is made, it is quite difficult to reverse - good or bad.

Within the decision making process in the Japanese system, there are unwritten - but strongly adhered to - rules which preserve the sense of balance. Take the purchasing of equipment, for example. In the U.S., the purchase of capital equipment is often determined by a bidding process, where the lowest bidder wins the contract, independent of past or future purchases. In Japan, it's somewhat different. The government appears to go through great effort not to play favorites. This strict sense of balance, in effect, allots to each major manufacturer a 'piece of the pie'. Around the laboratories, one can see many examples of this. For instance, if one electron microscope is purchased from JEOL corporation, then the next is purchased from Hitachi and so on. Another example is how Toshiba and Hitachi are equally represented in the superconducting magnet lab at NRIM.

Outside the group framework, a strict hierarchy of positions exists which inhibits most scientists from questioning the decisions of their superiors as well as making any kind of decisions on their own without the explicit okay of their superiors. The Japanese have a strong respect for their elders. Examples illustrating this are often seen at conferences, where a junior scientist will refrain from commenting on something until he or she has discussed it with his or her superior. It takes a good boss to be able to bring out the confidence, opinions and ideas of his subordinates in Japan.

As in the U.S., communication between groups varies widely with the group. Intergroup communication of the discovery of the Sr-Bi-Ca-Cu-O superconductor, for example, took months when it may have been more beneficial for NRIM as a whole if this information had been disseminated sooner. On the other hand, I was also aware of a number of strong inter-group, inter-laboratory, lab-university, and international collaborations in the Nuclear Materials division. These appeared to be more deep-rooted relationships dating back primarily to university days, but some went as far back as childhood. (Due perhaps to the nature of the research, there is a strong international network in the nuclear materials area, within which the Japanese have played an integral part.)

The biggest hurdle in the cultural barrier that I encountered was communication. The art of communication is very interesting in Japan. It varies drastically from the American way. We, for example, generally say what is on our mind at (or near) the moment the thought enters into our head. We are generally direct; primary emphasis being placed on the spoken word. No credit is given to that which is left unsaid. The Japanese way is almost the opposite. Thoughts and ideas are pondered long and hard before ever reaching the lips. Communication is very indirect, with at least as much emphasis placed on what is left unsaid as on the spoken word. Gesturing and intonation play an integral part of the final act of communication. (In fact, a popular stereotypical view that the Japanese have of Americans is that we often talk before we think and do so loudly.)

There is an entire philosophy in Japan dedicated to the art of gesture and the unspoken word. It is called Haragei. Haragei is literally translated to 'art of the belly'. In practice, it is the act of dealing with people or situations through ritual formalities and accumulated experience.

Quoting from a lecture given by Dr. S. Ieno of Tsukuba University to the Summer Institute participants and from the collected works of Mr. Michihiro Matsumoto, a specialist in the area of Haragei research:

> Ruth Benedict,[1] in the book The Chrysanthemum and the Sword [1], observed the Japanese nature as follows: 'The Japanese are, to the highest degree, both aggressive and unaggressive, both militaristic and aesthetic, both insolent and polite, rigid and adaptable, submissive and resentful of being pushed around, loyal and treacherous, brave and timid, conservative and hospitable to new ways. They are terribly concerned about what other people will think of their behavior, and they are also overcome by guilt when other people know nothing of their misstep. Their soldiers are disciplined to the hilt but also insubordinate'. In this view the Japanese are illogical. However, if they are observed from within, they are simply Hara-logical. The Hara-logical accepts either "yes" or "no" as it comes, and rejects nothing. Hara-logic and Haragei truth are in a category of their own. Particularly in politics, where Haragei is standard operating procedures, the truth not only hurts but can make one bleed to death. What argument is to Westerners, Haragei is to Japanese.

The Haragei concept can also be illustrated in other comparisons of Japan and the U.S. For example, from Saeki Shoichi, professor of American literature at Chuo University, in "Rediscovering America's Dynamic Society", Japan Echo, Spring 1988:

> In America, college students bring all their gripes and grievances right out into the open and confront the instructor with them. The direct approach allows for a quick solution - provided, that is, that the instructor's explanation is prompt and clear.
> The marked difference between the look and feel of classrooms in Japan and

---

[1]Benedict, Ruth, "The Chrysanthemum and the Sword: Patterns of Japanese Culture", Tuttle Publishing, Tokyo, 1965.

those in the United States comes to mind when I hear complaints of Japan-bashing. Washington says whatever it pleases with brash assertiveness, while Tokyo responds with passive, defensive, evasive mumbling. That the behavior of college students should echo the rhythms of Japan-U.S. trade friction is perhaps less amusing than frightening.

Although most Japanese won't admit to it, varying degrees of Haragei logic are used even in their daily interactions. In contrast to the U.S. (where people use the word "no" too liberally in my opinion), rarely will a Japanese person actually say 'no' when he or she means 'no', even when not agreeing with an opinion or a statement. To them it's more polite to counter a point without having to say outright "no, you're wrong". I personally encountered many shades of grey between 'yes' and 'no'. These were generally manifested in the phrase 'yes, but . . . '. The longer the clause behind the 'but', the closer to 'no' the statement actually was. Eventually I began to recognize this communication trait; and also learned to phrase my questions in such a way that an unambiguous answer was required. Despite Haragei, communication within each research group is open; and there is a strong sense of respect for each other. It is often said, "The nail that sticks up gets pounded down". I found this to be true in a good sense. There is no room for big egos in the Japanese system. More value is placed on how well one works within the group framework. Nobel prize winning biologist, Susumu Tonegawa (from the Chicago Tribune Magazine, April 29, 1990) compares the contrasting styles as follows:

> The key to America's high-powered intellectual performance is motivation . . . The [American] society encourages and rewards outstanding effort. In Japan, by contrast, the social ideal is the hard-working, low-key craftsman. To maintain group harmony, we are encouraged to downplay individualism. Japanese society frowns on people who stand out from the crowd, who disrupt the status quo.
> In the U.S., individuals are expected to take charge of their own lives. Self-expression is highly valued. And Americans, for better or worse, are aggressive. Quiet, self-effacing Japanese cut a poor figure in U.S. labs.
> Recently, team efforts in both countries have made major break-throughs in basic research. The team approach [of the Japanese] often utilizes facilities more efficiently than individuals can. But individuals, not groups, come up with new ideas, the sine qua non of research. Group work - the norm in Japan - stimulates good minds, but most team members exert a leveling influence on their members and thus stifle individual creativity.

Viewed from a different perspective, Americans can be perceived as aggressive and overbearing - cutting a poor figure in the Japanese labs as well. It's clear that not only do Japan and the United States, but Japanese and Americans have much to gain through collaboration with each other. The cultural differences, for the most part, can be viewed as complementary rather than conflicting. The stifling group mentality of the Japanese can be loosened by the Americans; whereas the Americans can benefit not only from the steadiness of the Japanese pace and their meticulous eye for detail, but learn to capitalize on the value of the group versus the individual ego.

In conclusion, my experiences indicate that the cultural barriers, once identified and accepted in their own context, are not insurmountable; and that personal friendship can break down the highest of barriers. The more I learn about the Japanese, the more I learn about myself. The time is ripe for Japanese - American collaboration. Lastly, there is one very good rule of thumb on how to proceed in Japan: In America, often forgiveness is easier to receive than permission. In Japan, the opposite is true. Even so, the Japanese will put up with quite a bit from the foreign scientist and will refrain from confrontation if at all possible. The best thing to remember is: if you wouldn't be able to get away with it back at home, don't try it with the Japanese. Courtesy and politeness need no translation.

A Comparison of Cultures

Sumo Wrestling	American Football
Much skill, years of training, dedication required.	Much skill, years of training, dedication required.
Highly competitive.	Highly competitive.
Passive show of strength.	Active show of strength.
Show of emotion looked down on, control important.	Lots of emotion allowed, sometimes affects game.
Deep respect for opponent.	Healthy disrespect for opponent.
Emphasis on balance and harmony, how game is played.	Emphasis is simple and direct: the bottom line is the final score.
Individual effort sees rewards with time, rank.	Individual efforts immediately rewarded.
Judgements not questioned.	Frequent questioning of judgments.
Performed under strict rules, rigid protocol.	Within the framework of the rules, anything goes.
Winning technique: simple, elegant, straight forward, steady strategy.	Winning technique: complex, lots of theories proposed, rules and strategy always changing.
Pyramid system: bottom up.	'Brightest Stars' stand out regardless of age, rank.
Relationships close, cooperative, within stable. Stablemates don't fight each other in official competition (except for tiebreaker situations).	Competition exists within team as well as between team.
As much (if not more) emphasis on events outside the match itself.	Main emphasis is on game itself, little regard for 'pre' and 'post' game fluff.
Very few foreigners allowed in the sport.	Foreigners welcome, especially if they are good.

JAPANESE RESEARCH STUDENTS IN THE UNITED STATES

Richard C. Bradt, Mackay School of Mines, University of Nevada, Reno

Although I've never had a jointly funded research program with either a Japanese company or government agency, I have had the opportunity to advise nearly twenty Japanese graduate students and laboratories, and a number of Japanese industrial scientists as well. Many of these contacts were made during a sabbatical in Japan in 1978, or during numerous subsequent visits to Japan. These cooperative research efforts have yielded nearly fifty journal manuscripts and currently another dozen or so scientific papers that are in progress. I've thoroughly enjoyed the interactions. The only problem is for me to keep up the pace. Although it hasn't always been easy, it has always been rewarding!

In addressing the various research aspects, I'll try to first discuss a few general observations that I have made while interacting with Japanese scientists, then I'll try to be more specific in terms of the graduate students, visiting faculty and industrial scientists. In the cooperative efforts in which I've participated, I've always found the Japanese scientists and engineers to be very enthusiastic, inquisitive and willing to work. Their approach to experimental studies is a very systematic one, thorough and methodical. Data gathering and analysis is meticulous. Even with the graduate students, I've always been able to eventually develop a colleague-like relationship so that they've been very interactive, making valuable suggestions and being critical, as well as responsive to criticism. Relative to Americans, myself included, I've found the Japanese scientists to be much more thorough with literature reviews and to use those references very wisely in planning experiments. There's absolutely no doubt in my mind that Japanese scientists are quite disciplined in their pursuit of literature searches before starting experiments. Often, I found them to identify peripheral references that are valuable. Very frankly, I've always been puzzled by this dedication and ability and have wondered as to its origins. Once the research is complete, as co-authors, they will invariably want the paper written in English.

The graduate students with whom I've been fortunate to conduct research have either been Monbusho scholarship awardees or industrial employees fully supported by their companies. For the most part, they have been very serious in their studies and research. If I did not know them prior to their joining my research group, they were usually far too respectful and a bit difficult to work with at first. However, after some time, a more collegial atmosphere developed and the research relationship substantially improved. I've generally found my Japanese graduate students to be delightful and most rewarding to do research with and to publish with. They develop a strong life long relationship with you, much closer than most American students. Whenever I go to Japan, a number of my former students meet me in Tokyo for a welcome unagi dinner celebration. They are almost like members of my family.

Japanese universities send their best young faculty abroad to study for a year or two, and I've been fortunate to have several of them choose to study and do research with me. Like the students, they too are initially not as collegial as they become later. It has been my experience that they are very serious and hard working. After they return to Japan, the cooperation and interaction continues for many years. In several instances, they have continued to send me their manuscripts written in English for correction and editing, for they continue to wish to publish in English.

Cooperative research with the Japanese scientists at the GIRI laboratories and in industry is equally challenging and rewarding. They set a pace that is difficult to match and will always contribute more than their fair share. Again, the collaboration never ends and a near family-like relationship develops.

Overall I've found it a pleasure to cooperate with Japanese scientists at all levels in various research endeavors. I feel that I always receive more than I am able to give to the joint efforts and I have never had any disappointments. It's been some of the most enjoyable and rewarding research in which I've been able to participate.

CROSS-CULTURAL STUDIES OF MANAGEMENT AND ORGANIZATIONAL BEHAVIOR

Toshimasa Kii, Georgia State University

Various, almost countless, comparisons have been made between the U.S. and Japan in terms of management style in the private sector, the role of the public sector, the role of education, human behaviors, values, and scientific innovation and creativity. What has come out of both the intellectual and popular discussions in these areas over the past ten years seems to be more political in connection with international business and trade, namely, that Japan is different, unique, and even enigmatic. However, there have been several informative writings on the issue surrounding innovation and creativity of Japanese industries, such as Sheridan Tatsuno's THE CREATED IN JAPAN (1990), Stephen Kline's INNOVATION STYLES IN JAPAN AND THE UNITED STATES (1990), Don Kash's PERPETUAL INNOVATION (1989), Eleanor Westney's IMITATION AND INNOVATION (1987), Gene Gregory's JAPANESE ELECTRONICS TECHNOLOGY (1985), James Abegglen and George Stalk's KAISHA: THE JAPANESE CORPORATION (1985), Rosabeth Kanter's THE CHANGE MASTERS (1983), Robert Cole's WORK, MOBILITY, AND PARTICIPATION (1979) among others. At the risk of being redundant about presenting some of the issues concerning management and organizational behavior as they relate to creativity and innovation, I would like to summarize the major issues from one of two interrelated perspectives. The two perspectives are structural and interactional. Since I will be exploring the Japanese organizational behavior from the interactional perspective in this presentation, let me simply state what it means by the structural perspective.

A human society has five major institutions for its survival which make up a macro level social structure of the society -- economic, political, family, educational, and religious. And it appears that all human societies have historically dealt with these institutions collectively or separately. Certainly, in more economically advanced nations such as the U.S. and Japan, these institutions are distinctively separated. And they have been examined and analyzed both independently as well as collectively for their interrelatedness in various disciplines, particularly in anthropology and sociology.

I use the structural perspective to mean that all these institutions will make up the macro structures of a society which impinge upon human behavior. Or, put another way, the social structures delimit as well as expand human activities in that society. But I would like to emphasize that the structure puts constraints upon human activities, although I must admit that I could not imagine what unconstrained human activities might look like. Since we feel constrained by the structures, we are knowingly or unknowingly negotiating every day with and within these institutions. But since the structures of the society are built on long historical currents, we seldom objectify them in order to transform them. Rather, they have become the environment within which we breath. Of course, many reforms and even revolutions have taken place which changed or overhauled existing institutions for ideological and practical reasons. But on the whole they have remained stable.

Now, let me turn to my use of the interactional perspective before I attempt to discuss what these perspectives have to do with cross-cultural management and organizational behavior. The interactional perspective means simply that individual humans do and must interact through symbolic communication with other human beings if they are to exist in a social organization. It involves negotiation. From birth on we are taught how to negotiate with others, either positively or negatively through the use of cues and guidelines which are provided by the society within which we interact. Because of this, individual perceptions of reality are variable and constantly changing. I would like to emphasize that this interaction puts constraints on human activity because it involves negotiation, symbolic or otherwise.

The reason I bring up this perspective is because we have a tendency to rely more on the structural perspective when attempting to understand another culture, namely the collective behaviors of the individuals in another culture. In some ways, it is easier and more comforting to do so since structures are seen in a more static fashion, thus leading to more or less dichotomous comparisons of American and Japanese institutional arrangements. For example, the MITI's roles in directing policies regarding high tech R&D and in orchestrating coordination of its policies with other ministries and the private sector are relatively easily understood. However, if we want to know how the policies are made and what processes are used to implement them, we must explore the behaviors of individuals. But there are a variety of individuals within a society whose interpretations of that culture, and of other cultures for that matter, are vastly different depending on who they are, what they are, and where they stand in that society.

We often speak of cultural differences between the U.S. and Japan when we conduct business, although I must admit that word, "culture," often becomes a whipping board when we fail to understand THEIR behavior as opposed to OUR behavior. We often say their values are different. Robin Williams, an American sociologist, has stated that the ten most central values to American culture are: 1) Equal opportunity, 2) Achievement and success, 3) Activity and work, 4) Material comfort, 5) Practicality and efficiency, 6) Progress, 7) Science, 8) Democracy, 9) Individualism, 10) Freedom. And I suggest to you that all these values are also regarded highly among the Japanese. Yet, we speak of value differences. So, from the interaction perspective, Japanese and Americans communicate symbolically in English in most cases with rather limited concepts of what these values really address. The real meaning of these values are understood only within the structures of each society. Certainly, language is an obstacle in communication. But it is only one obstacle. Democracy, freedom, individualism may mean different things to Americans and to Japanese. Yet, when they are spoken in English, or translated into English, the word, d-e-m-o-c-r-a-c-y, etc. are used but the concepts are not communicated because the interactional constraints under which these values are behaviorally expressed are not understood.

So, the Japanese and the Americans negotiate these and other issues symbolically and often end up more than frustrated because it takes too much energy to be flexible enough to see individual behaviors of another culture from their perspective even though we understand the structural constraints of that society.

How can we organize varied behaviors among individuals and make them culturally comparable? What follows is not exhaustive by any means. But it is an attempt to understand culturally sanctioned normative behaviors of individuals in the Japanese society.

Several years ago I attended an international scientific meeting in Tokyo. In that meeting I witnessed contrasting behaviors of a Japanese and an American scientist. A young American researcher rose up to the podium with full confidence, appeared relaxed, and introduced his paper by saying that his was a bit different from the previous one. He continued by explaining that his method of data collection was more elaborate and that his experiment was more tightly controlled. His presentation was straightforward and direct, although my impression was that his findings were not that much different from the previous presentation. A not-so-young Japanese scientist's turn came up. He profusely apologized for his poor methodology, praised the previous presenter and apologetically concluded his findings which were, I thought, quite different from the previous ones.

Certainly, one can dismiss these presentations as extreme cases. But I actually thought both did a good job. But my colleagues sitting next to me did not think so. The Japanese colleague thought the presentation by the young American scientist too self aggrandized.

The American colleague on the other hand thought the Japanese presentation was weak and did not understand why he apologized.

The more I thought about the incident, the more I became interested in the underlying cultural assumptions between the two societies regarding normative behaviors which are generally considered acceptable by the public. What are those underlying cultural assumptions?

Views toward Outside World

Although it is difficult to trace the origin, historically or geographically, the Japanese tendency is to view the outside world as unfriendly, if not hostile, uncomfortable, awesome. Americans on the other hand, probably for historical and geographical reasons, view the outside world as curious, worth stepping into to see how the other side lives, as a place to be explored, or even conquered. (One such contrasting behavior is found in the practice of child discipline in the family. The Japanese parent socializes the child to fear the outside world, thus threatening to throw him out of the house when he does not conform to the parent. In the U.S. the practice of "grounding" is to keep the child in his room, thus depriving him of freedom of exploring outside world.) Or, put another way, Japanese tend to focus more on the immediate environment in which they find themselves. Both Americans and Japanese see themselves as expandable but the primary direction appears to be different. The Japanese seem to expand inwardly whereas the Americans expand outwardly. This is rather a bold expression for the two cultures. But if you accept this perspective for a moment, I would like to propose that the Japanese and the American views of people around them are different. Thus, it appears that Japanese perceive individuals around them in three categories: the inner, outer, and in between. The primary group members such as family, close friends, are found in the inner world. They are extremely informal, tied together with strong emotionality. Their relationship is often characterized as the "sweet" dependency, allowed to express raw emotions. The people in the outer world are those with whom the Japanese do not wish to deal, if possible. The overwhelming majority of Japanese are in this category as well as most non-Japanese from the view of the particular individual Japanese. Often, Japanese are oblivious to people in the outer world, including the Japanese in it. Thus, from the American perspective Japanese behavior in public often appears self-centered, indifferent to people around, and even rude in crowded places such as train stations, theaters, department stores and parks.

What is interesting and complicated is the world in between. This is a secondary group of people who are related because they happen to work together, go to school together, see each other regularly at social clubs, etc. But it also includes business clients and those who provide services such as doctors, police, and neighborhood merchants. Interactional forms in this world vary tremendously depending on where the interacting individual is perceived to be on the continuum of the inner world to the outer world. But there is a common thread binding the relationship in this world. That is the formalized responsibility and obligation exchanged among interacting individuals. What are the criteria for this behavioral interaction?

BA

Ba translated into English can mean several different things -- frame, location, place, circumstance, situation, or position. Japanese behavior is circumscribed by where and with whom the interaction takes place. This is also true for American behavior. But the Japanese exhibit considerably more control, thus appearing more formalized and stylized. From the Japanese behavioral standpoint, *ba* also involves the hierarchical relations of persons who happen to share physically the same situation. Japanese individuals are hierarchically ordered, partly because of their social status and partly because of their organizational affiliation. And organizations are socially rank ordered in the perception of the interacting Japanese individuals. An individual may be an electronics engineer from a small firm who is as successful and respectable as any other engineer in the field, but his behavior would vary depending on who else happened to share the physical place with him. If engineers from larger firms are present, he would behave in certain ways, not based on his personal status, but based on his perception of his firm's status in Japanese society.

TATEMAE and HONNE

Tatemae may be translated as professed intention (as opposed to real intention), circumstantial opinion, agreed-upon-statement (agreed by a group or groups), or simply front. *Honne* may be interpreted as real intention, true feeling, and rationally derived opinion. The use of *tatemae* and *honne* are not opposed to each other as the above English translation may imply. They almost always compliment each other. Often Japanese will say things which are not what they really mean, particularly in formal meetings. But the purpose of *tatemae* is not to deceive or to lie. It is neither an excuse nor insincerity. Indeed, Japanese will speak from *tatemae* because they want to be sincere in the sense that they do not wish to hurt the feelings of the interacting others. Americans exhibit this type of behavior, also, but with the Japanese the behavior is more conscious and is led by *ba, soto,* and *uchi.*

SOTO and UCHI

This pair of concepts relates to the previous description of how the Japanese perceive the outside world. The inner world is *uchi* and the outer world *soto.* Or put in other terms, they are inside and outside, respectively. I have mentioned that the world between the inner and the outer worlds is the world in which Japanese work life is observed. I also mentioned that this particular world is best understood in terms of a continuum from the inner world to the outer world. Japanese use *uchi* and *soto* (and persons of the outside are called *tanin*) often depending on their group affiliation and their emotional attachment to the groups. The organization one works for is *uchi* in relation to other organizations but may be characterized as *soto* in relation to his or her family. Within the organization the department one works in is referred to as *uchi* and other sections may be *soto.* A subsidiary of a company one works for may be presented as an *uchi* company to his or her clients but a *soto* organization in relation to his or her own company. The university or the research outfit to which one belongs is *uchi* vis-a-vis other such organizations.

What would be the perception of Japanese individuals working on a team whose members are recruited from various organizations? If they have another team to refer to, perhaps to compete against, the team may become *uchi,* but probably never more *uchi* than the ones from which the team members originate. One condition for being an *uchi* member is the group identification based on some type of emotional attachment to the group. But this is a partial condition. In fact, the depth of emotionality may not be a crucial condition for an individual to feel an *uchi* member although it can be certainly acquired as a result of intense interaction among the members of the group. A more plausible interpretation for identifying *uchi* or *soto* is found in the way Japanese define the interpersonal relationship with other members in a given group in a given environment.

In the organizational context, therefore, the sense of *uchi* and *soto* consciousness is heightened or lessened depending on the immediacy of the interpersonal relations one feels toward others through obligations and responsibilities even though one may not feel an emotional attachment to the organization. The more obligated and responsible one feels toward members of a group in terms of task accomplishment, the more he or she identifies the group as *uchi.* As this sense of obligation and responsibility diminishes toward the members of the group, the perception of the group as *uchi* will fade into *soto.*

What about the group of participants in this conference? Do the concepts of *ba, tatemae, honne, uchi,* and *soto* explain behavior of the Japanese? The Japanese participants in the conference were invited from various organizations. They may have seen each other in the past and might have heard of the names of the participants through scientific publications. But except for a few members who worked in the same outfits most of them were only acquaintances or had never met before. Certainly, there were varied personalities and individualities among them. Yet, their behaviors vis-a-vis Americans at the conference were more uniform, more controlled, less direct, and less inquisitive. While the American participants more or less freely exchanged opinions, argued and confronted each other on occasion, the Japanese participants did not vehemently discuss, nor argue, and certainly did not confront other participants, particularly if the discussions concerned Japanese issues.

Ba was self explanatory at the conference. The Japanese were invited guests. Some of them had been students of the conference organizer. It was only appropriate for them to behave with great care and sensitivity toward the American participants. Even if they had had strong opinions, they had to control their expressions given what they perceived to be the frame of the conference, the positions of themselves, and the environment in which the conference proceeded. *Honne* did not come out, because it was inappropriate given the *ba*. But it was their perception of the *uchi-soto* framework of the participants that controlled the Japanese behavior. The Japanese participants became people of *uchi* and the Americans those of *soto*. Or, at least the former became more *uchi* than the latter. Members of *uchi* will consciously present what appears to be a unified front (or *tatemae*) in the face of *soto*. If some individuals of *uchi* disagree among themselves, they will control their urge to make it known in order to avoid the embarrassment of appearing disorganized in the view of *soto* members. This behavior is also reinforced by the desire to stay in *uchi*.

Views toward Uchi and Innovation

Given the aforementioned view of the outer world, how would the Japanese view the inner world? If the *soto* is perceived as unfriendly, uncomfortable, and awesome, a place to be avoided, then the *uchi* world becomes one of primary interest concerning both instrumental and expressive endeavors. Since Japanese perceive *uchi* and *soto* in relation to each other, a plausible way to comprehend what takes place in *uchi* is to see *soto* as consisting of various reference groups against which *uchi* measures its own relative worth. Two motivations appear to drive activities within *uchi*. The first is the motivation to enhance one's own group status vis-a-vis that of *soto* groups. It involves a symbolic desire to be accepted in the community of organizations of one's endeavor. Organizations are hierarchically arranged, and interactions among them are controlled by *ba* with normatively desirable behavior of *honne* and *tatemae*. Thus, when an organization perceives itself to be lower in the hierarchy in the community of the organizations, its activities will focus on emulating the activities of the organizations ranked higher. This is more than simply competition. The desire to compete to win (*honne*) is only a partial motivation; winning the acceptance of the community of the organizations is the other part.

The second is the motivation to constantly improve the organizational alignment of *uchi* in order to achieve the first motivation. It involves rather disciplined activities in terms of examining, experimenting, modifying, and refining the organizational arrangement so as to effectively manufacture and deliver products acceptable in the community of the organizations. It appears that this concentration of *uchi* activities is the impetus for Japanese style innovation.

The inner world contains two seemingly contradictory characteristics. One is the perceived limitation of physical space. Physical space can hold only so much unless the objects to be held are modified and refined on a smaller scale. The other is the perceptual expansion inward through simplicity and inconclusiveness. Here simplicity means the ability to transform complex issues into a simplified form. The appreciation of simplicity is only possible when complexity is comprehended. Inconclusiveness suggests that the inner world is limitless. Or put another way, anybody can fill in a portion of that inconclusiveness to come closer to completeness.

The above two characteristics of viewing *uchi* are related but I would suggest that the first is the domain of engineering and the second that of social relations. Modifying, reshaping, and refining material objects or human organizations, for that matter, can be accomplished through technical capacity. However, technical capacity alone will not achieve desired outcomes as they must be approved and accepted by the community of organizations. If the domain of engineering is the efficiency of task accomplishment, the domain of social relations is the effectiveness of task accomplishment. Neither products nor ideas can exist in a vacuum. They must be shared and considered by humans. Inward expansion is a perpetual process of feedback activity. And the larger the number of humans involved, the higher the degree of feedback activity. This activity is labor intensive. And the collective aspect of Japanese organizational activity by design promotes a high degree of feedback. Thus, innovations are ubiquitous but there are few individual innovators in Japan.

It is often implied that discovery and invention are the objectives of scientific endeavor and that emulation and refinement are the objectives engineering prowess. Japanese are popularly said to be good at the latter but weak in the former, but if we are to understand the Japanese-style innovation, the Japanese view of *uchi* and *soto* might hold a key.

JAPANESE RURAL HOUSEHOLD ORGANIZATION AND PATTERNS OF SCIENTIFIC/INDUSTRIAL GROUP MANAGEMENT

Robert McC. Netting, University of Arizona

As an ecological anthropologist, the particular technical innovations that interest me are not the state-of-the-art science of oxide superconductivity or other advanced materials but the knowledge and practice of intensive agricultural techniques by smallholder cultivators. Insofar as a farming system, like a modern factory, laboratory, or university department, is intimately associated with a particular system of social organization, they are all proper objects of ethnographic inquiry. In the context of a nation like Japan that was for centuries an agrarian civilization built on peasant wet-rice cultivation, it is also legitimate to ask just how much the structure and function of the farm household have been significantly carried over into the institutions of the industrial state.

The fact that Japanese agriculture, like that of China and other Asian peoples, was distinctly different from that of the Euro-American West, is often ignored. In fact, the adaptation to scarce land resources and high population density throughout this region was one of intensive, permanent agriculture with crop production that was relatively high, reliable, and sustainable (Bray 1976). High agricultural production per hectare was achieved largely by heavy inputs of human labor, in contrast to the lower land productivity and very high output per worker characteristic of large-scale, mechanized farming in the U.S. and Canada (Figs. 1 and 2). The tools of intensive cultivation, such as hoes, animal-drawn plows, garden tractors, and portable water pumps, may be simple and relatively cheap, but the techniques and accompanying knowledge are often extremely complex. Traditional Japanese farming involved transplanted rice grown on levelled, diked, and often terraced fields, with irrigation systems of dams, channels, and drainage ditches for the distribution and control of water. Practices of double-cropping, manuring, composting, and the use of domestic wastes insured dependable yields and restored soil nutrients. In addition to rice, farmers grew diverse crops of vegetables, sweet potatoes, fruit, tobacco, indigo, and mulberry leaves for silk worms (Smith 1959), serving both subsistence and cash needs. Domestic livestock and the aquaculture of fish further supplemented household production. Similar systems of intensive agrarian production by smallholders are found in areas of high population density around the world (Netting 1990).

The substantial labor costs of intensive cultivation are borne largely by members of the farm household. Adult men and women in Japan worked 1800 to 2500 hours per year (Clark and Haswell 1967), in contrast to the 600 to 800 hours that an African shifting cultivator may expend annually (Stone et al. 1990). Moreover the tasks of farming demand skilled, dedicated labor. There are a multitude of specialized jobs that must be independently scheduled and carried out responsibly. Returns to labor reflect the quality as well as the quantity of effort. For the smallholder, decision-making is a daily necessity. Every farmer may make somewhat different management decisions, depending on his particular resources of land, livestock, and labor, as further modified by weather conditions, cooperation with neighbors, and market prices.

The social unit of production most frequently found with intensive agriculture is the smallholder household (Netting 1989). It is not the feudal manor with serfs, the slave plantation, the big estate with wage labor, the mechanized agribusiness, or the socialist collective. Such a household is based on family members (in the Japanese case, often a stem family of two married couples related as father to son), who share a residence and both produce and consume in common. Other relatives, servants, and dependents may also

FIGURE 1.
Historical Growth Paths of Agricultural Productivity in the USA,
Japan, Germany, Denmark, France, and the UK, 1880-1970

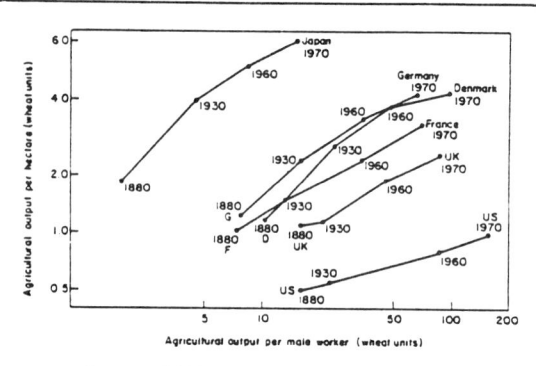

Source: Ruttan 1984

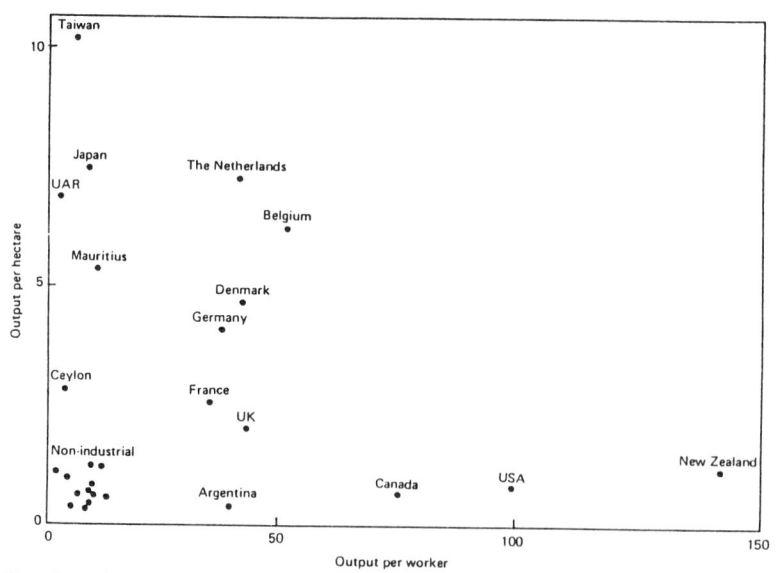

Figure 2. Agricultural outputs per hectare and per male worker

Source: Hayami and Ruttan 1971

be members of the household. The household supplies much of the farm labor and manages its own enterprise. Household members have continuing rights to land, equipment, and buildings either as full owners or tenants with long-term leases. Children or other heirs inherit these rights, including the accumulated investments in increased production that the household makes. The extra work and long-term care and diligence that increase production bring rewards, at least in part, to the farm family household.

Living together for years, sharing farming tasks and decisions, training new members in the duties appropriate to their age and gender position, and eating from a common pot foster interdependency and responsibility in household members. They must be able to work independently and without supervision in widely separated tasks, as well as cooperating efficiently in group efforts such as transplanting rice seedlings or harvesting and storing the crop. The effectiveness of the household labor unit depends on attitudes of trust, loyalty, and life-time allegiance that are not usually present in a group of wage workers who must be recruited, supervised, and paid to insure acceptable performance on the job. The incentives for household members are also of a different kind. They are directly producing their own livelihood on the farm, and even though their immediate returns on labor may be low, they have a claim on valuable property, either as present owners or eventual heirs. The farm household also provides long-term security in the form of care during childhood, sickness, and old age. There *is* no other insurance for most peasants. The emotional rewards of love, appreciation, and honor for the individual come largely from within the household primary kin group. Smallholder families are by no means necessarily harmonious groups of equals. Japanese cultural ideals of patriarchal headship, reverence for ancestors and the elderly, parental authority, and female dependence mean that there will be inequality and power differentials within the household. But the interdependency and diversity of farming and domestic activities and the importance of voluntary cooperation mean that household operations require a high degree of mutual understanding, discussion, and concensus. Though members are ranked by age and gender, the assertion of authority or dominance should not interfere with individual participation or responsibility.

Rural farm family households are obviously part of larger networks. In the Japanese village, households cooperated in the managing of common property resources like irrigation water, forests for fuel and timber, and rough grazing lands in the hills (McKean 1982). The corporate community of autonomous households also had to allocate such resources among its members, make and enforce rules for their use, and defend the resources against incursions by outsiders. Village assemblies of household heads governed the commons and guarded against overuse and environmental degradation. The households making up such a body were not equal in numbers, wealth, or prestige, and their relative rankings may have changed over the life course, but their long-range economic success was bound up with successfully protecting and utilizing common resources. They also had to have effective mechanisms for resolving conflict among their constituent member households without threatening communal unity. Household cooperation and adherence to village norms over the long term could not be coerced.

What could be farther from the rural household in its rice paddy than the lab of a giant corporation or a scientific institute. Isn't scientific innovation the exact opposite of traditional rural conservatism and a technology based on manual labor? Right? Wrong! I have always thought that intensive cultivation might be a nursery for certain desirable virtues--long disciplined work hours in the field may be the groundwork for scholar's burning the midnight oil. Postponement of gratification can lead to saving and investment. Inner directed motivation to do a good job is after all a peasant as well as a Protestant ethic. But, I thought, industry must have its own overriding organization, necessarily large-scale, hierarchical (especially management and labor), and with an assembly line-like breakdown into specialized tasks and roles. Isn't there an overwhelming logic of modern mechanization controlling human activity. Imagine my amazement to discover in Chie Nakane's brilliant *Japanese Society* (1970) that the farm household and its values lives on as the model structure in the firm, the lab, and the department, and that the organization is fundamentally different from that of the U.S., where a frontier, an abundance of natural resources, and a shortage of labor led farming and household organization in a different direction. Nakane (1970:1) contends that the frame (in Japanese *ba*), the institution or relationships that binds a set of individuals into a group, is of primary importance, while the attributes of individuals are a secondary matter.

The *ie*, the corporate farm family household, is a social group constructed on an established frame of residence and management organization (Nakane 1970:5). The vocabulary of terms (what Brian Pfaffenberger calls the "root paradigm") for organization and relationship comes from a traditional, rural setting. Like the household, the company or corporate enterprise, provides the whole social existence of a person and has authority over all aspects of his life; he is deeply *emotionally involved* in the association (Nakane 1970:3-4). Relations within the household are thought of as more important than all other human relationships, including kin who have married out, as compared to servants or clerks who have become household members. Households maintain considerable independence. Siblings in separate households can be of different occupation, status, and wealth.

In the modern business firm, as in the household, there is "a personalized relation to a corporate group based on work, in which the major aspects of social and economic life are involved" (Nakane 1970:7). "A company is conceived as an *ie*, all of its employees qualifying as members of the household, with the employer at its head" (Nakane 1970:7-8). The place of work is like a village community in which the sphere of living is concentrated. Individual discussion of domestic concerns goes on in the company context. Marriage may be preferentially with another employee, and there is company housing, group vacations, and even a common grave in some firms (Rohlen 1974). The corporation is also characterized by a high degree of closeness among its workers and an internal law or code of conduct that is binding on them. The ideal of life-time employment reflects an obligation by the company to care for its workers and a commitment by the workers to devote their entire careers to one firm. Companies display an individuality and continuity as social groups in competition with one another like autonomous households or distinct villages.

The "household" in these respects is more than an analogy. The social units of home and firm have fundamental similarities based on the inclusiveness of the group and its functional salience. Both carry on a variety of economic and social activities, and they share a commitment to work, corporate unity, and mutual support. The achievement of concensus is a significant goal. The loyalty of members and long term adherence to the group is expected. In both, a strict division of labor is absent, but there is emotional bonding within and considerable competition with other groups. Household-type groups are evidently extremely effective in mobilizing large amounts of voluntary, directed effort toward solving practical problems. In technology application, such groups encourage individuals to gain and use necessary skills. They offer incentives, both social and psychological, for quality of performance and for teamwork. They also justify the sacrifice of individual immediate gain to the long term good of the group. Even within a large Japanese corporation, the functionally effective work group is small, having no more than one or two dozen members in direct contact with the leader. Though this is perhaps larger than even a multiple family farm household, the personalistic, face-to-face nature of the group makes for comparable social relationships.

But how is the potential contradiction between authority and individual achievement, superior competence, and innovation resolved? The traditional authority of the household head is like that of the parent (*oya*) over the child (*ko*). This is the basis of the *oyabun/kobun* landowner/tenant or master/disciple relationships. Subordinates were supposed to accept unequivocally the opinion of the head (Nakane 1970:13). But given the very real seniority system in rural Japan, there also remained the possibility of life-time climbing in rank, both within households and in the village arena of status and influence. In the normal life course, children would become parents, and junior couples would graduate to the senior position. Rankings were clear, as emphasized by the rigid seating plan of a traditional house, but they had little to do with merit or achievement. One might imagine that the power of a leader and the members' fear of disrupting the harmony and order of the group might interfere with the open expression of opinion in a vertically organized Japanese group.

The archetypal American success stories of inventors like Thomas Edison or industrial innovators like Eli Whitney or Henry Ford seem to deny authority or group control in favor of unbridled individualism and iconoclastic personal genius. Our culture heroes are gifted leaders who bucked established order, rose from humble beginnings, and amassed private wealth. Japanese leaders, on the other hand, seem to have risen more gradually on the basis of training and seniority. They are said not to be generally despotic, and the authority of their positions in the group is checked by emotional sympathy, paternalism, and the consensus system in which junior members of the group lay their opinions before the supervisor. Public deference is balanced with greater freedom in private, like that between husband and wife (Nakane 1970:68). Subordinate status is therefore not intolerable. The main goal of leadership is to maintain happy and productive relationships among group members whose contributions and responsible participation must be recognized. As in the farm family household, clearly defined, ranked positions do not necessarily reflect in practice a rigid structure of command and obedience or dominance and subordination.

Even in traditional Japanese society, socio-economic strata did not decisively separate the interests of individuals. The feudal lord and his retainers formed a single household economy, and when there was a flood or poor harvest, the leader and followers shared the hard times (Nakane 1970:70). Within the rural community, there were no gentry or permanent status groups, and even the poorer landless segments of the population were not permanently confined to a lower class. Contemporary large-scale groups like unions or professional associations are characteristically divided into smaller, more personal groups or factions called *mura* (village communities). The name reflects the desired close-knit ties within an enduring social and economic entity. Such personal groups within large associations are equivalent and potentially opposed on some issues.

Both the literature and the testimony of our symposium participants suggest that a Japanese leader, regardless of his personal brilliance and achievements, does not generally administer a temporarily constituted research group in pursuit of his own unique vision as in the West. He remains subject to the social and emotional needs of the group members. Personal relations are the group's driving force, as in a classic Gemeinschaft rather than Gesellschaft entity. Nineteenth century social scientists agreed that the Gemeinschaft of the peasant household and community had been lost in Europe by the end of the 19th century. Wouldn't Tönnies and his evolutionist cohorts be amazed to find the distinctive structure of the rural Gemeinschaft alive and well in the organizational structure of modern Japanese high-tech society! In a culture of overt, formal hierarchy and delicately graded rankings, the groups of scientists that conduct innovative research show elements of an organizational pattern based on the traditional smallholder farm household and emphasizing "basic equality and communal rights" (Nakane 1970:143).

References
Bray, Francesca
    1986   The Rice Economies. Cambridge University Press.
Clark, Colin, and M. R. Haswell
    1967   The Economies of Subsistence Agriculture. London: Macmillan.

Hayami, Yujiro, and Vernon Ruttan
    1971   Agricultural Development. Baltimore: Johns Hopkins University Press.

McKean, M.
    1982   The Japanese Experience with Scarcity: Management of Traditional Common Lands. *Environmental Review* 6, No. 2:63-88.

Nakane, Chie
    1970   Japanese Society. Berkeley: University of California Press.

Netting, R. McC.
    1989   Smallholders, Householders, Freeholders: Why the Family Farm Works Well Worldwide. *In* The Household Economy: The Domestic Mode of Production Reconsidered. Richard R. Wilk, ed. Boulder: Westview, pp. 221-244.
    1990   Population Permanent Agriculture and Politics: Unpacking the Evolutionary Portmanteau. *In* The Evolution of Political Systems. Steadman Upham, ed. Cambridge: Cambridge University Press, pp. 21-71.

Rohlen, Thomas D.
    1974   For Harmony and Strength: Japanese White-Collar Organization in Anthropological Perspective. Berkeley: University of California Press.

Ruttan, V. W.
    1984   Induced Innovations and Agricultural Development. *In* Agricultural Sustainability in a Changing World Order. G. K. Douglas, ed. Boulder: Westview.

Stone, G. D., R. M. Netting, and M. P. Stone
    1990   Seasonality, Labor Scheduling, and Agricultural Intensification in the Nigerian Savanna. *American Anthropologist* 92:7-23.

Smith, Thomas C.
    1959   The Agrarian Origins of Modern Japan. Stanford: Stanford University Press.

JAPANESE CULTURE AND INNOVATION

Shin-Pei Matsuda, Hitachi Research Laboratory, Hitachi, Ibaraki, Japan

When examining relationships between Japanese culture and innovation, it is helpful to look both at the origin of Japanese culture and also at the employment system in Japanese corporations.

The Origin Of Japanese Culture

The origin of Japanese culture may be described in relation to four main subheadings: genetic heritage, surroundings, religion, and climate. Japan's genetic heritage involves its prehistoric racial experience and tradition dating back to 20,000 years B.C. Japan's surroundings include the location of the territory and the influence of its neighboring countries. The discussion of Japan's religions concerns Shinto (Japan's native religion) and Buddhism (a foreign religion brought in around 500 A.D.). Finally, climate is important in its relation to sources of agricultural production and life style.

Fundamental to the upcoming discussion is the fact that the culture of any nation depends strongly on the religion in which the people believe. Japanese culture is based heavily on Buddhism, and the fundamental Buddhist philosophy influencing Japanese behavior is the "sharing of responsibility".

Genetic Heritage

The origins of the Japanese people arise from three main sources which will be referred to here as the Native Japanese, the New Japanese, and the Immigrants and Refugees. Native Japanese populations allude to hunting and collecting groups (called Jomon-Jin) which inhabited Japan between 4000-200 B.C. and numbered approximately 0.2-0.5 million. The second source of people were the New Japanese (Wa-Jin, Yayoi-Jin). These groups came over as immigrants or invaders from mainland Asia over the course of seven centuries between 200 B.C. and 500 A.D. They numbered between 2 and 5 million and initiated such important activities as rice farming and iron production. These newcomers conquered the native population and over time mixed completely with them. The third source included immigrants and refugees from Korea who arrived between 500 and 700 A.D. Historical events of import at this time included the collapse of the Three Dynasty Age in Korea, the unification of the Korean peninsula by the Shiragi dynasty (circa 660 A.D.), and the unification of the Japanese Islands by the Emperor's family. Nippon in the seventh century was like the United States when it was still considered the "New World". Japanese culture has been influenced by the fact that the Japanese Islands were the dead-end of the Far East and have been the final homeland for refugees from Korea, China and Manchuria. Two other points relative to the origin of Japanese culture are of note. Buddhism was introduced to Japan through the Korean kingdom (Kudara-Kingdom) around 500 A.D., and up until 600 A.D., the Japanese still shared a common language with Korea.

Surroundings of Japan

Japan's surroundings have influenced its culture primarily as a result of its proximity to China, Japan's superpower neighbor. However, any influence from China or Korea was tempered by the body of water which separated the Japanese Islands from the mainland.

The general policy of the Japanese ruling power from 600 A.D. to 1600 A.D. was two-fold: 1) to keep up with advanced Chinese culture and technology, and 2) to defend its territory and tradition. The Japanese considered China to be both a father and teacher of culture. During this period, many aspects of Chinese culture were introduced in Japan and then modified and refined by the Japanese. Thousands of students were sent to China to learn and import elements of Chinese culture. They brought back with them knowledge of many cultural traits such as religion, political systems, education, characters (i.e. language), poetry, paintings, and chess. Korea was considered to be an elder brother of Japan. Much of the Chinese influence on Japan was received indirectly through Korea (e.g. Buddhism).

Religion

Religion in Japan also emerges from varied roots. Originally, there were the traditional religions of the hunting and collecting peoples. Later, there was Shinto, the religion of the New Japanese (the rice farming peoples). Imbedded in Shinto was the history of the Emperor's family and the worship of nature. Still later, and as mentioned above, there was Buddhism, a religion of foreign origin, but one which was considered to be new, fresh, sophisticated, and universal.

Buddhism was readily accepted by the Japanese people and went on to become Japan's largest religion. At present, approximately 80% of the Japanese population consider themselves Buddhists. The Buddhism adopted by the Japanese was based on several major tenets. The first is that everything in the universe has life. From this perspective, human beings are considered to be a part of nature. Furthermore, every living thing follows a path of cyclical existence, and all are interdependent. Finally, there is a natural law of cause and effect. In this sense, a man's life descends from, and is influenced by, all of his ancestors' doings.

Out of Buddhism also came a fundamental philosophy of group action. An individual is not only a integral part of his own environment but of the environments of all others as well. From this emerges the essential philosophy of shared responsibility. All group members are responsible for all of the group's actions. Certain behavior patterns arise as a result. Among these is a permissiveness toward other's failures, a stress on group harmony, a preference for shared common feelings, and ultimately, a strong loyalty to the group (e.g. loyalty to one's company).

Climate of Japan

The final subheading of import to the origin of Japanese culture is Japan's environment. This provides context for the development of culture. The climate of Japan is moderately warm with lots of rain, and Japan is located far enough north that it experiences four seasons. Food production arises primarily from rice farming, fish, and other sea foods.

The Employment System in Japanese Corporations

The company has multiple meanings in Japanese culture. It serves as a place to make money, a place to show one's abilities, and a place to participate in human activities. The meaning will ultimately depend on the stage of development of the standard of living.

The employment system in the Japanese corporation is based primarily on lifetime employment. In this system, salaries increase with the length of employment as do retirement pay and promotions. The company is "owned" by all employees, both management and labor. Managing officers are evaluated by the corporate society, not by the stockholders. Layoffs are accepted only if the company goes bankrupt. In short, the Japanese employment system of permanent employment insures the stability of life.

The evaluation of achievement and ability is performed by an employee's senior officer. A worker's reputation outside the company (i.e. among his peers) is secondary. Hence, it is the vertical relationship that exists between workers in a group which is of primary importance.

Embedded in this lifetime employment system are inherent disadvantages to changing places of employment. When an employee leaves one company for another, he loses the human connections and relations previously established in the first company. It takes time to become accepted as a member of a new group and to adapt oneself to a new company's culture. Also lost are one's reputation and achievements accumulated over time in the former company. Finally, one must be willing to accept negative impacts on salary, retirement pay, and rank.

EMPLOYMENT SYSTEM IN JAPANESE CORPORATION

Company means :

① Place to make money

② Place to show one's ability

③ Place to participate in human activities

The meaning depends on the development stage of the standard of living.

EMPLOYMENT SYSTEM   (Principally Life Employment)

- Insure the stability of life

- Salary increases with the period of employment.
    (Retirement Pay, Promotion)
- Company is owned by all employees, manager or laborer.

- Managing officers are evaluated by the society, not by the stock holders.
    (Layoff is accepted only if the company goes bankrapt.)

DISADVANTAGE OF CHANGING COMPANY

- Lose human connection and relation

- Takes time to become a member of a new group
    (company's culture)
- Lose reputation and achievements in former company

- Salary, Retirement Pay, Position etc

EVALUATION OF ACHIEVEMENT AND ABILITY

- Achievements are evaluated by his senior officer.

- Reputation outside corporation is secondary.
    (among peers)
- Promotion depends on his senior officer.

- Vertical relationship is more important.

JAPANESE CULTURE & INNOVATION

ORIGIN OF JAPANESE CULTURE

① GENETIC HERITAGE	----	Prehistoric Racial Experience Tradition (Language) 20,000B.C.~
② SURROUNDINGS	----	Location of Territory Influence from Neighboring Countries
③ RELIGION	----	Shinto (Native religion) Buddism (Foreign religion, 500A.D.~)
④ CLIMATE	----	Source of Production (Agriculture) Life Style

Culture of any nation depends strongly on the religion which peoples believe. Japanese culture is based on Buddism. Fundamental philosophy derived from Buddism on Japanese behavior is **"SHARE OF RESPONSIBILITY"**.

(1) GENETIC HERITAGE (ORIGIN OF JAPANESE PEOPLE)

① Native Japanese (Jomon-Jin)	4000 BC ~ 200 BC
Hunting & Collecting	Population 0.2~0.5 Million
② New Japanese (Wa-Jin, Yayoi-Jin)	Immigration or Invasion
	200 BC ~ 500 AD
Rice Farming, Iron Production	Population 2 ~ 5 Million
Conquer and mixing of two peoples.	
③ Immigrants and Refugees from Korea	500 AD ~ 700 AD
Collapse of Three Dynasty Age in Korea.	
Unification of Korean Peninsula by Shiragi-Dynasty ~660AD	
Unification of Japan Islands by Emperor' Family.	

○ Nippon in 7th century was like a USA (New World).

● Japanese could communicate with Korean without an interpreter before 600AD.

● Buddism was introduced to Japan through Korean Kingdom (Kudara-Kingdom) (~500AD).

● Japanese islands were the dead-end of Far East and final homeland for refugees from Korea, China and Manchuria. (Coalition Government)

(2) SURROUNDINGS OF JAPAN

① Located near China (Super Power)

② Islands are separated by sea from China and Korea.

③ Far East (Dead-end of Asia)

Policy of Japanese Ruling Power
(600 AD ~ 1600 AD)

● Keep up with Advanced Chinese Culture & Technology

● Defend Territory and Tradition

Thousands of students were sent to China to learn and import Chinese cultures. ---- Political System, Education, Poem, Characters, Paintings, Chess, Religion etc.

Chinese cultures were introduced and modified and refined in Japan.

○ China is a father and teacher of culture.

○ Korea is an elder brother of culture.
(Korea gets direct influence from China.
Japan gets indirect influence.)

(3) RELIGION

① Religion of Native Japanese (Hunting & Collecting Peoples)

② Religion of New Japanese (Rice Farming Peoples)
　Shinto--- History of Emperor's Family
　　　　　　Worship of Nature
③ Foreign Religion = Buddism
　　　　(New, Fresh, Universal, Sophisticated etc)

Buddism was accepted by Japanese and became the largest religion (80% at present).

Buddism

① Everything in universe has life.
　　Human being is a part of Nature.
② All Livings cycles.
　　A Living depends on other Livings.
③ The Law of Cause and Effect
　　A man's life is descended from all ancestors' doing.

```
┌─────────── Fundamental Philosophy of Group Action ───────────┐
│ ● A man is an environment of others.                          │
│   A man is an environment of himself.                         │
│             SHARE OF RESPONSIBILITY                           │
│ ● A member of a group is partly responsible for everything    │
│   done by any member.                                         │
└───────────────────────────────────────────────────────────────┘
```

┌─────────── BEHAVIOUR ───────────┐
│ ○ Permissive to other's failure │
│ ○ Harmony in group   │
│ ○ Prefer to have common feeling │
│ ○ Royalty to company   │
└─────────────────────────────────┘

(4) CLIMATE OF JAPAN

  ◎ Moderately warm, Lots of rain

  ◎ Rice Farming, Fishes and Foods from Sea

  ◎ Four Seasons
       × Sound of worm   ○ Voice of worm

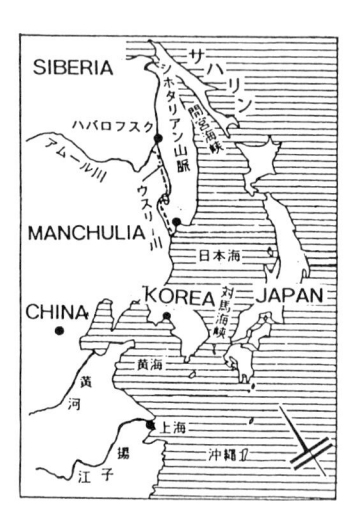

HIGH TEMPERATURE OXIDE SUPERCONDUCTORS

High temperature oxide superconductors are a new class of materials discovered in 1986 They are the subject of intense international research efforts and included here for discussion because they represent the earliest stage of innovation, that is research and discovery. We are at the starting point of an advanced materials development about which much optimism exists for important commercial developments, but as of now, there is no significant commercial production of these materials. The level of anticipation within the scientific community was made evident in March 1987 when a high temperature superconductivity symposium at the American Physical Society meeting was attended by more than a thousand physicists who extended the discussions long into the night in what has sometimes been called the Woodstock of Solid State Physics. In July 1987, more than a thousand people convened at a federal conference on superconductivity at which President Reagan presented an 11-point agenda to promote cooperative research, to transfer scientific developments more rapidly into commercial products and to protect the intellectual property rights of scientists involved in superconductivity research. In 1987, hundreds of research groups worldwide began a search for new materials and there was tremendous excitement. Resources, prestige and public interest all combined to generate a vigorous and competitive search for new discoveries. Newspaper and television reports have imagined breakthroughs in cheaper and more effective power transmission, magnetic levitation, ultra fast computers and medical imaging. We are led to believe that a new era of technology is approaching.

In normal metal conductors, the flow of electricity occurs by the motion of electrons within the metal. There are frequent collisions with the lattice such that electrons are scattered and require a constant source of energy to maintain a flow of current in the face of this resistance. In April 1911, Heike Kammerlingh Onnes was studying the resistance of mercury at liquid helium temperatures and was surprised to discover that its electrical resistance vanished at a temperature of about 4.2° kelvin. Within a year, he had found that tin became a superconductor at 3.7°K and lead became superconducting at temperatures below 7.2°K. These temperatures are referred to as the critical temperature for superconductivity. Onnes and his research group also discovered that the presence of a magnetic field decreases the transition temperature. Electrical currents generate magnetic fields so that an increased electrical current also lowers the transition temperature. It turns out that there is a maximum critical current for superconductivity and a maximum critical magnetic field. These define a critical temperature—magnetic field—electrical current surface within which a material is superconducting as shown in Fig. 1. The critical current and critical field for mercury, tin and lead are too small for any practical applications.

Onnes received a Nobel prize for his discovery of superconductivity and it remained a phenomenon of interest to solid state physics that seemed to have no commercial potential. After World War II, interest increased with the discovery of more than a thousand new superconducting materials, most of which were inter-metallic compounds or alloys rather than pure elements. Some of these had higher transition temperatures and many were found to have high critical field and critical current values. In the 1960's inter-metallic compounds such as niobium tin and niobium gallium were discovered. By 1973 a transition temperature of 23.2°K was found for niobium germanium.

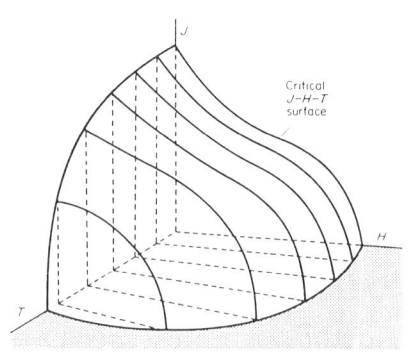

Fig. 1. Critical surface of a superductor in temperature (T)—magnetic field (H)—current density (J) space. for points below the $J_c(H,T)$ surface, the material is superconducting

Development of practical applications of high fuel superconductors required more than a decade of metallurgical research and development in order to obtain practical materials. At present, superconducting niobium titanium alloys and niobium tin are available as multifilament conductors in a number of configurations. One application is for nuclear magnetic resonance medical imaging for which the high cost of liquid helium refrigeration can be accepted. This technology requires a strong and very uniform magnetic field with a solenoid surrounding the patient. In the magnetic field which is established, atomic nuclei act like tiny magnets and respond to radio frequencies which allow the mapping and imaging of the body's chemical composition. Another important application of the development of very high magnetic fields has been superconducting particle detector magnets and beam guiding magnets for high energy physics research. Because of the high magnetic flux densities which can be achieved with superconducting magnets, smaller magnets are possible with very large savings in the electrical energy required. The superconducting super collider project scheduled for construction in Texas would be impossible without the use of high power superconducting magnets. These magnets will account for about one billion dollars of the total ten billion dollar cost of the SSC.

A number of other potentially exciting applications for superconducting magnets have been proposed. These include magnetic separation and filtration devices which might be applied to mineral purification, water purification and desulfurization of coal. There has been continued interest in magnetically levitated railway systems. In Japan the Japanese National Railway has constructed a test-track 7 kilometers long on which test vehicles have been successfully operated. Superconducting power station turbogenerators have been proposed which would take advantage of the high magnetic flux density which can be achieved with superconductors. High power transmission systems using superconductors have been proposed as well as magnetic systems for fusion reactors and for energy storage systems which would efficiently provide a way of load balancing for electrical generating systems. None of these applications has proved practical, in part because of the high cost of helium refrigeration required by the low critical temperature of available materials.

There are also a number of potential applications for superconductors in the form of thin films for electronic devices. During the 1960's the Josephson effect was discovered in which a superconductor-insulator-superconductor configuration could be used for ultra fast low power switching. This made conceivable a new class of high performance computers. In addition to active Josephson junction elements, a complete technology would require a host of passive circuit elements including insulators, capacitors, resistors and so forth. As a result, this development is a very difficult technological challenge. Efforts in this direction in the United States were pretty much abandoned in the face of rapidly improving competitive semiconductor technology; research continued in Japan. A variety of other applications such as delay-line signal processors, microwave circuit filters, high frequency antennas, high efficiency wave guides and microwave resonators seem more likely applications.

The large majority of oxide compounds are insulators but substantial group are semiconductors and a somewhat smaller group have long been known to be metallic conductors. Understanding the crystal structures responsible for this kind of conductivity has occupied a number of researchers and it seems reasonable that the search for new superconducting materials should embrace these compounds. Beginning in 1964, with the discovery that titanium oxide and the niobium oxide are superconductors a number of superconducting oxide materials have been discovered as illustrated in Table 1. None had very high critical temper-

Table I. Superconducting oxides known prior to 1986.

Compound	T_c	Discovery
TiO, NbO	~1 K	'64 - J.K. Hulm et al.
$SrTiO_{3-x}$	~0.7 K	'64 - J.J. Schooley et al.
A_xWO_3	~7 K	'64 - Ch. J. Raub et al.
A_xTO_3 (T = Mo, Re)	~4 K	'69 - A.W. Sleight et al.
Ag_7O_8X	~1 K	'66 - M.B. Robin et al.
$Li_{1+x}Ti_{2-x}O_4$	~14 K	'73 - D.C. Johnston et al.
$Ba(Pb_{1-x}Bi_x)_3$	~14 K	'75 - A.W. Sleight et al.

From the 1989 JTEC Panel Report on High Temperature Superconductivity in Japan, M.S. Dresselhaus, et al. National Technical Information Services, U.S. Dept. of Commerce.

tures but barium bismuth lead oxide had a low carrier density and consequently was of considerable research interest. Barium lanthanum copper oxide was known to be a metallic conductor and in early 1986, Johannas G. Bednorz and Karl A. Müller discovered that it was a superconducting oxide with a transition temperature of about 35°K, almost twice the value that had seemed to be an asymptotic limit for superconductivity. Their discovery was published in *Zeitschrift für Physic* in September 1986. The publication itself and its dissemination did not excite an enormous amount of interest but at the University of Tokyo a superconductivity research effort under the direction of S. Tanaka included studies of oxide superconductors led by Koichi Kitazawa. In that laboratory the superconducting phase was identified as having the K_2NiF_4 structure and a substantial Meissner signal confirmed that the superconducting state had been reached. Kitazawa attended a meeting of the Material Research Society at Boston in December and in a hastily scheduled presentation publicly confirmed the Bednorz-Müller discovery.

Kitazawa's confirmation of the superconductivity in $Ba_xLa_{2-x}CuO_4$ and determination of the structure transformed the research accomplishment of Bednorz and Müller into a research innovation widely recognized as having a short time potential for scientific discovery and a longer term potential for technological applications. The potential for scientific recognition was realized when Müller and Bednorz received the 1989 Nobel prize in physics. There was a race on to discover analogous compositions to the barium lanthanum copper oxide. Kitazawa had informed Paul C. W. Chu of the University of Houston about the composition and structure of this material. Chu telephoned a former student, Mau Khuen Wu, at the University of Alabama who began an empirical search for analogous compounds. Chu held a press conference on February 16 announcing a new material which was a superconductor having a critical temperature above 90°K. On February 21, S. Hikami, at the University of Tokyo announced the discovery of a similar material. On February 25, Professor Z.X. Zhao of the Peking Institute of Physics described that yttrium barium copper oxide was a liquid nitrogen temperature superconductor and the *People's Daily of China* reported it on February 27. By March 2, when Chu had announced the composition of his material, it was clear that there had been the simultaneous discovery of this liquid nitrogen temperature superconductor by independent groups in China, Japan and the United States. These discoveries turned up the heat for many research programs and the public interest because they evoked the anticipation that superconducting devices could be designed which would operate at liquid nitrogen temperatures which would substantially decrease the costs of refrigeration. In addition the rapid increase from a critical temperature of 23°K to 35°K to 95°K led to the anticipation that still higher critical temperatures would be reached, perhaps even a superconducting material at room temperature.

H. Maeda of the National Research Institute for Metals (NIRM) at Tsukuba, Japan, found that the compound $Bi_2Sr_2Ca_{n-1}Cu_nO_{2n+4}$ was a high T_c superconductor ($T_c \sim 110°K$). Soon thereafter, Z.Z. Sheng and A.M. Herman in the United States and Kondah in Japan found superconductivity in the system $TlBa_2Ca_{n-1}Cu_nO_{2n+3}$ ($T_c \sim 125°K$). A number of other cuprate superconductors have been discovered. In analogy with $Ba(Bi,Pb)O_3$, R.J. Cava et al found $Ba_{1-x}K_xBiO_3$ with a critical temperature of about 40°K. Most oxide superconductors are electron deficient and, prior to 1989, theories were being developed on the assumption that superconductivity in these materials resulted from holes moving through CuO_2 planes in the crystal structures. Then in 1989, Y. Tokura, H. Takagi and S. Uchida in Japan found $Nd_{2-x}Ce_xCuO_{4-y}$ to be an electron doped conductor, creating chaos with emerging theories; Tokura's group and other groups in the United States have found other related compositions. Table II shows the host of new superconducting oxides that have been discovered subsequent to 1986.

Table II. Superconducting oxides discovered after 1986.

Compound	T_c	Discovery
$(La_{2-x}M_x)CuO_4$;		
M = Ba	~ 30 K	'86 J. G. Bednorz & K. A. Müller
M = Sr	~ 40 K	'86 K. Kishio et al
	~ 40 K	'87 R. J. Cava et al.
M = Ca	~ 20 K	'87 K. Kishio et al.
$YBa_2Cu_3O_7$	~ 95 K	'87 M. K. Wu et al.
$LnBa_2Cu_3O_7$	~ 95 K	'87 Various laboratories
$(La_{2-x}Na_x)CuO_4$	~ 20 K	'87 J. T. Markert et al.
$Bi_2Sr_2CuO_6$	~ 22 K	'87 C. Michel et al.
$Bi_2Sr_2Ca_{n-1}Cu_nO_{2n+4}$	~ 110 K	'88 H. Maeda et al.
$Tl_2Ba_2Ca_{n-1}Cu_nO_{2n+4}$	~ 125 K	'88 Z. Z. Sheng & A. M. Herman
$(Ba_{1-x}K_x)BiO_3$	~ 30 K	'88 R. J. Cava et al.
$Nd_{2-x+y}Ce_xSr_yCuO_4$	~ 20 K	'89 J. Akimitsu et al.
$RBa_2Cu_4O_8$	~ 80 K	'88 D. E. Morris et al.
$Pb_2Sr_2(Ca,R)Cu_3O_{8+y}$	~ 77 K	'88 R. J. Cava et al.
$(Ln_{2-x}Ce_x)CuO_4$	~ 25 K	'89 Y. Tokura et al.
$(Ln_{2-x}Th_x)CuO_4$	~ 20 K	'89 J. T. Markert & M. B. Maple
$Ln_2CuO_{4-x-y}F_x$	~ 25 K	'89 A. C. W. P. James et al.
$Nd_{2-x+y}Ce_xBa_yCu_3O_{10-z}$	~ 30 K	'89 H. Sawa et al.

Every scientist who has conducted and published research knows very well that the rationally organized published presentation has little to do with the halting missteps and confused understandings that actually occurred in the laboratory and mind of the researcher as the work progressed. Rustum Roy, an outstanding material scientist and a leader in the Science-Technology-Society Movement has provided the following interpretation of the history of the search for high T_c superconducting oxides:

THE PAST AND FUTURE IN HIGH T_C SUPERCONDUCTORS
THE METAPHOR OF PROGRESS

Rustum Roy, The Pennsylvania State University

Introduction
The viewpoint expressed below has two major biases. First I was very well informed on the subject as head of the only university group in the U.S. that had published on (high T_c perovskite) oxide superconductors before 1986. Second, I am not involved in the present high T_c research. However, because we were so distressed by the gross inaccuracies appearing in the scientific and public press, Professor A.S. Bhalla and I have created an archival record of the *publicly available* information on the high T_c discovery. In addition, under the auspices of the Materials Research Society, I created a "video history" taped record of interviews with the eight major research groups involved in the discovery of the 90K material. This video history forced *Science* magazine to, finally, present a radically revised history of the discovery (Ref. 1) showing unequivocally that J.W. Ashburn and his professor, M.K. Wu, at the University of Alabama made the first YBaCu materials in the U.S. although, there is no doubt whatsoever that the first public announcement of the YBaCu was by Zhao in Peking on February 19, 1987—a fact known to and consistently ignored by all scientists in the U.S. and Japan. In the following figure I have summarized our analysis of this history (Ref. 2).

On the basis of this detailed familiarity with the early record, the following conclusions can drawn regarding the *cultural* and *social* influences on attempting to shape the history. This can perhaps be most helpful for analysis of cultural factors if the facts are presented in tabular form, with an analytical opinion on the right.

HISTORY OF HIGH T_C SUPERCONDUCTORS

The Record	Cultural/Social Factors
1. Bednorz and Müller did no novel crystal chemistry at all. The rare earth cuprates had been thoroughly studied as ceramic metals by four of the best synthetic groups in the world: Raveau and Hagenmuller in France; C.N.R. Rao in India; B. Lazarev in Moscow. B&M *reasoned* that such compounds might be superconducting.	Bednorz and Müller did read and know the literature and attempted to work with Raveau. It was a success of their reasoning, measurement and interpretation, not of making a new material.

	Explicitly on La(BaSr)CuO₄ *Metallic*			*Spinel + Ba(BiPb)O₃* *Superconducting*

Precursor Work (On Perovskite)

France	USSR	India	Japan	USA
Careful crystal chemistry			National team effort at 5 universities + some industries NTT Suzuki	SC work winding down Matthias, Sleight, Roy
Hagenmuller Raveau	Lazarev	RAO		

↓
Switzerland

Discovery of ~36K Material (1986)

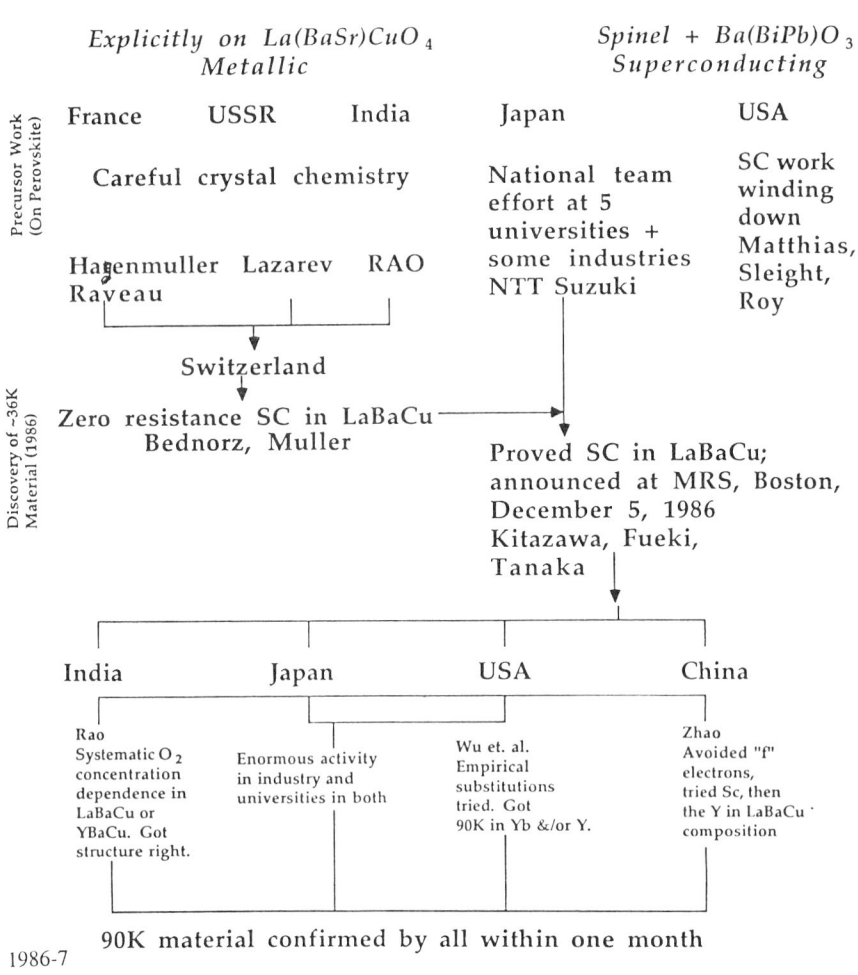

Zero resistance SC in LaBaCu ──→
Bednorz, Muller

Proved SC in LaBaCu; announced at MRS, Boston, December 5, 1986
Kitazawa, Fueki, Tanaka

India	Japan	USA	China
Rao Systematic O₂ concentration dependence in LaBaCu or YBaCu. Got structure right.	Enormous activity in industry and universities in both	Wu et. al. Empirical substitutions tried. Got 90K in Yb &/or Y.	Zhao Avoided "f" electrons, tried Sc, then the Y in LaBaCu composition

1986-7 90K material confirmed by all within one month

Worldwide: 500 person-years of research on new materials

1987-8   CaBiSrCu₂   107K   Maeda (NRIM)
1988   CaTlBaCu   125K   Sheng & Hermann (Ark.)

1500 person-years yielded only these materials from closely related families.

Fig. 1. A short history of oxide superconductor materials

2. Bednorz and Müller report (sent to IBM HQ) and papers elicited very, very little interest on the part of anyone (IBM V.P.; IBM patent section; general science or even superconductor community). Müller in his *Science* paper explicitly says he did not expect any great attention. And in fact taking T_c from 23 → 36 was hardly earth shaking.

The lack of interest correctly represented the community's viewpoint—"no big deal," indeed a hardly noticeable advance.

3. The Tokyo group was fully geared up to look for new materials. However, even there it was Kitazawa's *student* who first broke ranks to confirm the effect in LaBaCu.

This proves that maintaining long-term centers of excellence in key fields allows one to jump in quickly. Senior researchers in U.S. and Japan did not think much of the work.

4. Kitazawa's *late* paper at M.R.S. confirming the Meissner effect is historically the key event in triggering the other experienced groups, *which were already funded* to get into the game (Bell, IBM, Houston, Peking, Bangalore).

Personal presentation and excitement conveyed by Kitazawa was the "starting gun." I'm not sure why. Again a few long-term funded groups were fast off the mark.

5. There is absolutely no argument that the **scientific credit for public disclosure** of the "90K" materials goes to Zhao Zhang in Peking. The announcement appeared in *People's Daily* on February 25, 1987 and in the *N.Y. Times* two days later, giving the KEY DISCLOSURE that the composition lay in the system Y-Ba-Cu-O.

China is not a mainstream science country and hence rarely gets credit. Very few give credit to Zhao, although it is certain that IBM, Bell, etc., used his results for their starting point.

6. In the meantime verbal accounts by the participants agree that J.W. Ashburn, a graduate student in physics at the University of Alabama, with no training in or understanding of structural chemistry, was "mixing up a variety of batches" and, it is claimed, that on January 31 noticed that a "mix" of Y-Ba-Cu gave a high T_c discontinuity near 90K and reported this to his professor, M.K. Wu. No conceptual guidance for selecting the ions or stoichiometry was evidenced. Indeed the Alabama and the Houston group which conducted their Meissner measurement did not know till well into March whether it was the "green phase or black phase" which was superconducting. Nor could they use a simple phase diagram to estimate the composition.

This is a striking example that even in 1987 the most simple-minded empiricism gets the result fastest (not best).

7. There is no evidence at all in the published literature that C.W. Chu (then located in Washington, DC) or the University of Houston had before or after (till months after January 31) ever done any synthesis of materials. Yet on the basis of claims to reporters, the public and magazines such as *Science* continued to credit C.W. Chu for discovering "1:2:3," even after repeated challenges to produce evidence.

Culturally, this story illustrates the value of bypassing the scientific literature which nobody reads anyway and going to the press. C.W. Chu, who *on the record,* had nothing *whatsoever* to do with the *synthesis* held at least two *press conferences* to make claims for which *no* evidence was available and which drew attention to himself, as though he had made the material.

8. The empirical fact is that within a few days to 3 weeks of the *N.Y. Times* report of Zhao's YBaCu discovery several groups at least including Bell Labs, Bellcore, IBM Yorktown, IBM Almaden and Bangalore had obtained the composition and structure of the $YBa_2Cu_3O_{7\pm}$ phase.

1:2:3 was a success waiting to happen once the cuprates were looked at. Very poor empirical cut and try work succeeded first; good crystal chemistry and analysis did the cleanup work.

9. The enormous "hype" in both the lay press (and in the *United States* scientific press) make for a policy nightmare— forcing funding decisions by Congress and agencies long before sober analysis of the opportunity. By and large the *science press* is at least as much to blame for the excesses. The reporting of any claim by any laboratory became a weekly affair, with the editors serving as cheerleaders.

The lay press can easily be manipulated with exaggerations. The cultural issue is who will monitor and control the excesses of scientists in making unjustified claims. The behavior of the science press is much more serious cause for concern (see below).

World Cultural Bias Today: Science as Metaphor of Progress

The history of high T_c superconductors is one, but perhaps the defining, example that Western culture which has flirted for a long time with the "Idea of Progress," is increasingly committed to believing that progress comes mainly or exclusively from science and technology. The science establishment, although aware of the dangers of such exaggeration has fully and shamelessly participated in inserting science and technology as the veritable metaphors of hope in our society. This ranges from utterly ludicrous statements from the world's most distinguished scientists that this or that experiment into this year's fashionable hadron, lepton or boson "will yield the secrets of the universe," or help us "know the mind of God," and similar religious verbiage. Carl Sagan, explicitly arguing against traditional religions, intones in appropriately sepulchral voice about the wonder of "billions upon billions" of stars, and the hope in finding extraterrestrial intelligence.

Faced with the obvious negative correlation between the success of U.S. science and the decline of technology, the science establishment rushes to pronounce the wonders of the latest discovery however trivial, unconfirmed, or of dubious significance it may be. *Science* magazine's treatment of superconductors, and *Nature*'s dealing with "quasicrystals" are excellent examples. Indeed once such editors become committed to the advocacy role, becoming salespersons for "more money for science," it becomes nearly impossible to go back to honest journalism, critical evaluation of claims and admitting mistakes.

The key cultural issue illuminated by the high T_c story is how society, and the community of science, will enforce some standards on the reporting and exaggeration of scientific advances (breakthroughs!). The R/D policy and funding management is grossly distorted by the laissez faire system now in existence. High T_c superconductors constitute physical science's latest exemplar to bolster the metaphor of "progress through science and technology." The evidence shows that the history has been strongly distorted by manipulation of the public media while ignoring the scientific record.

References
1. Report by R. Pool, *Science 241,* 655-657 (1988).
2. R. Roy, "HTSC, Restoring Scientific and Policy Perspective." *Proceedings of the World Congress on Superconductivity,* C.G. Burnham and R.D. Kane, eds. World Scientific, Singapore, 1988, pp. 27-42.

A major figure in the transformation of Bednorz' and Müller's research accomplishment into a research *innovation* was Professor Koichi Kitazawa, who publicly confirmed the discovery and described the composition and structure of the superconducting phase. He continues to be a leading figure in high T_c superconducting research. He writes:

THE FIRST FIVE YEARS OF THE HIGH TEMPERATURE SUPERCONDUCTIVITY: CULTURAL DIFFERENCES BETWEEN THE U.S. AND JAPAN

Koichi Kitazawa, University of Tokyo Hongo, Tokyo, 113 Japan

Although the history of superconductivity dates back to 1911, the practical usage of it began only in the last two decades. The scale of the commercial application is still minimal in comparison with its counterpart the semiconductor. The limited usage of superconductivity has been attributed to its extremely low critical temperature which requires liquid helium as a coolant. Suddenly, in 1986 the critical temperature started rising so rapidly that scientists lost their confidence in the basic theoretical scheme of solid state physics to describe superconductivity. Industries were trapped in a fear that they might be left behind in its application unless they started immediate R&D efforts. The early chaotic situation has calmed down as it has become clear that long and hard efforts will be needed before the new superconducting materials can be put into practical utilization.

In the meantime, there have been various research projects proposed worldwide. In this paper some personal views are presented about the development of high temperature superconductivity and about the relative efforts put into this field by the U.S. and Japan.

Years before 1986
Until 1983, there were many researchers in the U.S. in the field of superconductivity. As it was gradually recognized that no large scale application of superconductivity would

be expected soon, they left the field one after another. Especially the decision of IBM in 1983 to terminate the large scale project of developing the Josephson computer was a symbolic event that cast dark clouds over the superconductivity society. Much smaller scale efforts survived in academia as well as in industry. Superconducting wires have been fabricated by a few small scale companies.

On the other hand, since about 1960 in Japan there has been a medium but constant number of researchers in this field supported rather constantly by successive national projects, such as MHD power generation, nuclear fusion reactors, superconducting power generators, computers for large scale scientific computation, the fifth generation computer, magnetic levitation trains, etc. Superconducting wires have been fabricated by many large-size companies such as Toshiba, Hitachi, Mitsubishi Electric, Furukawa, Sumitomo Electric, Fujikura Wire, Kobe Steel, etc. These research groups have been allowed to survive as a tiny group in each company without producing profits but as the nucleus for future possible developments. Technological competition developed among them in producing high performance superconducting wires. By 1986, they could fabricate a wide variety of wires for differing purposes but on a rather small scale. The accumulation of technical knowledge, however, has become large in each company.

The Ministry of Education ran a three year special project in 1983 "New superconducting Materials" organizing about 150 university researchers, encouraging them to go into nonconventional materials including oxides and sulfides in addition to the conventional metals and to elaborate on the superconducting mechanisms. Owing to this project, various institutes such as the University of Tokyo could get equipped for superconductivity research (e.g., SQUID susceptometer) and could establish the methodology to prove, identify and characterize superconducting materials. This is recognized to have helped them a lot in the earliest stage of the high temperature superconductor research since 1986.

Discovery of High Temperature Superconductors

In spite of many previous efforts, the highest critical temperature had remained at 23K from 1973 until 1986. Bednorz and Müller of IBM Zurich started a search for new superconducting materials in 1983, being interested in the oxide superconductors although the highest T_c of these was only 13K. After a couple of years of effort in vain, Bednorz found an article written by Michel and Raveau of Caen University describing the metallic conductivity in the Ba-La-Cu-O oxide solid solution system. He synthesized the material and measured the resistivity down to the low temperature which had not been attempted by Michel and Raveau. He found an indication of superconductivity up to 35K. On April 17, 1986 Bednorz and Müller submitted a paper to the *Zeitschrift fur Physik* on the title "Possible High T_c Superconductivity in Ba-La-Cu-O system." It is said that their sample was tested at an IBM research laboratory in the U.S. but a negative judgment was given back to them. In September, they obtained the SQUID susceptometer to test the Meissner effect which is the most powerful tool to prove superconductivity. They did find the Meissner effect and submitted a second paper to the *Europhysics Letters* on October 19th. They announced their results in a few academic conferences in Europe. Their announcements, however, were not taken up seriously by the superconductivity community in Europe or in their company.

Reconfirmation and Identification of the Superconducting Material

Bednorz' and Müller's first paper was seen by researchers at the Electrotechnical Laboratory in Japan in September. They tried to reproduce it but did not get positive results. That this paper had appeared was communicated by Professor Sekizawa of Nippon University to Kitazawa of the University of Tokyo at a workshop of the Ministry of Education's special project on October 4, 1986. This was because the University of Tokyo group was known to be working on oxide superconductors. Kitazawa, however, neglected this information for about a month and only told his collaborators about it. He knew that there were several similar materials reported to be superconductive, some with T_c at room temperature but that none of them had ever been reconfirmed. In the beginning of November, a research associate Takagi suggested to Professors Uchida and Kitazawa that the reconfirmation work could be a good project for an undergraduate student who was about to start his thesis experiments in their group. Kitazawa, although still reluctant, suggested Takagi's different synthesis procedure of the material, which was just to mix the raw powders with agate mortar and pestle and then to bake, a more primitive but much quicker method than that followed by Bednorz and Müller. The student prepared a dozen samples. On November 13, the very first sample exhibited the Meissner signal indicating that at least 6% of the total volume should be superconductive with a T_c about 23K.

This news was transmitted to Professor Tanaka, the principal investigator of the group, who then arranged additional reconfirmation measurements by another institute and decided to put several graduate students to work on this project. Takagi, with these students, worked to make the superconducting volume fraction as high as possible and to raise the T_c as high as possible by preparing dozens of samples with differing compositions and under different atmospheres and heat treatment. By December 5, they had identified the superconducting phase to be the so-called K_2NiF_4 structure mentioned as one of the three phases included in the sample of Bednorz and Müller. The zero resistance was obtained above 23K for the first time and the largest Meissner signal reached 30%. By then, therefore, they had become confident that the material of this crystal structure was really the highest T_c superconducting material ever known.

Announcement of the Results

The earliest report of the reconfirmation was presented by Tanaka in a workshop of the theoretical group in the Ministry of Education project on November 21 and the *Asahi* daily newspaper reported about it on November 28. Kitazawa attended the Materials Research Society (MRS) fall meeting in Boston and was supposed to present an invited lecture on the Ba-Pb-Bi-O superconductor system on which they had been working for several years. When he arrived at the conference site, there were several people who had already heard about the news of reconfirmation and asked him about the details. He told them what he had learned before he left Japan.

On December 4, he gave a scheduled lecture but did not mention the new developments because his colleagues were reluctant for him to disclose it in public. This was mainly because they had not identified the material. Two talks later, Professor C.W. Chu of Houston University reported on the same material as Kitazawa did. At the end he mentioned that his group had worked on Bednorz' and Müller's material. They found a similar resistivity drop below about 30K, and concluded that the material was an interesting candidate as a possible superconductor. Replying to a question raised from the floor about the Meissner effect, Chu said that the Meissner signal was less than their detection limit about 0.5%. Kitazawa made a comment saying that they observed a Meissner signal of 14%, indicating that the superconducting fraction was more than 14%. Then the organizer of the symposium on superconductivity, Dr. Braginski of Westinghouse requested Kitazawa to describe details of the Tokyo studies on December 5 after the scheduled sessions were over. In the late evening of the 4th, he talked with Tanaka on the phone and learned that they had identified the superconducting phase and reached 23K as the zero resistance T_c with the maximum Meissner signal of 30%. Therefore in the talk Kitazawa gave details of sample preparation, crystal structure and heat treatment conditions. This is said to have nucleated the enthusiastic research efforts around the world.

Kitazawa then visited AT&T Bell Laboratories on December 9 and Stanford University of the 12th for special seminars on the new material. Tanaka sent him at Stanford a second manuscript submitted to the *Japanese Journal of Applied Physics* on December 8 about the identification of the superconducting material. Following a suggested mailing list by

Professor T.H. Geballe, Kitazawa sent copies from Stanford to about 20 researchers who might be interested in the results and one to the editor of *Physics Today*.

Because the University of Tokyo group had had experience in oxide superconductivity research for more than several years, they were the largest single group in the field and relatively well equipped. Assisted by the special project of the Ministry of Education, they could proceed relatively rapidly to elucidate the unique properties of the new superconductor. Also their group was a hybrid group of physicists and chemists, which enabled them to prepare many well characterized materials of various compositions in a short period. They also discovered the higher T_c system Sr-La-Cu-O reported by Kishio et al. on December 18. They filed a patent on this material system on December 23 in the name of the senior Professor Fueki. That was the world's first patent on the HTSC materials. They were invited to lecture about the new superconductors and their unique properties both in Japan and abroad because the speed of information transmittance was not enough by academic journals or even by the newspapers. Kitazawa talked more than 40 times abroad during the first three years of development. Therefore, during the early stages the University of Tokyo group played a publicity service section role for the HTSC community.

Discoveries of Further High T_c Superconductors

By this time the critical temperature was still only 37K, which was much lower than liquid nitrogen temperature, 77K. Therefore, the new developments did not create a large enough incentive for the industrial world to initialize R&D efforts on a large scale.

Professor M.K. Wu of Alabama University got a phone call from Professor Chu, his supervisor while he had been a student, from Boston immediately after the MRS meeting and was asked to join in collaborative work on HTSC materials. In January 1987, during the course of efforts to find new oxide systems, he and his student Ashburn found the Y-Ba-Cu-O system to show T_c even higher than liquid nitrogen temperature. This was immediately communicated to Chu in Houston and in collaboration the two groups proved that it was a real superconductor of zero resistivity with T_c higher than 90K. This news was disclosed on February 16 at a press conference held at NSF by Chu but the composition of the material was not disclosed until March 2, the date of printing of an article in *Physical Review* Letters, because of patent considerations.

In the meantime Professor S. Hikami of another section of the University of Tokyo who used to be a theorist also announced the discovery of a liquid nitrogen temperature superconductor on February 21 in a workshop of the Ministry of Education project, but without disclosing the composition of the material. On the 27th, the newspaper *Peoples Daily of China* reported the discovery of a liquid nitrogen temperature superconductor by Professor Z.X. Zhao's group at the Beijing Institute of Physics. Zhao described the material for the first time to be Y-Ba-Cu-O. On March 2, it became clear that all these materials were the same. During the month of March, the crystal structure of this material was determined by Chu's group and several other groups around the world independently and nearly simultaneously.

By mid-March there were several groups around the world (including the author's group) who announced that not only yttrium (Y) but also most of the other rare earth metal elements can create superconductors with T_c about 90K. This period was the most enthusiastic period in the search for new materials and now involved many researchers in industry laboratories.

One year later (January 10, 1988) group leader Dr. H. Maeda of the National Research Institute for Metals (NRIM) in Japan announced the discovery that the Bi-Sr-Ca-Cu-O system was a 110K superconductor. He had been searching for new materials independent of other researchers in his institute. Several days later, Professor A.M. Hermann of Arkansas University announced the even higher T_c with the Tl-Ba-Ca-Cu-O system. The T_c was optimized later to 125K in this material by researchers at IBM Almaden, and this is the record high T_c up till now. From a practical point of view Y, Bi and Tl-cuprates are currently the three most promising materials on which research is focused.

From a scientific point of view, an important discovery was the material now known as an electron-doped superconductor, the Nd-Ce-Cu-O system, which should be mentioned. This was discovered by Professor Y. Tokura of the University of Tokyo in January 1989. This discovery has had a great impact on the theoretical aspects of high T_c superconductivity. Tokura also succeeded in finding a material rule governing HTSC materials and was able to discover new ones according to his prediction. His rule, in a sense, has nearly terminated the first phase of the material research.

Comparison of R&D Efforts on HTSC

Based on the author's personal view, after the discovery by Bednorz and Müller, the basic science of HTSC was at first dominated by Japanese contributions. As time elapsed, more high quality research results came from the U.S. than from Japan, except in some particular fields like nuclear magnetic field resonance. This was because there are more than twice as many physicists in the U.S. in the field and on the average they are much better equipped. As far as basic physics is concerned, the American community has a very high potential. On the other hand, there are perhaps as many or even more materials scientists in Japan who are engaged in searching for new materials, improving the material quality and processing methods, growing crystals, making thin films, etc. From the viewpoint of practical results, the role of materials scientists seems to be important.

There have been a few polls taken to predict who is contributing the most science and who may reach practical applications first. Polls in both countries agree that the U.S. is contributing more to basic science but Japan is ahead for applications. I believe that this is due to the fact that the role of materials scientists is better appreciated in Japan but physicists are more dominant in the U.S. One typical and frequent mode of U.S.-Japan collaboration is that U.S. physicists with better testing equipment wait for Japanese samples. No opposite cases are known to the author.

As for research efforts in industry, Japanese companies have been leading in application oriented fields except the ones aimed at military purposes such as high frequency passive devices. Several tens of Japanese companies have started R&D with their own funds while many fewer companies in the U.S. have started on a comparable scale. There are many small scale U.S. venture companies funded by the government on a project basis. The unwillingness of American companies to invest in long term R&D is not a long standing tradition but seems to have developed during these last two decades.

Another factor affecting this difference may be the fact that in the U.S. during 1987 and 1988 a pessimistic view was spread both in scientific journals and in newspapers about the practical usage of HTSC materials. This was the time when industries were seriously considering starting research projects. This view, I believe, was partly true but exaggerated. In Japan optimism prevailed over pessimism. People said that nobody could have foreseen the present day of semiconductor integrated chips during the early days of highly unreliable transistors. Indeed, Japanese materials scientists have so improved the performance of the HTSC wires that renewed R&D efforts have been initiated in American companies.

The author's personal view is that American researchers avoid difficult materials problems more than their Japanese counterparts. They seem to be afraid that their efforts and capability may not be appreciated in the company if they cannot get immediate success. In Japan, even if a project has to be terminated without success, there are mechanisms in the company to appreciate researchers who tackle a difficult subject. And besides, a project has a better chance to survive when the researchers are enthusiastic.

During this period many Japanese companies started research programs in HTSC as if they could not stand the pressure from enthusiastic researchers in the company. The managers also got pressure from personnel sections who felt that recruiting good students required a good image as a company carrying out high profile research such as HTSC. On the other hand, U.S. companies in general maintained rather skeptical attitudes unless they were provided with government funds. U.S. researchers seem to have less power to influence top management. The managers' eyes seem to be more directed toward stockholders who prefer immediate profit. Therefore, the author would like to stress that this difference arises from a very basic question: which is stronger; the stockholders of the company or the people who work for the company?

The New Superconducting Materials Forum and the International Multi-Core Project

The Science and Technology Agency (STA) of Japan has a branch for promotion of materials R&D. An officer in charge of this section, Mr. S. Hattori, was perhaps the first to have realized the importance of new developments in the HTSC field. The agency has three national laboratories, but those laboratories had problems in association with budget rmductions during a period when metal industries were hit by the recession. One laboratory had been playing an important role in developing conventional metallic superconducting wires. Mr. Hattori thought that the HTSC research could be a very suitable topic to activate R&D in these laboratories and that it should be initiated as soon as possible for the greatest effectiveness. So he called for a tentative planning committee meeting a couple of weeks after the reconfirmation of HTSC, inviting the author and group leaders from those laboratories and industries. The committee was headed by Professor K. Tachikawa of Tokai University who just retired from one of the laboratories where he had been in charge of the superconductivity section.

Committee meetings were held frequently, initially discussing current HTSC research and then good ways to promote information exchange. In the meantime research was started in the groups of committee members. The committee decided to establish the New Superconducting Materials Forum (NSMF) which was supposed to sponsor four one day symposiums and four smaller scale workshops each year and to publish a list of preprints as frequently as possible. Both types of meetings were planned to be instructional. The agency did not have a budget for this and hence recruited companies as members for nominal cost. They got more than 100 members immediately and the first symposium was held on May 1, 1987, collecting about 700 researchers mainly from industries and national laboratories under the agency.

The exciting discovery of the 110K new material by Maeda in January 1988 came from one of these laboratories (NRIM). Another laboratory, NIRIM, became active in investigating the crystal structures of the HTSC materials by electron microscopy. The officer Hattori then planned a rather big project for this agency making use of these achievements, the International Multi-Core Project. This was intended to promote joint research projects at the several core sites at the three laboratories by inviting both Japanese and foreign companies to send researchers. As a result of this project the three laboratories were able to get the highest voltage electron microscope and the highest field magnet as well as other advanced equipment. More than 20 researchers from industry and some postdocs from abroad participated. The NSMF has functioned as a major source for industry to acquire the latest information about HTSC research.

The International Superconductivity Technology Center and The Superconductivity Research Laboratory

The Ministry of International Trade and Industry (MITI) is a larger organization than STA, and usually supports much larger projects. In general their consortium projects require private member companies to pay more than half of the total expenses associated with each project. The superficial reason for this is said to be that in this way only serious companies join as members. The real reason is that the government simply does not have a large enough budget. Because of this, an officer who is in charge of planning must pay careful attention to opinions of the private sector. He has to create such an atmosphere that companies are cooperative with his project. At the very least he has to avoid strong opposition from expected members. A project therefore is better not too clearly defined in advance but should rather be brewed gradually while it is discussed with expected members.

Within a month after the reconfirmation of HTSC, an officer of MITI, Mr. M. Urashima, started frequent visits to Professor Tanaka at the University of Tokyo, at first just getting information and then gradually talking about the possible style of a new project on HTSC. Then they started meeting influential technical managers of private companies. During this process they desired to establish a new research laboratory rather than just fund researches in national laboratories and companies.

There had already been two examples: the VLSI research lab and the optoelectronics lab. These laboratories had been formed as consortium laboratories under MITI and terminated after five years. The most difficult part of managing them was collecting good people in a short period and then being able to terminate their employment so as to terminate the project. Because of life-span employment in Japan, it is not possible to hire good researchers for just five years. Therefore, they devised a clever idea — requesting each of the member companies to dispatch researchers or administrative personnel. The salary is paid by the company and they go back to the original company on termination of the project research. The office and building is built on land which one of the companies can lend for the period. Land is usually a big problem in Tokyo. But the close location of the lab can be attractive to the company which lends the land because better communications can be secured in rapidly developing fields. NEC and Fujitsu were the two companies that lent land close to their own labs for the above two consortia.

By the end of February 1987, the MITI officer and Professor Tanaka had reached the conclusion that at least some new project should be possible and started various efforts to get consensus among other MITI officers and among influential Diet members. The officer arranged seminars by Tanaka and some from industries for the MITI community and for some Diet members. In order for the authorization, a round table committee was formed consisting of executive managers of influential companies in the field of superconductivity, managers of national laboratories under MITI and university professors. The committee chairman was a senior and well known retired professor who did not have any direct interest in the possible project. Under this committee, an investigation committee was formed with Professor Tanaka as the chairman and technical managers of companies and active scientists from national laboratories and universities as members. They made a purely technological report within two months based on a presentation of each member. The round table committee concluded rather vaguely that a positive governmental support and collaborative researches should be needed.

In parallel the officer and Tanaka arranged for filming of a committee meeting, university laboratory and the laboratories of member companies. This was quite effective and created widespread interest in the new technology. MITI in collaboration with the two committees even published commercial books; one for readers with technical background and the other for non-technical readers. On May 26 the Science and Technology Committee of the Diet held a hearing with HTSC specialists including Tanaka. The MITI officer and Tanaka, taking advantage of the surrounding enthusiastic atmosphere, started to organize companies one by one asking to what extent they would be willing to pay for the consortium and whether they would be able to send researchers. The response was more than expected. They decided to have two kinds of members: one just for information and the other for the consortium laboratory.

After intensive discussions an agreement was reached to open the consortium to foreign members. Tanaka advocated the words "HTSC for the future of humankind", which was in accordance with ideas that Japan should contribute more to the international research community. They knew that one of the most difficult problems in the U.S. when thinking of a consortium is to reach an agreement on the issue of how to share R&D results among the members. In Japan the members agreed that they would decide on that question later when necessary. What would happen is that they would discuss the problem for a certain period of time and then choose some neutral and experienced person whom they all trusted and follow his decision if they cannot decide by themselves. But the officer and Tanaka worried that this time it was possible that foreign members might not be satisfied with this rather Japanese style rule. So they collected information about the rules adopted for this question in the U.S. But it took a long time until they reached a conclusion (in March 1989) and hence they could not give a clearcut reply to inquires from abroad in the initial period.

The MITI established the International Superconducting Technology Center (ISTEC) in Tokyo October 1, 1987. It was also decided the Superconductivity Research Laboratory should be formed under ISTEC. At first they expected about 30 members to join ISTEC and a dozen to join the laboratory. It turned out that about 100 companies applied to join ISTEC and 40 to join the laboratory by paying 1 million dollars as the admission fee and sending two researchers each. As the site for the lab, Nagoya offered land and a building. Also a gas company in Tokyo offered to lend land. Hence, it was decided that the main lab was to be built in Tokyo and a branch to be set in Nagoya. Because Nagoya has been the largest center for ceramic industries, it was decided that the branch would concentrate on ceramic processes. As Tanaka became confident of success, he collected a dozen researchers from industry to begin researches immediately as visiting scientists at the university. They would be shifted to the new laboratory as soon as the lab was built. He also negotiated with some companies so that they would dispatch his former students who once worked on superconductivity. He later got five of them.

Tanaka reached the retirement age of 60 at the University of Tokyo and was selected as the head of SRL. The construction of the new building started in April 1988 and it was opened on October 25. In the meantime, the Nagoya branch opened on July 8 with a group leader and six researchers from member companies. Those who were staying at the University of Tokyo as visiting scientists started planning and negotiating to purchase equipment in order for them to arrive soon after the opening of the lab. The lab started up amazingly quickly and the researchers increased to nearly 100 within a year. The member companies increased to 111 with 46 of them members of the lab by November 1989. Success resulted from the preceding thoughtful efforts of Tanaka and others before things had been completely decided. Plans were developed and accepted in a flexible and cooperative manner just as when a village is going to hold a festival. I would like to stress that this was done neither by strong power nor by money.

A Personal View on the Consortium and Free Competition

It does not seem likely that a consortium can be easily created in the U.S., considering the discontinuous spectra of the distribution of companies. In the the automobile industry, for instance, there are ten companies producing passenger cars in Japan, while there are just three in the U.S. for a much larger market. In the field of computers there are five large Japanese companies fabricating main frame computers while in the U.S. there is one giant and some others. There are at least five iron and steel companies with blast furnaces in Japan, while there are two large ones in the U.S. That is, there are many companies in each industrial field in Japan competing hard domestically while one or just a few giant companies enjoy monopolized markets in the U.S., unless they get international competition. Until 30 years ago, there were many more companies also in the U.S. U.S. capitalism seems to have outgrown the period of free competition which is supposed to be the essence of capitalism. In this sense, Japan may be still in an early period of capitalism or there may be some social mechanism which works to prevent the decrease of the number of competing companies. (A typical exception may be beer, with four companies in Japan, but many in the U.S.; banking may be another exception.) True cooperation can be set among equal partners but not so easily among giants and dwarfs. The essence of a consortium is cooperation among member companies. One reason that Japanese companies can join and become deeply involved in a consortium is the continuous spectrum of rival domestic firms.

While a high critical temperature and high critical field set the framework for the potential of oxide high temperature superconductors, obtaining a sufficient critical current density and developing satisfactory methods of fabrication, mechanical properties and price are equally important considerations. With regard to superconducting oxides for thin film and electronic device applications, critical research innovations were the achievement of high current density thin films at liquid nitrogen temperature by P. Chaudari et al. at IBM and T. Morakami et al. at NTT in May and July of 1987. This and subsequent highlights of thin film and device achievements (Table 3) indicate that thin film high critical

Table 3. High-T_c superconducting thin film and device highlights.

Achievement	Organization	Date
First high-J_c film at 77K (YBaCuO)	IBM	5/87
First $J_c > 10^6$ A/cm^2 at 77K (YBaCuO)	NTT	7/87
First in-situ growth (YBaCuO)	Cornell	8/87
First ultra-thin (100 Å) film (YBaCuO, T_c = 82K)	Kyoto Univ.	6/88
First high-J_c in-situ laser-deposited film	Bellcore	6/88
High J_c in all high-T_c film materials (Tl...,Bi...,Y...)	Sumitomo	8/88
New perovskite substrates (LaGaO$_3$, LaAlO$_3$)	IBM & TRW	9/88
Synthesis of $n = 3, 4, 5$ BiSrCaCuO films	Matsushita	9/88
First low noise SQUID (Tl...)	IBM	11/88
First high-J_c CVD film	Tohuku Univ.	11/88
Film with low microwave losses (86 GHz, 77K)	Siemens & Wuppertal	11/88
First high-J_c film at 77 K on silicon with buffer layer	Bellcore & NEC	12/88
Picosecond pulse propagation	AT&T	3/89
First two-level high-T_c device (microstrip resonator)	Stanford Univ. & HP	4/89
High-Q coplanar transmission line resonator (Q 14× higher that for Cu at 9 GHz, 77K)	Siemens & Tech. U. of Munich	5/89

Source: 1989 JTEC Panel Report on High Temperature Superconductivity in Japan, M.S. Dresselhaus, et al. National Technical Information Services, U.S. Dept. of Commerce.

temperature superconductivity research in Japan and the United States are on a par. Thin film accomplishments are mostly related to learning how to grow high quality films; high current density reflects the development of single crystals or oriented crystal alignment in these materials. As shown in Table 4, the current density achieved in $ReBa_2Cu_3O_7$ films

Table 4. Thin film $ReBa_2Cu_3O_7$ critical current density achievements at 77K

J_c (A/cm^2)	Organization	Date
1.0 ×10^5	IBM	5/87
1.8 ×10^6	NTT	6/87
2.5 ×10^6	Sumitomo Electric	2/88
4 ×10^6	U. of Kyoto	6/88
4-5 ×10^6	Bellcore-Rutgers	8/88
5.5 ×10^6	Karlsruhe	11/88

Source: 1989 JTEC Panel Report on High Temperature Superconductivity in Japan, M.S. Dresselhaus, et al. National Technical Information Services, U.S. Dept. of Commerce.

suggests that the rate of development in Japan and the United States is pretty much equivalent.

The thin film and device highlights illustrated in Table 3, illustrate differences in the Japanese and American approaches. While the overall thin film growth activities of Japanese and American researchers have been roughly equal, Japanese laboratories have tended to make long term commitments without sharply defined application goals, stressing the expectation of long development cycles. Even without short-term device applications, there has been a strong Japanese commitment to materials synthesis projects. In contrast, there is an appreciable American commitment to the high-frequency properties of superconducting film devices. A number of American laboratories have constructed resonators with superconducting layers. Researchers have constructed filters with both superconducting and metal films with low losses at liquid nitrogen temperature. Investigators have shown that very short pulse propagation is more effective with superconductor films than with metallic conductors. Studies have proceeded with the development of superconducting infrared detectors. Several American companies are working on microwave device components. These device applications in the field of electronics for space and defense have been spurred on by DARPA and DoD contracts in the United States for which there is no Japanese equivalent.

For the present, large-scale applications of high field superconductivity remain in the realm of metallurgically well developed low temperature materials such as niobium titanium and niobium tin. In both Japan and the United States large-scale application studies are mostly supported by government funds and coordinated by national laboratories with a comparable scale of activities. In Japan two major projects are superconducting magnetic levitation for trains and superconducting electric generators, both aimed at specific commercial markets. In the United States the superconducting super collider is aimed at high energy physics while the superconducting magnetic energy storage system is intended for a combined defense and commercial objective. With regard to the development of high current density bulk oxide high temperature superconductors substantial achievements have been reported in yttrium barium copper oxide by melt-textured growth techniques which led to oriented crystal structure (S. Jin and Bell Telephone Laboratories). This general approach has been elaborated by M. Murikami and others at Nippon Steel as quench and melt-growth and further as a melt-processed melt-growth technique developed by Murikami after moving to the superconductivity research laboratory of ISTEC. A number of similar techniques have been developed by other laboratories in Japan, the United States and China. High critical current wire and tapes have been mostly demonstrated in Japan by Sumitomo Electric and Furukawa Electric. American researchers have generally focused more on thin and thick film materials. Representative achievements in Japan and the United States are illustrated in Table 5 and Table 6.

Table 5. Representative benchmark data for wires, tapes, and thick films of high-T_c superconductors in the U.S.

Laboratory	Process	Conductor Properties
U. Houston, Texas Center for Superconductivity	Controlled cooling of melt through $YBa_2Cu_3O_{7-x}$ peritectic (1°C/hr).	$J_c = 15,000 - 18,500 A/cm^2$ @ 77K, 0T; $J_c = 75,000 A/cm^2$ (pulsed) @ 77K, 0T; $J_c = 37,000 A/cm^2$ (pulsed) @ 77K, 0.6T
AT&T Bell	Melt-textured growth. Directional solidification of $YBa_2Cu_3O_{7-x}$ melt	$J_c = 17,000 A/cm^2$ @ 77K, 0T; $J_c = 4000 A/cm^2$ @ 77K, 1T
AT&T Bell	Hot forging of $YBa_2Cu_3O_{7-x}$ powder at 1000°C, 26 MPa, 6 hr.	$J_c \sim 3000 A/cm^2$, T unspecified
Stanford U.	Laser heated pedestal growth of $Bi\text{-}Sr\text{-}Ca\text{-}CuO_x$ fibers	$R = 0$ @ 80-85K J_c(pulsed) $= 60,000 A/cm^2$ @ 68K
Argonne National Laboratory	Tape cast $YBa_2Cu_3O_{7-x}$ powder with and without Ag powder. Tape placed onto Ag foil substrate, followed by sintering	$R = 0$ @ 86-90K; $J_c = 300 A/cm^2$ @ 77K
Superconductor Technologies, Inc.	Spin-on composition of Tl-Ca-Ba-Cu 2-ethylhexanoates deposited onto substrates	$R = 0$ @ 100K, $R_s = 250$ mΩ @ 77K, 150 GHz. No J_c reported
Massachusetts Institute of Technology	Spin-on process of Pechini-citrate/ethylene glycol polymerization mixture for Bi-"4334" films on (100) $SrTiO_3$	$R = 0$ @ 70-75K $J_c = 5 \times 10^5 A/cm^2$ @ 4K, 0 T
IBM-Yorktown Heights	Spin-on composition of Y, Ba, Cu trifluoroacetates dissolved in methanol. Various substrates.	$R = 0$ @ 91K; $J_c = 10^4 A/cm^2$ @ 77K, 0T on $LaGaO_3$
Microelectronics and Computer Technology Corporation	Spray pyrolysis of Bi-Sr-Ca-Cu nitrates onto (100) MgO and BeO. Post deposition melt-quench-anneal to densify film. Bi-"2212", "4334" stoichiometry.	$R = 0$ @ 81K; $J_c = 4000 A/cm^2$ @ 77K on (100) MgO
Sandia National Laboratory	Screen printing of $YBa_2Cu_3O_{7-x}$ powder (5μm) dispersed in an alcohol. Printed onto substrates.	$R = 0$ @ 91K; $J_c = 93 A/cm^2$ @ 76K
Los Alamos National Laboratory	High-rate magnetron sputtering from single target Tl-"2212" and "2223" targets	$R = 0$ @ 90K; $R_s = 6 - 7$ mΩ @ 15K, 22GHz; 25 μm thick films. No J_c reported

Source: 1989 JTEC Panel Report on High Temperature Superconductivity in Japan, M.S. Dresselhaus, et al. National Technical Information Services, U.S. Dept. of Commerce.

Table 6. Representative benchmark data for wires, tapes and thick films of high-T_c superconductors in Japan

Laboratory	Process	Conductor Properties *
Sumitomo	Ag-sheathed tubes drawn to wire, then cold rolled to tapes	(Bi,Pb)SCCO-0.14×4 mm² tape: $J_c = 1.7 \times 10^4 \text{A/cm}^2$, 0T; $J_c = 1700 \text{ A/cm}^2$ at 0.1T; YBa$_2$Cu$_3$O$_{7-x}$ $J_c = 4 \times 10^3 \text{A/cm}^2$
	Multifilamentary HTSC wires/Ag sheath	36 (Bi,Pb)SCCO filaments; 0.16 mm wire; $J_c = 1050 \text{A/cm}^2$
Hitachi	Ag-sheathed, HTSC powders drawn and rolled into 0.5 mm tapes	YBa$_2$Cu$_3$O$_{7-x}$ $J_c = 3.3 \times 10^3 \text{A/cm}^2$; TBCCO-$J_c = 6 \times 10^3 \text{A/cm}^2$; (Tl,Bi)SCCO-$J_c = 10^4 \text{A/cm}^2$
NRIM	Tape cast, sintered, and rolled to high density	(Bi,Pb)SCCO; $J_c = 1850 \text{A/cm}^2$ 0T; 30μm thick, 3 mm wide, 100 mm long
Nippon Steel	Quench-melt-growth to form monolithic conductor	YBa$_2$Cu$_3$O$_{7-x}$ $J_c > 10^4 \text{A/cm}^2$, 1T
Mitsubishi	Aerosol particle deposition 1μm/hr onto substrate, melt textured, and annealed	BSCCO-1μm thick on (100) MgO: $J_c = 8000 \text{A/cm}^2$, 0T, $J_c = 100 \text{ A/cm}^2$, 0.4T
NRIM	Magnetron sputtering on Hastelloy X substrate with MgO buffer layer	YBa$_2$Cu$_3$O$_{7-x}$ - 1-2μm thick: $R = 0$ at 80K, $J_c = 200 \text{A/cm}^2$

*All Jc values at 77K.

Source: 1989 JTEC Panel Report on High Temperature Superconductivity in Japan, M.S. Dresselhaus, et al. National Technical Information Services, U.S. Dept. of Commerce.

One of the leading producers of low temperature superconducting large-scale magnets in Japan has been Hitachi. They have provided magnets for magnetic levitation devices, for fusion reactors, for high energy physics applications, high field generators and synchrotron radiation. Dr. Shin-Pei Matsuda is Director of the Superconductor Research Center of the Hitachi Research Laboratory. His comments about the Hitachi research program on high critical temperature superconductors represents the long time frame within which Hitachi research is conducted:

HIGH Tc SUPERCONDUCTIVITY IN JAPAN

Shin-Pei Matsuda, Hitachi Research Laboratory, Hitachi, Ibaraki, Japan

This brief report examines the status of High T_c Superconductivity research in Japan. The first part looks at some of the comprehensive High T_c Superconductivity research programs which have been established in the recent past, and the second part looks specifically at Hitachi's research program as a case study.

Applications of High T_c Superconductors

For every potential application of high T_c superconductors we must realize that there is competition from metal superconductors and also from normal conductors. From an economical point of view the high T_c superconductors have not reached a state of commercial application but are still the subject of basic materials research. We see as *possible* applications: magnets operable at liquid nitrogen temperature (MRI, MAGLEV, etc.), magnets operable at very high magnetic fields at low temperature (fields above 20T at 4.2°K), magnetic shields and also for electronic switching devices (Josephson devices, transistors) where there is very strong competition from semiconductor devices.

High T_c Superconducting Research Programs in Japan

The high T_c superconducting research program in Japan has been spearheaded by two specially created research organizations. Each will be discussed briefly below.

ISTEC

The International Superconductor Technology Center (ISTEC) was established in 1987 by both the private sector and government. A "Superconductivity Research Laboratory" supporting 100 researchers was created. Two researchers are typically sent to the Superconductor Research Laboratory by each company member. Membership in ISTEC includes 46 companies. Among these are representatives from the electric power, electronics, machinery, chemical, and iron and steel industries. Initial membership costs are $700,000 (U.S.) with annual fees set at $80,000 (U.S.).

SUPER-GM

The Superconducting Generator and Materials Program (SUPER-GM) was established in 1988 by funds from MITI as part of an eight year project. Its goals include the development of a 70 MW superconducting generator and high Tc superconducting cables.

Hitachi's R&D Structure on High T_c Superconductors

Hitachi's program for the research and development of High T_c Superconductors is designed to take place in two phases. Phase I runs from early 1987 to early 1990, and phase II runs from mid-1990 to early 1993. The program is classified as basic research. Within Hitachi, the participating departments include the Central Research Lab, Hitachi Research Lab, the Advanced Research Lab, Hitachi Chemical Lab, and Hitachi Cable Lab. Their respective missions are thin film electronic devices, materials and wire, mechanisms, materials synthesis, and wire fabrication.

In regards to the technical backgrounds of the participating researchers, one-half have experience in Low Temperature Superconductors (including superconducting magnets and Josephson Devices), and the other half are new-comers with experience in metallurgy, materials chemistry (catalysis), and ceramics.

Hitachi's company policy regarding superconductors is two-fold. First, Hitachi's goal is to lead the industrial and scientific worlds in superconducting technologies in general. This acknowledges the fact that a market presently exists in Low Temperature Superconductors. Second, Hitachi wants to develop long range projects. This includes participating in national projects associated with high temperature superconductors and cooperating with foreign corporations.

Innovations from the Hitachi group have been in low temperature film formation by plasma assisted oxidation of Y-Ba-Cu-O films ($T_c = 80°K$) on a 450°C substrate (Dr. Takagi, CRL) and tape-shaped wire formed by drawing and rolling (M. Okada, HRL). With densification and crystal orientation developed with rolling process, the critical current J_c has exceeded 10,000 A/cm^2 at 77°K,0T.

Comments

In thinking about the future of high T_c materials we have to keep in mind that even the nearest application is at least five years away, and that the future market for these materials will be limited. The application of any *new material* usually requires 10 years.

A key point in new materials application is reliability which requires not only discovery but also development of fabrication methods and extensive testing. Finally, revolutionary new systems in which there is *substitution of a conventional system* by a new system is extremely difficult.

David Larbalestier is L.V. Shubnikoff Professor in the Department of Materials Science and Engineering at the University of Wisconsin and Director of the Applied Superconductivity Program. Prior to coming to Wisconsin in 1976 he had worked with superconducting magnet research in the British High Energy Physics Research Laboratory, (the Rutherford Laboratory) and developed filamentary magnet conductors from the brittle compound Nb$_3$Sn. Working with niobium tin, niobium titanium alloys and high critical temperature oxides, he has assembled at Wisconsin an extensive fabrication facility, electromagnetic characterization facility and has focused on critical current density as a key parameter in advancing the technology of superconducting magnets. In discussions of the development of superconductivity he proposed that key inventions are those of K. Onnes, who discovered superconductivity and L. Shubnikoff, who discovered that alloying a pure metal extended the critical field of a superconductor in 1935. Shubnikoff, a Soviet Jew, was subsequently sent to prison and there are relatively few references to his work. Then in 1960 A. Kunstler discovered that niobium tin was able to conduct very high current densities in very high fields. This made practical extremely high strength magnets. Then, of course, Müller and Bednorz discovered high temperature superconductivity in 1986, the seventy-fifth anniversary of Onnes' discovery. At the present time, having gotten over the invention hurdle, many other factors are required to develop the technology. The critical current density is most important but it's also essential to have satisfactory fabricability, mechanical properties and price-performance behavior.

Professor Larbalestier emphasized the role of the universities in the development of superconductivity. First of all, universities tend to define and also diffuse the scientific culture. Universities have integrated processing and characterization in departments of Materials Science and Engineering and have demonstrated that interdisciplinary approaches are necessary in the formative years of a new technology. The feedback processes of a university culture enforce rigorous thinking and experimentation. Second, a number of start-up companies have developed out of university programs. In the United States there is Conductus which came out of Stanford/Berkeley, American Superconductor Corp. which came out of MIT and Superconducting Technologies, Inc. from the University of California at Santa Barbara. The small scale of university research allows the initiation of new programs which can multiply by networking with larger organizations. Because superconductivity is a stimulating scientific problem, and also a complex one, new problems can be welcomed, solved and then new ones attacked. However, there's a very serious problem that relates to the criteria for tenure at universities. Increasingly, problems in advanced materials are more complex and inappropriate for a single investigator. There needs to be a process for evaluating the contribution that individuals make to interdisciplinary programs. Focusing only on individual recognition and tenure evaluation fragments the capability of attacking important complex problems.

In the innovation process managers who understand the company, the technical community and the national objectives are essential. It is also important that long time horizons be kept in mind. Looking at the development of low temperature high performance magnets, the basic niobium titanium alloy now being used was developed in 1962. By 1966 wire consisting of 55 niobium titanium filaments dispersed in copper had been developed and in subsequent years improved upon such that a magnetic resonance imaging device one meter in diameter with a one tesla magnetic field was achieved in 1982, a development period of about twenty years. In this development previous innovations served as models and also as training grounds for fresh innovation. With regard to superconductor technology the principle driving force in the engine of innovation has been the need of user communities, principally high energy physics. Superconducting magnets clearly are an exemplar of science driven technology. In this long time frame the presence of people having a long time commitment has made a significant difference in the development of superconductors. Some who come to mind are John Hulm at Westinghouse, David Sutter, the Department of Energy, Ted Gabelle and Mac Beasley at Stanford, K. Tachikawa at NIRM/Toki University, K. Yashukochi at Neihan University, and S. Tanaka at the Tokyo University, now ISTEC. These committed managers have maintained a long time horizon and been important in driving the field. People do make a difference.

Larbalestier suggested that it was particularly striking that perceived different approaches are being taken in the United States and Japan:

	National Commitment	Level of Activity		
		Industry	Nat'l Labs	Universities
Japan	High/long term	Strong and very large	Strong but small overall	Widespread at moderate level
US	Fickle/medium term	Variable and limited in scale	Largest and best funded	Widespread but moderate level

It seems clear that industry is driving the field in Japan while the situation in the United States is more fluid with defense and space objectives and national laboratories having the largest share of available resources.

There have been a number of thoughtful reviews of high temperature superconductivity in the last few years: U.S. Congress, Office of Technology Assessment, *Commercializing High-Temperatures Superconductivity,* OTA-ITE-388, U.S. Government Printing Office, Washington, DC, June 1988; Brendan Barker, *Superconductivity,* Elsevier Science Publishers,

Ltd., Oxford, 1989; Alan M. Wolsky, Robert F. Giese and Edward J. Daniels, "The New Superconductors: Prospects for Applications", *Scientific American,* p. 61-69, February 1990; U.S. Congress, Office of Technology Assessment, *High-Temperature Superconductivity in Perspective,* OTA-E-440, U.S. Government Printing Office, Washington, DC, April 1990; M.S. Dresselhaus, et al., *High Temperature Superconductivity in Japan,* Japanese Technology Evaluation Center, Loyola College in Maryland, Baltimore, MD, November 1989; and others. These document the development of high temperature superconductivity inventions and innovations in Japan and the United States and come to the conclusion that, starting from the same point in 1986, progress with discovery and invention has been approximately equivalent. In a field of advanced materials where there has been euphoria about the development program and dedication to advancing research, we see no evidence of a greater creativity on either side, a conclusion with which our conference participants concurred. It remains to be seen how the development of invention and research innovation into commercial innovation and manufacturing effectiveness comes to be played out.

There may be some indicators of long term development in the history of large-scale superconductor projects. John K. Hulm (pg. 66 in JTEC Panel Report) points out "It is quite remarkable that large-scale superconducting technology offers so many examples of project failure in the United States (i.e., the Isabel accelerator at Brookhaven which was abandoned after much of the tunnel was constructed and a helium liquifier installed, the mirror fusion test facility at Livermore which was abandoned after construction of a large superconducting magnet, the superconducting transmission line project at Brookhaven, and the ERPI-Westinghouse superconductor generator which was abandoned after partial rotor construction) and almost no similar incidence in Japan". As described above by Professor Kitazawa, there were extensive informal discussions and consensus building in Japan before a government-industry consortium and government programs for the development of magnetic levitation and of superconductor generators were put in place with long range planning. In contrast the 1990 report of the National Commission on Superconductivity in the United States suggests "A disadvantageous business environment in the United States, rooted in high interest rates, unfavorable tax policy, antiquated antitrust laws, and other factors, discourages U.S. industry from investing in high risk technologies such as superconductivity....The Commission supports the efforts of the Defense Advance Research Products Agency (DARPA) to develop conductor technologies because of the convincing advantage that superconductors provide for military requirements and the beneficial spin-off they will have on the commercial sector". DARPA has been the principal U.S. agency sponsoring superconductivity research which in the United States is focused on electronic defense and space applications. Depending on beneficial spin-off for commercial innovations and manufacturing know-how would seem to be a risky proportion.

LOW PRESSURE DIAMOND SYNTHESIS

The high hardness and the high refractive index, which gives the diamond an extraordinary brilliance together with a high dispersion that gives their fire, have made diamonds a popular and valuable gemstone. Because of their extreme hardness, the hardest material known, there are many important industrial uses for diamonds as abrasive grit and cutting tools with a world market of about one billion dollars. Diamond is not only the hardest material known, but has the highest thermal conductivity at room temperature, is an excellent electrical insulator, is totally inert and is transparent to ultraviolet, infrared and x-rays as well as visible light. Diamond has the highest elastic modulus of any known material, it is the most incompressible substance known and it has a relatively low thermal expansion coefficient which gives it a good resistance to thermal shock. As a doped semiconductor material, it may have several advantages over silicon for electronic devices. With respect to many of its properties diamond is superior to all other materials.

Because of its unique properties, there have been efforts to produce synthetic diamonds for more than a century. A Scottish chemist, James Ballantyne Hannay, reported that he had made diamonds in 1880 by heating paraffin, bone oil and lithium in sealed iron tubes. In 1893 the French chemist, Henri Moissan claimed that he had made diamonds from a mixture of carbon and iron heated to a very high temperature and quenched into a water bath. However, it was not until 1955 that F.P. Bundy, H.T. Hall, H.M. Strong and R. H. Wentorf, Jr. announced successful growth of diamonds at high pressure and high temperature at the General Electric Company. Graphite in the presence of a liquid metal solvent such as iron or nickel was brought to a temperature and pressure range where diamond is the thermodynamically stable phase, where crystals nucleate and grow. The method is used commercially; diamonds are formed at a temperature of about 1500°C and a pressure of about 60,000 atmospheres. More recently, diamond films have been produced at low pressures by chemical vapor deposition. This low pressure process has been used to make commercial products such as wear resistant coatings for cutting tools, heat sinks for sensitive electronic devices, high elasticity films for tweeters in stereo speakers and as windows in scientific instruments. It can properly be described as a nascent industry, one with great potential for the future.

The story of low pressure diamond synthesis clearly illustrates that discovery, invention and innovation are socio-technical processes strongly influenced by extratechnical social, cultural and personal influences. As reported by Angus below, several people conceived of producing diamonds by deposition from the vapor phase during the 1950's. William G. Eversole at the Union Carbide Corporation began his work in 1949, achieved successful growth of diamond on diamond seed crystals in 1952 and applied for patents that were issued in 1962. The rate of growth was very low (less than 0.1 micrometer per hour). There is an 1956 author's certificate on Russian work by B.V. Derjaguin and B. V. Spitsyn for similar results. (But this did not see the light of day until 1980 and thus had no influence on the developing story outside Moscow). Eversole's work was stopped primarily because of the success of the General Electric Company high pressure high temperature synthesis announced in 1955. Later, John C. Angus took up the Eversole work, meeting personally with Eversole to learn details of his work, and improved the process with the use of atomic hydrogen to remove graphite and prepare the surface for subsequent diamond growth. This work was published in the *Journal of Applied Physics* in 1968 and placed the innovative research accomplishment squarely in the public domain. Angus has been a continuing participant in the development of CVD diamond synthesis and his perceptions of the innovation follow:

INNOVATIONS IN THE CHEMICAL VAPOR DEPOSITION OF DIAMOND: PERCEPTIONS OF A PARTICIPANT

John C. Angus, Case Western Reserve University

Introduction

The chemical vapor deposition of crystalline diamond at atmospheric and subatmospheric pressures has become one of the most active fields in materials science. This great interest is driven by the extreme properties of diamond, which can now be obtained as a coating on diverse substrates by using relatively simple, low pressure processes. In this paper I will present the impressions of a participant in the development of this technology. I will focus primarily on the early barriers to innovation in the field. The paper will reflect my own experience and is not meant to be a comprehensive history of the field of low pressure diamond growth.

Earliest Work on Diamond Chemical Vapor Deposition: 1940's and 1950's

The most significant early, sustained effort at growing diamonds at low pressures was that of William G. Eversole at the Union Carbide Corporation. This work was started in 1949 and the first successful growth of diamond seed crystals was achieved in November of 1952 (1). It is remarkable that Eversole successfully synthesized diamond before either of the high pressure syntheses by ASEA in Sweden and by General Electric in the United States.

During the 1940's a group at the General Electric Corporation also attempted to grow diamond at low pressures. Little has been published of these efforts (2).

In 1956 Boris Spitsyn proposed the growth of diamond at low pressures through the thermal decomposition of carbon tetraiodide (3). Apparently, this proposal was made while he was a student at the Physical Chemistry Institute in Moscow. In 1959, this writer, while a graduate student at the University of Michigan, also proposed that diamond could be grown at low pressures where it is the metastable phase (4).

As far as known, each of these four above proposals for growing diamond at low pressures arose independently, without knowledge of the other efforts. This writer came upon the idea after studying the theory of supercooling and metastable phases during a graduate materials science course.

With the current intense level of activity in diamond chemical vapor deposition, it is somewhat difficult to imagine the situation as it existed in the early 1960's. There was no known activity in the field anywhere. No publications had arisen from the General Electric effort and the Soviet group under the direction of Derjaguin had not yet published. There were, however, numerous patents on low pressure diamond growth that had been issued in the United States and abroad. In retrospect, the most important were clearly those issued to Eversole in 1962 (5). Other patents of interest were issued to Brinkman (6) and

Hibshman (7). In addition to these serious attempts at diamond growth, there were some obviously absurd patents as well. Perhaps the most amusing is the claim that diamonds can be grown by converting the carbon in a bicycle tire by spinning it sufficiently rapidly that the centrifugal forces spontaneously convert the graphite into diamond.

Barriers to Innovation

There were several very significant barriers which held back the development of low pressure diamond growth. The first, and not often appreciated barrier, was the enormous success of the synthesis of diamond at high pressures, i.e., where it is the thermodynamically stable phase. The success of the high pressure program had the effect of killing off the nascent low pressure programs in development at Union Carbide and General Electric.

The second major barrier facing workers in the field was the misapplication of thermodynamic theory to the problem of diamond growth. There was a widespread feeling that the growth of diamond at pressures where it is the metastable phase was somehow thermodynamically forbidden. This is certainly a strange argument coming from beings who themselves are metastable objects. The successful application of equilibrium thermodynamic theory to the high pressure process enhanced the belief that diamond could only be grown at high pressures. It should be pointed out, however, that the most sophisticated observers were under no illusions on this point. Professor Percy Bridgman, the father of high pressure science, was quite clear in noting that diamond or any metastable phase could be made if the kinetic barriers to more stable phases were sufficiently high (8). Also, J.J. Lander of the Bell Telephone Laboratories in 1966, based on his careful low energy electron diffraction measurements on diamond, concluded that epitaxial growth of diamond at low pressures was a distinct possibility (9). Despite the observations of Bridgman, Lander and others, the strong perception remained that it would only be possible to grow diamond where it was the thermodynamically stable phase.

A third serious barrier which had to be overcome was the very low growth rates achieved by the early workers. It was held by many observers that, even though it was possible to grow diamond at low pressures, it would never be possible to increase the growth rates to the point to where a practical diamond synthesis process could be achieved. They argued that at carbon supersaturations sufficient for practical diamond growth rates, the nucleation and growth rates of graphite would be even greater. This was not an argument that could be dismissed by theoretical arguments. One had to, in fact, find conditions at which graphite nucleation and growth was suppressed sufficiently to permit the growth of diamond at reasonable rates.

There were some unfortunate institutional barriers which hindered development of diamond CVD during the 1960's and 1970's. The political situation in the Soviet Union was such that the leading Soviet workers in Professor Derjaguin's group could not openly discuss their results. This writer visited the Soviet Union in 1971 and presented a paper at an international diamond conference in Kiev (10). In this paper we described the use of a hot tungsten filament to generate atomic hydrogen for the removal of graphite from a diamond-graphite mixture and for the subsequent surface treatment of the diamond for further growth. At the time, the Soviet group was using atomic oxygen for removing graphitic deposits. The Soviet workers eventually used atomic hydrogen, not only in a cyclic etching process, but also during the growth part of the process as well. This was a critical discovery leading to the chemical vapor deposition of diamond at high rates (11). Unfortunately, the details of this research were not divulged. It is highly probable that if open communication of results had been possible at that time by the Soviet group, the field of low pressure diamond growth would have developed about a decade earlier than it actually did.

Another problem which influenced the rate of development of low pressure diamond growth was the unfortunate polywater episode which occurred about 1970. The announcement and subsequent retraction of the discovery of polywater by the same research group that was simultaneously announcing low pressure diamond growth, generated an enormous amount of skepticism and doubt about the validity of any low pressure diamond claims.

Finally, the reluctance of U.S. corporations to commit resources to high risk, but potentially high payoff projects was a very serious barrier. No effort of the scale or length of either the Japanese or Soviet efforts was ever mounted in the U.S.

Overcoming the Barriers

In the early 1960's the field was moribund. The efforts at Union Carbide and General Electric had been abandoned. The only literature on the subject was a short description of the General Electric effort by Guy Suits (2), the electron diffraction experiments of Lander (9) and patents including those of Eversole (5), Brinkman (6), Hibshman (7) and others. After reviewing the earlier work, and meeting with some of the principals, in particular Eversole and Hibshman, we concluded that the chemical vapor deposition from hydrocarbons was the method most likely to succeed. We decided that in order to renew the field it would be necessary to obtain absolutely convincing experimental proof of low pressure diamond growth of the quality that could appear in a peer reviewed journal. Support for this effort, as a "blue-sky" project, was obtained from Richard F. Cornelissen of the Air Force Cambridge Research Laboratories. The first goal of reproducing Eversole's work was achieved by the mid 1960's (12). As a further means of proving that new diamond was being grown, we added diborane to the gas phase. P-type, semiconducting blue diamond was obtained during diamond growth, but no conductivity was achieved by annealing under diborane alone (10,13).

Our work also made it clear that hydrogen played a critical role in the diamond growth process (14). The presence of molecular hydrogen dramatically slowed down the spontaneous nucleation and co-deposition of graphite during diamond growth. However, molecular hydrogen by itself was not enough to completely suppress graphite growth. John Forgac, a graduate student in our group spent a very significant effort on performing the diamond CVD at super-atmospheric pressures, up to 10 atmospheres. This method of increasing the chemical potential of hydrogen did not result in increased diamond growth rates. Much of this work is unpublished. We used hot tungsten filaments for generating atomic hydrogen which was used to clean off graphitic deposits from the diamond after growth. The atomic hydrogen also rationalized and prepared the diamond surface for the next growth cycle. However, our group did not use atomic hydrogen during the growth process. A drawing of an early hot filament used in our research is shown in Fig. 1. Our use of a hot filament to produce atomic hydrogen arose from the experience of Nelson C. Gardner, who used the method for cleaning field emission tips during his graduate work at Iowa State University.

It was possible to obtain support for these high risk studies during the late 60's and early 70's. In addition to the grant from the Air Force Cambridge Research Laboratories our research on diamond was funded by the Advanced Research Projects Agency (ARPA) of the Defense Department and by the National Science Foundation under its hard materials program. We were, however, unable to develop a long term, large scale development effort to permit a more complete exploration of the parameter space. U.S. corporations that were approached were uninterested. The ARPA effort eventually was terminated, apparently because of a belief that the process would never become practical.

The great achievement of the Soviet group was to use atomic hydrogen during growth (11). This permitted much higher growth rates and, perhaps of equal importance, it permitted the nucleation of new diamond crystallites on non-diamond substrates. The use of atomic hydrogen was also mentioned in other papers from the Soviet group during the 1970's and early 1980's as well (15). The experimental method, however, was not revealed.

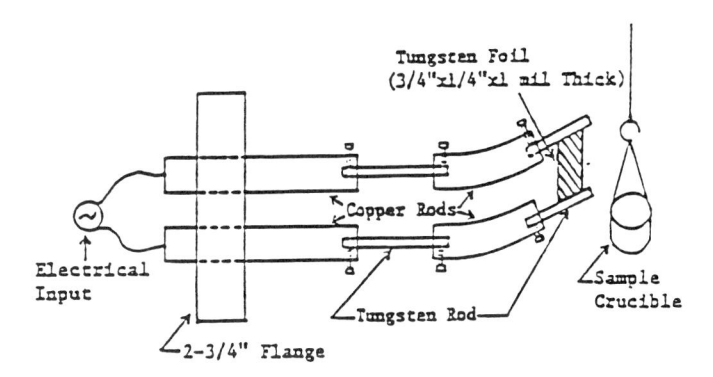

Fig. 1. Drawing of early apparatus for producing atomic hydrogen for cleaning co-deposited graphite from diamond.

It is not known with certainty which of the Soviet workers is primarily responsible for this development, but it is believed that B. Spitsyn, V. Varnin, L.L. Builov and D. Fedoseev all played roles in the discovery.

A Japanese group at the National Institute for Research in Inorganic Materials (NIRIM), under the direction of Nobuo Setaka, started a major effort on the chemical vapor deposition of diamond in 1974. They developed several methods for growing diamonds in the presence of atomic hydrogen at high growth rates (16). Their seminal papers on the use of hot tungsten filaments and microwave discharges for growing diamond films were sufficiently detailed to permit other workers to duplicate their results. These remarkable results mark the beginning of the current development of low pressure diamond technology. The majority of current research efforts in the world today can be traced directly to the NIRIM effort.

It is important to recognize that the Japanese effort went a full seven years, from 1974 to 1981, before any positive results were obtained. There was significant pressure to abandon the project (17). Several factors appear to have played a role in permitting such a long development effort. The first is the institutional climate in Japan, which does not require such immediate, short range payoffs as in this country. The second was the published results from the American and Soviet groups. Finally, the low pressure diamond project was embedded in a group which was achieving successful results in the high pressure synthesis of diamond.

Residue from the Past

Despite the enormous progress in the past decade since the announcement of the NIRIM results, there is still strong resistance to further innovations in this field. For example, virtually all research groups are still using vapor phase chemical vapor deposition from hydrocarbons. A great majority of workers are using simple variations of the methods revealed by Sato, Kamo, Matsumoto and Setaka in their earliest papers (16). Although the temperature, pressure and composition ranges have been expanded significantly, little effort has been expended on completely novel growth methods. The use of combustion methods for growing diamond (18) and the introduction of fluorine chemistry into the growth process (19) are exceptions to this rather negative assessment. Many of the possible processes for low pressure diamond deposition proposed during the 1950's and 1960's have yet to be

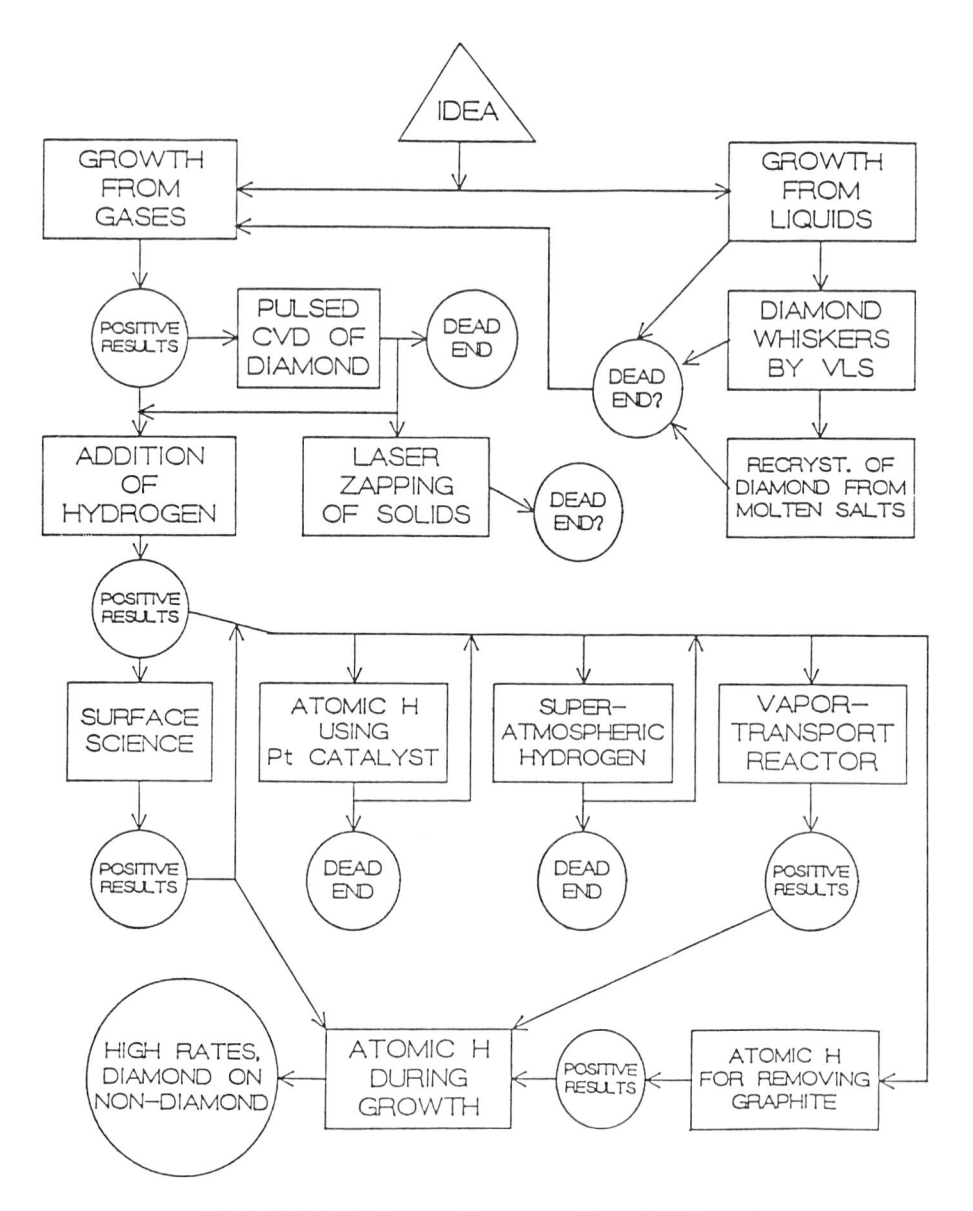

Fig. 2. Historical development of low pressure diamond CVD Technology

explored in depth. Perhaps the most obvious example is the growth of diamond whiskers by a vapor-liquid-solid process using molten iron. In fact, this was the method used by Derjaguin in his first publication on low pressure diamond growth (20). It certainly warrants further research. Also the intriguing results of Cherian and Patel on the recrystallization of diamond from nickel containing molten salts at atmospheric pressure have received little attention (21).

Summary

The development of low pressure diamond technology was not a simple linear progression from the original inception of the idea to the present situation. There were numerous false starts, dead-ends and feedback loops in the historical process. A somewhat oversimplified "road-map" of some of the major milestones in the evolution of low pressure diamond technology is shown in Fig. 2. It should be emphasized that the technology is still immature and some of the paths marked "dead end" surely warrant additional attention.

The major barriers that had to be overcome for the development of low pressure diamond technology were: 1) the success of diamond synthesis at high pressure, 2) the mistaken perception that it was thermodynamically forbidden to grow diamond at low pressures, 3) the belief that the simultaneous nucleation and growth of graphite would always preclude high rate growth of diamond at low pressures, 4) institutional barriers for communication by Soviet scientists during the 1960's and 1970's and 5) the association of low pressure diamond growth with the polywater episode.

Literature

1. J.C. Angus, "History and Current Status of Diamond Growth at Metastable Conditions," Proceedings of the First International Symposium on Diamond and Diamondlike Films, J.P. Dismukes, Editor, Volume 89-12, pp. 1-13, Electrochemical Society 175 Meeting, Los Angeles, CA, May 9, 1989.
2. C.G. Suits, "The Synthesis of Diamond—A Case History in Modern Science", A paper presented before the American Chemical Society, Rochester, New York, Nov. 3, 1960, Reference is made to an unpublished report by R.A. Oriani and W.A. Rocco.
3. B.V. Spitsyn and B.V. Derjaguin, author's certificate dated July 10, 1956; USSR patent 339,134, May 5, 1980.
4. J.C. Angus, "Synthesis of Diamonds," 3M Company Report, November, 1961.
5. A.D. Kiffer, Tonawanda Laboratories, Linde Air Products Co., "*Synthesis of Diamond from Carbon Monoxide,*" June 6, 1956; W.G. Eversole, U.S. Patents 3,030,187 and 3,030,188 (1962).
6. J.A. Brinkman, U.S. Patent 3,142,539, July 28, 1964.
7. H.J. Hibshman, U.S. Patent 3,371,996, March 5, 1968.
8. P.W. Bridgman, *Sci. Am.* 193, 42 (1955).
9. J.J. Lander and J. Morrison, *Surf. Sci.* 4, 241 (1966).
10. J.C. Angus, N.C. Gardner, D.J. Poferl, S.P. Chauhan, T.J. Dyble and P. Sung, *Sin. Almazy* 3, 38 (1971); presented at International Conference on Applications of Synthetic Diamonds in Industry, Kiev, 1971.
11. B.V. Derjaguin, B.V. Spitsyn, L.L. Builov, A.A. Klochkov, A.E. Gorodetski and A.V. Smolyaninov, *Dokl. Akad. Nauk SSSR* 231, 333 (1976).
12. J.C. Angus, H.A. Will, W.S. Stanko, *J. Appl. Phys.* 39, 2915 (1968); J.C. Angus, AFCRL Report 66-107, AD63-705, 1966.
13. D.J. Poferl, N.C. Gardner and J.C. Angus, *J. Appl. Phys.* 44, 1418 (1973).
14. S.P. Chauhan, J.C. Angus and N.C. Gardner, *J. Vac. Sci. Technol.* 11 (1), 423 Jan./Feb. 1974; S.P. Chauhan, J.C. Angus, N.C. Gardner, *J. Appl. Phys.* 47, 4746 (1976).
15. D.V. Fedoseev, S.P. Vnukov, B.V. Derjaguin, *Zh. Fiz. Khimii*, 51, 26 (1977); D.V. Fedoseev, K.S. Uspenskaya, V.P. Varnin and S.P. Vnukov, *Izvestiya Akad. Nauk SSSR, Seriya Khim* 6, 1252 (1978).
16. S. Matsumoto, Y. Sato, M. Kamo, N. Setaka, *Jpn. J. Appl. Phys.*, Part 2, 21 L183 (1982); S. Matsumoto, Y. Sato, M. Tsutsumi, N. Setaka, *J. Mat. Sci.* 17, 3106 (1982); M. Kamo, Y. Sato, S. Matsumoto, S. Setaka, *J. Cryst. Growth* 62, 642 (1983); Y. Matsui, S. Matsumoto, N. Setaka, *J. Mat. Sci. Lett.* 2, 532 (1983).
17. M. Kamo, personal communication.
18. Y. Hirose and M. Mitsuizumi, *New Diamond* 4, 34 (1988).
19. D. Patterson, B.J. Bai, C.J. Chu, R.H. Hauge, J.L. Margrave, Proc. Second Intl. Conf. on the New Diamond Science and Technology, Washington, D.C., September 24-27, 1990, to be published, Materials Research Society.

20. B.V. Derjaguin, D.V. Fedoseev, B.V. Spitsyn, D.V. Lukyanovich, B.V. Ryabov, A.V. Lavrentev, *J. Cryst. Growth* **2**, 380 (1968).
21. A.R. Patel and K.A. Cherian, *J. Cryst. Growth* **46**, 706 (1976); *Indian J. Pure Appl. Physics* **119**, 803 (1981).

Recreations of technological innovations tend to be much like recreations of research reported in technical journals. In both cases the activities are cast in a rational form that does not represent the stumbling, uncertain and often confused reality. Fig. 2 is important in its depiction of something closer to what actually happened. As Angus notes, Russian research in which atomic hydrogen was employed during the growth process allowed the successful deposition of diamond films on a variety of different substrates. This result was widely available in the West through publications of B.V. Spitsyn, L.L. Builov and B.V. Derjaguin (*J. Cryst. Growth* **52**, 219-226, 1981), but that report did not describe the experimental apparatus or conditions in sufficient detail that the work could be easily repeated. Acceptance of this achievement was also affected by Derjaguin's earlier role in announcing the discovery of a new form of water (polywater) which initially aroused much excitement and resulted in more than 200 publications before the purported "discovery" was shown to be an impurity effect. This episode caused a good deal of embarrassment for the early enthusiasts and certainly made the scientific community wary of further Derjaguin claims.

The research program on low pressure diamond synthesis at the Japanese National Institute for Research on Inorganic Materials at Tsukuba Science City was part of a larger program that also included high pressure high temperature diamond synthesis. It had been slogging along since 1974 and took up the atomic hydrogen growth process invented by the Russians. This resulted in very rapid progress using hot tungsten filaments and microwave discharges for growing diamonds on a variety of substrates at relatively rapid rates, about 1 micrometer per hour. These accomplishments were published in English language journals beginning in 1982 (*Jpn J. Appl. Physics* **21**, L183-185, 1982) and soon thereafter in widely read American journals (*J. Mat. Sci.* **17**, 3106, 1982; *J. Cryst. Growth* **62**, 642, 1983). The Japanese researchers described their apparatus techniques and results in sufficient detail that the work could readily be reproduced in other laboratories. These results and publications constituted the completion of the research innovation that had been commenced with the Russian work and opened the floodgates to subsequent events. Dr. Nobuo Setaka, now Director General of NIRIM and Drs. Yoichiro Sato and Mutsukazu Kamo describe NIRIM philosophy and its diamond CVD research:

CERAMIC INNOVATION AT NIRIM

Nobuo Setaka, Director General, National Institute for Research in Inorganic Materials

I think that we are required to carry out interdisciplinary studies and to study material synthesis in order to promote innovations in the field of ceramics. These studies are very important for developing materials science. The research organization and research attitude at NIRIM are particularly suited to these studies.

NIRIM has been organized with a special research system since its establishment. Researchers are organized within research groups in a way unique to NIRIM. In its organization NIRIM is different from other National Institutes in Japan.

The system has these features:

1) A materials name is given to each research group.

2) Each group consists of about seven researchers including different fields of specialization. For example: ceramist, chemist, physicist and mineralogist and so on.

3) The research period for a research group is set at 5 years. At the end of this research period, the group is dissolved and a new research group is organized with a new theme.

With these groups a flexible organizational matrix is formed. One axis is the organization of the director, the administration and the research group structure. The other axis is a special study field. It is thus easy to achieve cooperation among researchers beyond the limits of a particular group; such cooperation has naturally occurred between groups in NIRIM.

The research attitude in NIRIM is first to synthesize ceramics and then to develop new methods for synthesizing ceramics. The synthesis of advanced materials and the development of new methods are main themes in each research group as well as the evaluation of physical properties of materials. The researchers in NIRIM recognize that studies with relation to synthesis are of central importance for promoting materials science and innovation in the ceramic field. However, it is not easy to synthesize advanced materials; there is no guiding principle for synthesizing advanced materials. It takes a long time but the research period in NIRIM is around 5 years. Therefore, we are able to tackle risky problems with some confidence.

FROM DREAM TO TECHNOLOGY

Yoichiro Sato and Mutsukazu Kamo, National Institute for Research in Inorganic Materials

Introduction
At the initial stage of our research program, we did not necessarily expect that our program would have an outcome which might be considered as an innovation, i.e., an outcome of great practical importance. We wished for that, naturally, and we appreciated the possibility of electronic and optical applications. However, we would have felt very much rewarded if our results had been such that they could have had an impact on those who are interested in crystal growth, whether of experimental or theoretical background. To show unambiguous experimental evidence of diamond growth was the most essential part of our initial program.

In view of the central subject of the Conference, we will try first of all to point out what may be considered as innovation. Then, we will describe briefly our own view of our situation within the institution and how we could continue working in this field. A brief comment will also be given to the influence of some of the previous pioneering works on our own achievements.

Innovative Factors in Diamond CVD

First, innovation brought about by the currently adopted chemical vapor deposition (CVD) techniques will be discussed in a general sense, and then the contributions of NIRIM may be suggested.

Innovation in Diamond Synthesis

The most important features of the current CVD techniques may be summarized as:
1) Nucleation of diamond is feasible on various materials.
2) Growth rate is reasonably fast.

Because of 1), we may refer to the deposition of diamond by CVD as "synthesis" of diamond from the gas phase. Historically, most of the attempts to grow diamond from the gas phase have been directed to the "growth" of diamond on diamond seed crystals. If nucleation is not possible and only "growth" is feasible, CVD will find only limited applications. The point 2) is important for practical application and has been pointed out in an important paper by B.V. Spitsyn et al (1) in which the growth features anl properties of diamond are described in great detail together with discussion of the growth concept.

Other innovitive factors may be derived from a comparison with conventional high pressure techniques (3):
3) Most of the CVD reactors are simple in construction and operation.
4) In principle, there is no limitation to the size of deposition area.
5) The CVD techniques are supposed to have better capability for impurity control.

The feature 2) favors faster and wider spread of the technology and 4) should be a great advantage for the fabrication of electronic devices or light emitting devices.

Contribution of NIRIM

The principle contributions of our earlier works published in the years 1982-1983 (3,4) may be summarized as:
a) Use of CVD reactors equipped with gas flow systems.
b) Introduction of Raman spectroscopy as a means of characterization of the deposits.
c) Mechanical surface treatments of substrates to increase nucleation density in order to form films.

Explanation of the background may be necessary for a). Flow-type gas supplying systems are most common in various CVD reactors, but we had some discussion as to whether we should adopt the chemical transport system which B.V. Spitsyn et al had employed (1). They stressed the growth rate advantage of the chemical transport system over conventional CVD techniques. Not much time was spent, though, before concluding that flow systems would have an advantage over closed systems in maintaining gas purity and in controlling gas composition, gas supplying rate and impurity doping conditions.

Although b) is no innovation to the CVD method itself, it should be considered to form an important part of the overall innovation of diamond CVD technology as a whole. Application of Raman spectroscopy to the characterization of diamond, in the authors' view, has played a key role in attaining technical developments for synthesizing high quality diamond by CVD (5, 6) in a relatively short period of time. Raman spectroscopy proved to be a sensitive tool to detect non-diamond structures in the deposits, which are rather hard to detect by X-ray or electron diffraction. There are reasons to believe that it will remain an important tool. It should be noted that "what cannot be detected cannot be controlled."

Another by-player, important but clumsy, is the mechanical surface treatment given in c). Scratching of the substrate surface by hard abrasive powders, typically diamond, is the procedure we adopted. It can attain a nucleation density of about 10^9 nuclei/cm^2, which enables us to prepare polycrystalline films of submicron thickness but no less. Various modifications have been developed, but most of them are as clumsy as our own.

Breakthroughs which produce much higher nucleation density without mechanical treatments are looked for.

We have recognized the importance of various technical subjects which are more or less common to various applications. Most of the most recent work at NIRIM is directed toward: 1) Control of defects and impurities, 2) control of nucleation density, 3) low temperature synthesis, 4) high speed growth, 5) homogeneous film formation over a large area, 6) epitaxial growth on non-diamond substrates and 7) homogeneous nucleation.

Situation at NIRIM

The program of diamond CVD was considered seriously for the first time at NIRIM in early 1973 and the program was initiated in 1974, as a part of the program of Diamond Group under Nobuo Setaka, present director of NIRIM, to be continued for a period of five years 1974-1978. The subgroup working on the CVD program consisted of three members. In the first five years, substantial progress was achieved by the other subgroup for high pressure synthesis, in cooperation with High Pressure Station in NIRIM. Because of this achievement, the Diamond Group continued another five years, 1979-1983, with the same staff.

Successes in depositing diamond on non-diamond substrates at NIRIM were achieved in 1981-1982. Almost eight years had elapsed from the beginning of the program.

Our own understanding of the positive factors within the institution that have led to the successes may be summarized:

1) The subject seemed to be worth challenging: this recognition enabled continued efforts to proceed.

2) High-pressure and CVD staffs were in the same group:

This led, first of all, to the survival of the Group. Communication among co-workers helped all to have knowledge of diamond synthesis in general and to have a deeper understanding of the high pressure synthesis. For example, the ^{13}C enriched diamond single crystals necessary for CVD research were supplied by our high-pressure colleagues.

3) Budget condition was not good, but some of the useful apparatus was made accessible to the CVD staff thanks to the kindness of colleagues:

This permitted us to make studies on more basic sides of vapor deposition, which included studies on a) chemisorption and thermal desorption on diamond surnaces, b) surface graphitization under diffmrent ambient gases, c) Raman spectroscopy of ^{13}C isotope enriched diamond and graphite, as a part of identification techniques.

4) The staff was subject only to attenuated political pressures. Also, the staff was not clever enough to avoid undertaking a risky project. Determination of the leader is a primary requirement. Youth also is an important factor.

Some of the Pioneering Works Which Influenced Ours

It should be pointed out that it is highly probable that we would not have undertaken the program at all unless there had been pioneering works of American and Soviet groups. However, in 1973, when N. Setaka suggested the program, we had knowledge mostly of papers and patents related to growth of diamond on diamond seed crystals. Growth on seed crystals with reasonable growth rate or efficiency appeared extremely difficult, not to speak of growth on non-diamond substrates.

Here are a few lines from two papers which had great impact on our studies.

1) "The tendency to discuss diamond synthesis in the terms of equilibrium thermodynamics can obscure the fact that kinetic factors may provide the possibility of diamond synthesis in a temperature-pressure regime where diamond is actually thermodynamically unstable with respect to graphite." J. Angus et al (1968) (7).

For one of the authors, Y.S., theoretical reasoning started from this paper together with limited knowledge of the fundamentals of crystal growth, thermodynamics, and structural organic chemistry.

2) "Selective growth of diamond is ensured by introducing atomic hydrogen into the crystallization zone: this suppresses crystallization of graphite....Diamond crystals up to several tens of microns in thickness were also grown on non-diamond substrates." (B.V. Spitsyn et al (1981) (1).

It is fortunate that we were about to concentrate our efforts on C-H gas systems when we came to know of this paper. It was this paper that turned us to concentrate on non-diamond substrates immediately. Identification of the growth product is much easier than for homoepitaxial growth.

Concluding Remarks

We think that the CVD techniques which have been worked out at NIRIM over the past years are still at the stage of laboratory experiments. Our contribution, therefore, may be considered scientific or mental in nature. We conclude our paper by citing: "In the light of this history it is remarkable that the question has received so little serious attention." J. Angus et al (1968). For those interested in diamond CVD in general, several excellent reviews are available (8).

Diamond CVD today is on its way from dream to technology.

References
1. B.V. Spitsyn, L.L. Bouilov and B.V. Derjaguin, *J. Cryst. Growth,* 52, 219 (1981).
2. See, for example, a review work, F.P. Bundy, H.M. Strong and R.H. Wentorf, Jr., in *Chemistry and Physics*, Vol. 10, ed. P.L. Walker, Jr. and P.A. Thrower, Marcel Dekker, New York, 1973, p. 213.
3. S. Matsumoto, Y. Sato, M. Kamo and N. Setaka: *Jpn. J. Appl. Phys. 1*, L183 (1982); S. Matsumoto, Y. Sato, M. Tsutsumi and N. Setaka, *J. Mater. Sci.* 17, 3106 (1982).
4. M. Kamo, Y. Sato, S. Matsumoto and N. Setaka, *J. Cryst. Growth, 6:*, 642 (1983).
5. A.T. Collins, M. Kamo ind Y. Sato, *J. Phys. Condens. Matter 1*, 4029 (1989).
6. Y. Sato and M. Kamo, *Surf. Coatings Technol.;* 9/40, (1989).
7. J.C. Angus, H.A. Will and W.A. Stanko, *J. Appl. Phys.* 39, 2915 (1968).
8. For example:
    a) R.C. DeVries, *Annual Rev. Mater. Sci. 17, 161 (1987).*
    b) A.R. Badzian and R.C. DeVries, *Mater. Res. Bull. 23, 385 (1988).*
    c) J.C. Angus and C.C. Hayman, *Science, 241*, 913 (1988).
    d) K.E. Spear, *J. Am. Ceram. Soc.* 72, 171 (1989).

The most recent development in diamond synthesis has been the use of high energy plasmas and flames to achieve rapid rates of diamond deposition — growth rates approaching a millimeter per hour. Dr. Moriyoshi describes these developments:

THE ROAD TO HIGH RATE SYNTHESIS OF DIAMONDS

Yusuke Moriyoshi, National Institute for Research in Inorganic Materials
Tsukuba, Japan

The possibility of diamond synthesis at lower pressure than 1 atm was first considered by Derjaguin et al (1) who applied for a patent concerning the synthesis of diamond in 1956. Later Eversole took a patent about the synthesis of diamond by using thermal CVD at lower pressure than 1 atm in 1962 (2). Thereafter, Angus tried successfully to prepare diamonds by using Eversole's method (3-4). These researchers indicated that diamond could be grown on diamond seed crystals by vapor deposition from CH_4 at temperatures and pressures on the order of 1050°C and 0.3 torr. However, the crystalline quality of the diamond was not established. Also, the deposition rate was extremely low, about 1 Å/hour, and simultaneous code position of graphite was always a problem. Derjaguin et al carried out the synthesis of diamond by using an electron beam and thermal pulse CVD (5-6). Through these experiments, they clarified that the dissociation of H_2 begins at about 800°C and takes place rapidly at temperatures above 2000°C. The resultant atomic hydrogen retards effectively the codeposition of graphite. They also made diamond whiskers which were grown on diamond single-crystal substrates from the thermal pulse CVD by using a Xenon tube of 6k/V and pointed out the VLS method for growth of diamond whiskers.

In 1981, Spitsyn et al played an extremely important role in diamond synthesis at lower pressure than 1 atm; they showed that the synthesis of diamonds was ensured by introducing atomic hydrogen into the deposition zone. They clarified the synthesis conditions for diamond and successfully made diamonds with crystal habits by a CVD at lower atmosphere than 1 atm (7). The growth rate of the diamond films reached about 0.02 μm/min at 1000°C. Through the deposition of diamond, they pointed out an important role of atomic hydrogen, the growth temperature, and also indicated the possibility of p-type diamond doped with boron.

The National Institute for Research in Inorganic Materials (NIRIM) has studied the synthesis of diamond at lower than atmosphere pressure since 1975. They were thinking that the important point of diamond synthesis from gas phase was the presence of non-equilibrium chemical species in the gas phase and on the growing surface; that is, the presence of excess excited species in gas phase such as CH_3, C_2H, C and so on. Using this idea, NIRIM has tried to prepare diamonds by using various methods from gas phase of H_2 and CH_4. Matsumoto et al in NIRIM reported a new method of thermal CVD heating a filament just over a substrate, by which they successfully prepared diamonds with crystal habits without graphite at pressures below 1 atm to prepare diamond particles and films (8). The growth rate of the diamond was fast, about 0.1 μm/min. Kamo et al in NIRIM developed a microwave plasma to prepare diamond particles and films (9). Matsumoto et al also developed a radio frequency glow discharge (10). Thereafter, many scientists in the world began to use these methods for the synthesis of diamond particles and films.

Thermal plasmas have been used for the synthesis of diamond since 1987. At first it was thought that the higher temperatures, usually above 6000°K, and the high heat content of the plasma would be unsuitable for the synthesis of diamond. At these conditions, it was considered that diamond transforms into graphite at normal or reduced pressure. Therefore, sufficient cooling of the substrate is a particularly important key to solve the problem. The

Fig. 1. Schematical diagrams of thermal plasmas (a) a rf thermal plasma torch, (b) a D.C. arc, and (c) a D.C. jet.

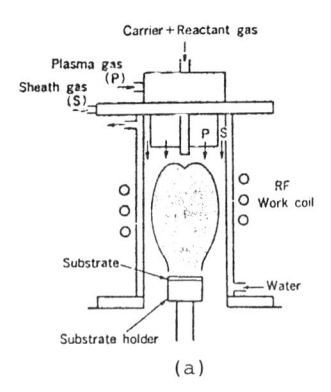

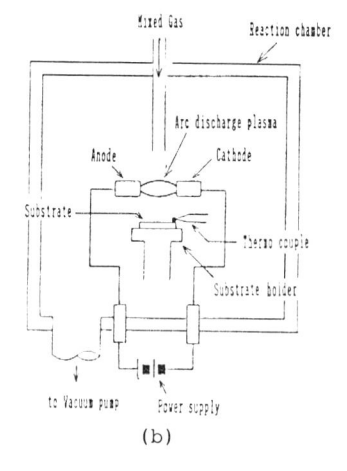

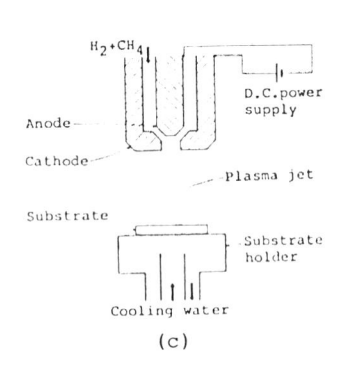

important advantage in the synthesis of diamond using a thermal plasma is the high speed of deposition resulting from a very high concentration of excited chemical species.

The first paper concerning high speed synthesis of diamond was reported by S. Matsumoto et al (11), who used a radio frequency induction (rf) heating to produce a thermal plasma. The apparatus used is illustrated in Fig. 1(a). The torch was a conventional one made of coaxial silica tubes operating at about 30 kW with 4MHz. The resultant film thickness was not so uniform, one of the films was 12 m thick near the edge and 6 m thick at the center of the substrate. The growth rate amounted to 3-5 μm/min in crystal diameter and 1 μm/min in film thickness for single crystals and films, respectively. These growth rates are about 100 times larger than those by the other methods reported previously.

Akatsuka et al thought to exchange the heating filament in low temperature plasma into an arc discharge plasma (12), since much more atomic hydrogen and hydrocarbon radicals could be generated with the arc discharge plasma than those with the heated filament. As

Fig. 2. The activity JRDC has been carrying out since its foundation is to promote commercialization of new technology that contributes to the improvement of national economy. JRDC collects information on the research outcome from universities and public research institutions as well as from individual(researchers across the nation to promote "Cooperative Development with Industry" or "Coordination for Licensing" depending on the degree of risk in developing the technology so that technology can be exploited by industry. JRDC is an intermediary that links inventors and companies.

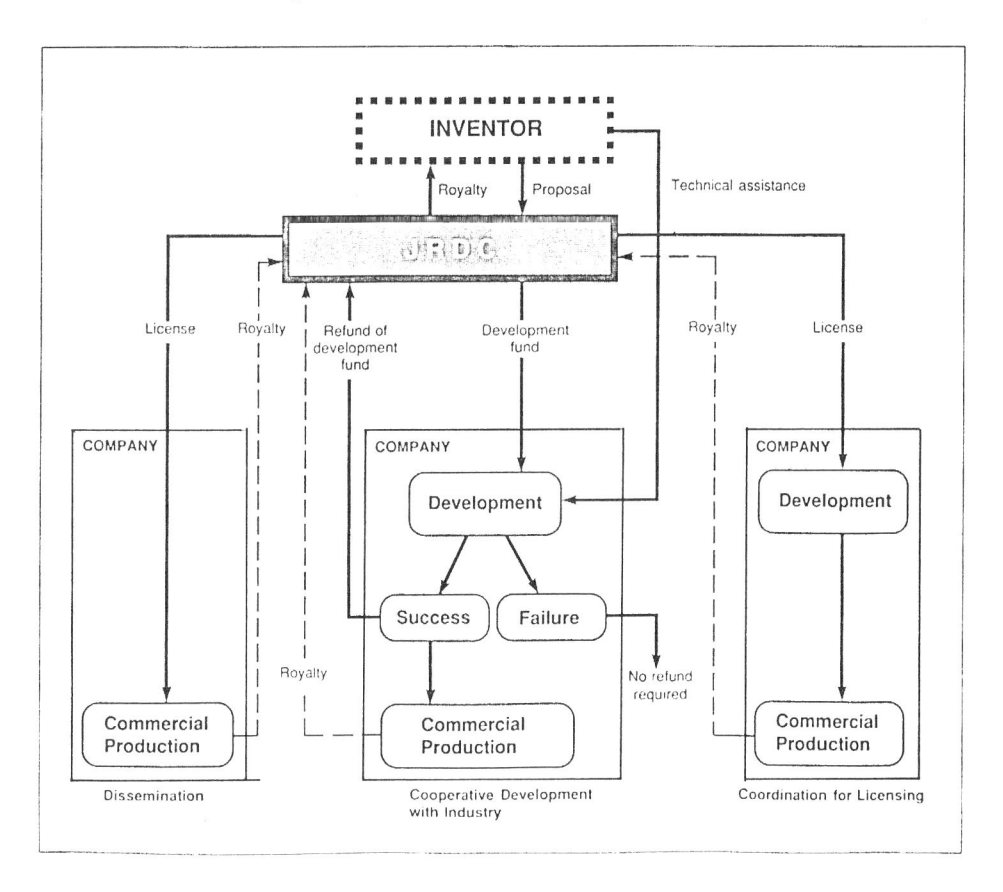

a result, high growth rate of diamond deposition could be expected. The experimental apparatus used in their study is shown in Fig. 1(b). The D.C. arc discharge plasma is between the anode electrode and the cathode. The reaction gas is decomposed by the arc discharge plasma just above the substrate and then the diamond is deposited onto a water-cooled substrate. The obtained growth rate of diamond films is about 3-4 μm/min. They reported that the growth rate is higher than those of any other CVD methods.

From a viewpoint of electrical power, a D.C. plasma jet has very high efficiency. It produces a temperature above 8000°K which is one of the most effective methods to obtain high concentration of excited chemical species. Therefore, it is a powerful tool to prepare diamond particles and films. Kurihara et al (13) successfully made diamond films on a substrate cooled with water by using the jet as schematically shown in Fig. 1(c). The growth rate is high about 1.5 μm/min at 6kW-18kW. It is possible to elevate the electric power used to MW, much more high rate deposition would be expected.

Other methods are possible for high speed production of diamond films and particles. For instance, a thermal plasma of hydrogen at 1 atm can be made at low power by using microwave discharge. The resultant thermal plasma was generated at 2-5 kW with 2.45 GHz, by which Mitsuta et al produced diamond films on a silicon substrate at a rate of 0.5 μm/min (14). Also, diamond synthesis by using a combustion flame is particularly important (15). Diamond films with good crystallinity on a substrate can be deposited in about 2 μm/min at 1 atm. An extremely large combustion flame would be possible to develop. At present, various combined methods are being tested in the US and Japan to produce high quality diamond films.

Diamond Technology Transfer

The diamond technology transfer from National Institutes and Universities to private companies is carried out in the following ways.

The National Institutes and Universities can transfer the technologies and results of research to private companies by way of the Research Development Corporation of Japan (JRDC). It is the intermediary organization belonging to STA. In some cases, patents are transferred to JRDC and then commercialized by suitable private companies. JRDC coordinates such systems in two ways: development by contract and coordination of licensing. Contract Development is applied in the case of patented inventions and technology which involve a greater financial risk. Contract Development supplies a considerable fund to the companies, in order to effectively develop the new technology for industrial use. Once production is established companies refund the JRDC. However, it is said that it is very difficult for industry to get these funds from the JRDC.

References
1. B.V. Derjaguin, USSR Patent, No. 399134, 7/10/56.
2. W.D. Eversole, US Patent, No. 3030187, No. 3030188, 4/17/62.
3. J.C. Angus, H.A. Will & W.S. Stanko, *J. Appl. Phys.*, *39* 2915 (1968).
4. S.P. Chauhau, J. C. Angus, and N.C. Gardner, ibid., *47,* 4746 (1976).
5. B.V. Derjaguin et al, *J. Crystal Growth,* *2* 380 (1968)
6. B.V. Derjaguin and D.V. Fedossev, *Sov. Phys. Dokl.,* *18,* 771 (1974).
7. B.V. Spitsyn et al. *J. Crystal Growth,* *52,* 219 (1981).
8. S. Matsumoto, Y. Sato, M. Kamo, and N. Setaka, *Jpn J. Appl. Phys.,* *21,* L183 (1982).
9. M. Kamo, Y. Sato, S. Matsumoto, and N. Setaka, ibid. *62,* 642 (1983).
10. S. Matsumoto and N. Setaka, NIRIM reports, *39, (1984).*
11. *S. Matsumoto, H. Hino, and T. Kobayashi, Appl. Phys. Lett., 51,* 737 (1987).
12. Y. Akatsuka and Y. Hirose, *Jpn. J. Appl. Phys.,* *27,* LI600 (1987).
13. K. Kurihara, et al, *Appl. Phys. Lett.,* *52,* 437 (1988).
14. Y. Mitsuda, T. Yoshida, and K. Akashi, *Rev. Sci, Instr.,* *60,* 249 (1989).
15. Y. Hirose, Abstract. 1st Inter. Conf. New Diamond Sci. Tech., (Japan New Diamond Forum, Tokyo 1988), p. 38.

The initial response to these developments opening the doors to what *Science* Magazine in 1990 described as a "glittering prize for materials science" (R.L. Guyer and D.E. Kochland, Jr., *Science,* **250,** 1640-1643, 21 December 1990) was not immediately recognized but soon developed. Beginning with corporate research programs in Japan (Kobe Steel, Sumitomo, NEC, Mitsubishi, Toshiba, Fugitsu, Edimitsu, Asahi, and others) the perception of possible commercial innovations based on the NIRIM successes initiated a self-catalyzing increase in low pressure diamond growth research. In the United States the Crystalume

Corporation was formed to exploit this possibility and General Electric Company, a manufacturer of diamonds by the high temperature high pressure process, inaugurated a substantial research effort in 1984. In that same year Professor Rustum Roy of Pennsylvania State University visited the NIRIM laboratory and was instrumental in obtaining research support by the Office of Naval Research and a corporate consortium. Professor Roy interprets this rather slow response as follows:

INFLUENCE OF POLICY AND CULTURE ON R/D: THE CASE OF DIAMOND FILM RESEARCH

Rustum Roy, The Pennsylvania State University

History

Early Work

For the present purposes a schematic presentation of the history of diamond synthesis as shown in Tables 1 and 2 will serve to make our points. There is virtually no disagreement on the facts and people involved in diamond film research although, no doubt, meetings such as the present one will help enrich the texture of the record by providing the viewpoints of different contributors, many of whom are present at this meeting.

Table 1 lists the 100-year history of scientists attacking the synthesis of diamond as a benchmark goal. From this table two points can be seen. Before WWII vapor liquid and solid state approaches were all tried. Since 1950, success via the high pressure route came rather rapidly and more or less independently at ASEA, Norton and, of course, GE. Yet the successful commercialization of high pressure diamonds by GE starting in 1955, clearly set back all vapor phase efforts. Moreover, the thermodynamics of Rossini and Jessup became firmly entrenched because GE showed that diamonds could (only) be made in the stable regime.

Yet vapor phase approaches persisted in scientific eco-niches. (Table 2). The earliest was at Union Carbide in parallel with their high pressure efforts where Eversole, starting in 1952, succeeded (1958) in depositing diamond epitaxially on diamond substrates by thermal pryolysis of hydrocarbons. In parallel work in 1956 at the Institute for Physical Chemistry in Moscow a graduate student, Boris Spitsyn, working under B.V. Derjaguin succeeded in the same goal. Excellent *technical* reviews of CVD diamond work are found in DeVries (1) Badzian and DeVries (2) and DeVries and Roy (3). Through the sixties, the only U.S. thermal pyrolysis work was by Angus who (1967) confirmed and extended the Eversole results. In 1970 the status of CVD thermal pyrolysis diamond work was that a substantial effort in Moscow and Angus' work had showed that *pyrolysis on a diamond substrate* led to a mixture of graphite and diamond, and that the graphite could be removed by cycling hydrogen only over the hot mixture. It is not difficult to see why this was guardedly of great interest to science or industry: growth of microscopic amounts of diamond on a diamond powder substrate with a cyclic process was not a very impressive technical achievement. In 1971 Angus at a paper in Kiev reported that he had used a heated filament in his H_2 during the cleaning cycle. The Soviet workers (Fedoseev, Derjaguin, Varnin) utilized this "hint" or others, and from the early seventies on became the world center for

TABLE 1. EARLY HISTORY OF DIAMOND SYNTHESIS*

1880	HANNAY	Bone Oil, Li, Iron Tube
1886	MOISSAN	Carbon in Molten Iron, Graphite Cruc.
1917	RUFF et al.	Pyrolyzing C_2H_2, CH_4, CO, etc., 790°C. Claimed NO, diamonds
1920	PARSONS	Tried duplicating earlier work. Concluded: No one had succeeded.
1921	TAMMANN	CCl_4, CBr_4, CI_4 at 800°C
1924 on	HERSHEY	Repeated Moissan v. carefully. Evidence good: patent fights with GE.
1938	ROSSINI, JESSUP	Thermodynamic calculations
1939	LIEPUNSKI	Calculations. Recommends use of Fe.
1941-50	BRIDGMAN	Supported by Norton, GE, Carborundum. Broke up in 1947.
1951	G.E. starts in-house	Hi-P and CVD.
1953	NORTON*	Synthesizes all Hi-P phases except diamonds.
1953	ASEA*	von Platen, Lenader, Lundblad
Dec. 1954, Feb. 1955	G.E. succeeds* and announces	
1957		Syn. diamonds, commercial CBN made.

*Given what we know today, some of these no doubt made diamonds BUT COULDN'T PROVE IT; NO RAMAN, NO XRD.

CVD diamond research with a long stream of papers—albeit short on detail and long on kinetics and chemistry. These papers stimulated literally no one in the world except the group in NIRIM in Japan (Matsumoto, Kamo, Sato and Setaka) who *with no official program* sustained a modest effort for many years.

CVD Goes Public

First, by the late seventies in Moscow and then by 1981 in Tokyo, the benchmark synthesis had been achieved: *diamond films could be grown at 1 atm ON MANY SUBSTRATES at a reasonable rate* $\approx 1u/hr$. Figure 1 shows the title and abstract of two key papers published in English (obviously some years after the original work and its location publication) which clearly announced this unexpected, indeed incredible, achievement with OBVIOUS TECHNOLOGICAL SIGNIFICANCE.

TABLE II. KNOWLEDGE FLOW IN CVD DIAMOND

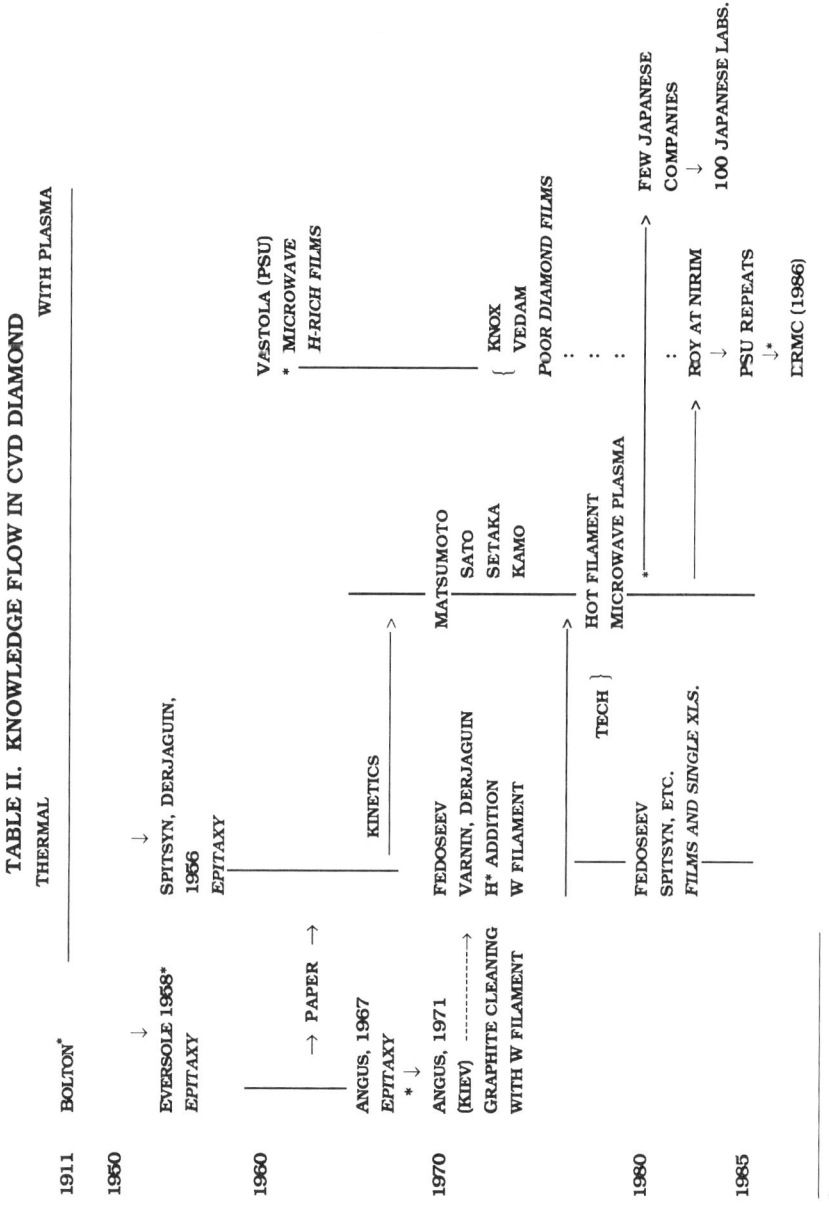

	THERMAL	WITH PLASMA
1911	BOLTON*	
1950	↓	
	EVERSOLE 1958* EPITAXY	SPITSYN, DERJAGUIN, 1956 EPITAXY
	→ PAPER →	
1960		KINETICS
	ANGUS, 1967 EPITAXY * ↓	
		VASTOLA (PSU) * MICROWAVE H-RICH FILMS
1970	ANGUS, 1971 (KIEV) GRAPHITE CLEANING WITH W FILAMENT	FEDOSEEV VARNIN, DERJAGUIN H* ADDITION W FILAMENT
		MATSUMOTO SATO SETAKA KAMO
	TECH }	HOT FILAMENT MICROWAVE PLASMA
		{ KNOX VEDAM POOR DIAMOND FILMS
1980	FEDOSEEV SPITSYN, ETC. FILMS AND SINGLE XLS.	* → ROY AT NIRIM → FEW JAPANESE COMPANIES → 100 JAPANESE LABS.
1985		PSU REPEATS →* LRMC (1986)

* From Gardner

From the viewpoint of CULTURAL FACTORS IN SCIENCE, this record will surely intrigue historians. For these papers were TOTALLY IGNORED by every laboratory in the world, even major companies (GE, De Beers, etc.) making synthetic diamonds and university groups such as the author's own, to which we turn.

Journal of Crystal Growth 52 (1981) 219-226
© North-Holland Publishing Company

VAPOR GROWTH OF DIAMOND ON DIAMOND AND OTHER SURFACES

B.V. SPITSYN, L.L. BOUILOV and B.V. DERJAGUIN

Institute of Physical Chemistry, Academy of Sciences of the USSR, Moscow, USSR

It is shown that diamond crystallization by chemical vapor deposition should preferably be carried out at reduced pressures. Selective growth of diamond is ensured by introducing atomic hydrogen into the crystallization zone; this suppresses crystallization of graphite. The growth rate of homoepitaxial diamond films reached 1 μm/h at 1000°C; film properties were identical to those of bulk crystals. The lattice parameter in boron-doped films (~0.1 at.%) decreased by 0.0009 Å; the film and substrate parameters coincide at dopant concentrations of ~1 at.%, and the semiconductor diamond film intergrows with the substrate without stress. Diamond crystals up to several tens of microns in thickness were grown also on non-diamond substrates. At large supersaturation, the crystal habit is octahedral and at low supersaturation, it is cubic. The linear growth rate is constant at the early stages of crystal growth but then it diminishes to a level typical for the homoepitaxial growth of diamond films.

Journal of Crystal Growth 62 (1983) 642-644
North-Holland Publishing Company

LETTER TO THE EDITORS

DIAMOND SYNTHESIS FROM GAS PHASE IN MICROWAVE PLASMA

Mutsukazu KAMO, Yoichiro SATO, Seiichiro MATSUMOTO and Nobuo SETAKA

National Institute for Research in Inorganic Materials, 1-1 Namiki, Sakura-mura, Niihari-gun, Ibaraki 305, Japan

Received 12 April 1983

Crystalline diamond predominantly composed of (100) and (111) faces was grown on a non-diamond substrate from a gaseous mixture of hydrogen and methane under microwave glow discharge conditions.

Fig. 1. It is inexplicable that these two papers, Soviet (in 1981) and Japanese (in 1983) in a major U.S. journal announcing unequivocal preparation of diamond films (including the key role of H_2) and on non-diamond substrates, were totally ignored by the world community.

TABLE III. HISTORY OF PENN STATE INVOLVEMENT IN DIAMOND SYNTHESIS

a. High Pressure

1957	**Tuttle and Roy** (ONR) Alternative (hydrothermal) high pressure route to diamonds (our ____ catalysis finally achieved by ___ in 1990).
1960	**Roy and Dachille** Anvil approach, long-term contact with <u>Vereschagin</u> USSR Academy of Science
1969	**Carborundum** diamond factory starts in State College - 1966 with PSU help.
1970	Joint diamond - boron nitride - boride work with **Niemyski-Badzian** in Warsaw

b. Vapor Phase

1962-67	**Vastola and Knox** Microwaves and hydrocarbons
1971-74	**Knox and Vedam** RF & microwave plasma decomposition of hydrocarbons - di-synthesis achieved but disbelieved
1971-73	**Messier and Roy** Hyperdense Ge achieved by sputtering
1984	**Roy** visits "NIRIM" in Tsukuba, sees diamonds on Si. Triggers major effort.

Penn State Research on Diamond Synthesis

At Penn State, which had by far the largest university high pressure research program in the U.S. (under Professors Tuttle, Roy, Wylie, Burnham, Harker, Boettcher, etc.) ONR provided funds in 1957 to try to make diamonds by *alternate routes:* mainly hydrothermal carbonate rich liquids imitating nature (Table 3). What is relevant here is that independent of the high pressure group, the VERY FIRST RECORD OF GROWING "carbon" films in $H_2 + CH_4$ mixtures in a MICROWAVE PLASMA were those by Vastola et al. This was not buried in an obscure journal but rated a full page spread in *Chemical and Engineering News* (May 7, 1962, p. 44). In today's technology these solid films were "diamond-like carbon" containing hydrogen. However that work was followed up in the early seventies by B.E. Knox and K. Vedam at Penn State's MRL who, in the Final Report (31 January 1975) to the Defense Department's Advanced Research Projects Agency, Contract 2415, provided electron diffraction evidence (obtained independently by J.J. Comer of the Air Force Cambridge Laboratories) *that in Ch_4-H_2 mixtures in a microwave plasma* they had produced diamond (plus some graphite materials) films on many substrates. The results were dismissed by most, including myself as Laboratory Director, as a quirk contrary to thermodynamics.

However, Penn State's MRL by the seventies had become prominent in thin film work by PVD and CVD. Thus in February 1984 when visiting colleagues at NIRIM the present author was shown a 3" Si wafer with a myriad sparkling points of light, and the x-ray and Raman evidence, for diamond, he was truly astounded that he had ignored the published work.

This "knowledge" of the success in Japan was received by most colleagues in the U.S. with very modest, if any, interest or even skepticism. What is important here is that I could not convince most university colleagues and those in the highest tech corporate labs that this was a very *significant* result. Obviously selling this to "peer reviewers" would be impossible. Fortunately agencies such as the Office of Naval Research exist—that understand deeply the question of SIGNIFICANCE of an invention. Hence by 1985 because of personal contacts with Dr. A.M. Diness of ONR, Penn State's MRL was launched in trying to confirm the Japanese work. I had by then gone to Moscow's Institute for Physical Chemistry again where I was a very frequent visitor of other groups, and been brought up to date on the enormous depth of the Soviet work.

We note here what is particularly relevant to this conference is that even when in 1986 we had succeeded in duplicating the Soviet-Japanese work that our unambiguous results were greeted with enormous skepticism. In every case the skepticism lessened if they physically saw a film and the data. Moreover, because we scrupulously insisted on crediting the Soviets and Japanese, the TV and newspaper media both underplayed that aspect, and tended to downgrade the significance of their achievement. At this point as a major university laboratory we were confronted with the question of how to manage this new knowledge and for whose benefit. Two major companies approached us with the idea of exclusive linkage to them. In spite of offers from two of the largest venture capital firms, we decided that that route was wholly inappropriate for a university in this case. Our judgment at PSU-MRL was that (a) diamond films were a "universal" enabling technology which would affect dozens of industries and (b) that our usual networked-consortium model would be very appropriate for a broad attack on synthesis and processing of such materials. Thus was born the Diamond and Related Materials Consortium which has had roughly 25± members from all over the world now for four years. It has achieved exactly what was hoped: widespread generation, collection and dissemination of knowledge on diamond films.

Policy and Culture Issues

By sheer coincidence, I have been deeply involved and intimately knowledgeable about both recent diamond and the superconductor developments, while being active in national science policy analysis and formulation. It has become increasingly clear to me that the weaknesses of the U.S. R/D system have nothing whatsoever to do with the (1) level of funding, (2) the number or quality of the scientists/engineers, or (3) the availability of facilities. Yet these are the only three issues which are ever addressed by the national science policy establishment.

From my more detailed writing in the field (see Refs. 4-6), I have therefore attempted to list what I regard as the most important technical factors which have and still contribute to the sorry state of U.S. science and technology. (I am omitting here "social" factors such as interest rates, patient capital, LBO's, education, labor relations, etc.).

1. The wholly unjustified view of the superiority of U.S. research, its position in the world, its methodologies for funding, prioritizing and conducting research.

This results in the utterly ludicrous attitudes and statements such as "Japanese researchers are not creative," "Japanese technological superiority depends on U.S. basic research" (which is "taken" or "stolen" without adequate payment, etc.). These are not casual remarks: even the President of the U.S. National Academy of Scientists believes and proposes such ideas.

2. The enormous confusion in the language and terms in science policy: science, technology, engineering, basic science, applied science; and the religious fundamentalism in the U.S. physics (and chemistry) community that believes, and tries to make a nation base its policy on the absurdity that if we just do more "basic science," all our technological problems would be solved.

This has led to inadequate contact with and learning from worldwide colleagues and laboratories.

3. The gross neglect of the scientific literature in the U.S. as compared to Japan, the USSR and Europe.

It is my considered judgment that if 10% of the research budget were specifically allocated to reading, analyzing and publishing summaries of the literature, the U.S. would learn more *new* science than by funding more research.

4. The short-term and often sub-critical nature of U.S. funding patterns is grossly counterproductive when it is absolutely certain that a few steadily supported groups with slowly mutating goals serve the national interest.

The Balkanized structure of the U.S. science establishment, and the absence of any political leadership, especially one with any competence in S/T dooms the country to continuing decline.

Summary

The history of diamond film research clearly illustrates three non-technical realities: (1) the published literature is almost totally ineffective in transferring knowledge; (2) sustained efforts in research areas are effective in rapid capitalizing on new opportunities; and (3) too much reliance on establishment paradigms (in this case thermodynamics in the C system) blocks innovation.

References

1. DeVries, R. *Annual Rev. Mat. Science 17*: 161 (1987).
2. Badzian, A. and R. DeVries, *Mat. Res. Bull.* 23:385-400 (1988).
3. DeVries, R. and R. Roy. *CVD Diamonds—CVD Diamond and Cubic BN: A Collection of Key Papers*, A.I.P., New York, NY 1991 (in press).
4. "The Nature and Nurture of Technological Health," *Ceramics and Civilization, Vol. II. High-Technology Ceramics--Past, Present, and Future*, W.D. Kingery (ed.), American Ceramic Society, pp. 351-370 (1987).
5. "University-Industry Knowledge Transfer: Exaggerated Expectations," *Forum for Applied Research in Public Policy 3* (4), pp. 32-36 (Winter 1988).
6. Roy, Rustum. Creating New Wealth: National Research and Development Funding in the 1990s. Testimony before the Committee on Science, Space and Technology, April 11-12, 1989, *Congressional Record 135* (42 & 43) U.S. Government Printing Office, Washington (1989).

A TRW-KOBE STEEL JOINT VENTURE

John Ogren, TRW

TRW has been involved in joint activities with Kobe Steel, Sony and Hitachi. All have followed a similar management pattern. The activity with Kobe Steel involves diamond film technology and I have served as the TRW interface person. Without getting into technical details I propose to describe the personal interactions that have led to this joint effort.

The story begins five or six years ago when our TRW CEO, Dr. Rubin Metler, was golfing with the president of Kobe Steel, Ltd. at Rancho Mirage in California. The president of Kobe Steel was explaining that his company must diversify and that a number of new activities had been initiated aimed at products with a "large value-added content". Together they agreed that there would be a meeting at which TRW would define their technological needs, Kobe would describe their new activities and together they would decide if there was a match warranting joint activities. Returning to Cleveland, Dr. Metler delegated responsibility for further action to a branch of the corporate office which might loosely be described as the "foreign office". That office arranged a meeting with Kobe Steel and people from TRW's automotive sector to describe their technology needs. The TRW participants were mostly accountants and mostly concerned with whether Kobe Steel might supply good quality parts at a lower cost. For its part the Kobe Steel participants described superconducting magnetics, robotics and other high technology. There seemed to be no match of interest but another meeting was set and then another with no results. A fourth meeting was decided on at Redondo Beach at which TRW personnel were not very anxious to attend because rumors were flying that the TRW-Kobe interface was falling apart. No one wanted to be associated with a memo to the CEO indicating that no common interest areas could be found anywhere between the six billion dollar per year TRW Corporation and Kobe Steel. I attended that meeting and discussed some needs about rocket propulsion. However, an important item on the program was a talk by Dr. Kogikobashi of Kobe Steel in which he described work he had just completed at NIRIM in Tsukuba City on diamond films. He described the deposits he had made, the plasma from which the films had been deposited, Langmuir probe data on the plasma, Raman spectra on the films and was very convincing that a real innovation had been achieved in making polycrystalline films from a methane hydrogen mixture in a resonant zone of a three hundred watt 2.45 GHz microwave discharge. As much as anything else, this provided a way in which the managers at TRW's executive headquarters could salvage the TRW/Kobe interaction and at the same time, find reasons for a trip to Japan.

That meeting was held in 1986 and in the subsequent four years there has been a continuous interaction in which TRW, a materials user, writes specifications and Kobe Steel has supplied prototype samples which are evaluated by TRW. No money has changed hands yet, and the joint venture will only have come to fruition in a business sense when that occurs.

Low pressure diamond synthesis is now a nascent industry in which straightforward substitutions of diamond films in simple applications such as coatings for cutting tools, heat sinks, high elasticity films for tweeters in stereo speakers and small windows have been accomplished. There are a number of potential applications which take advantage of the exceptional properties of diamonds. Mechanical and thermal applications include bearings, barrier coatings, surgical blades, medical implants and wire drawing dies. There are a number of electronic applications which would include high performance, high temperature radiation hard high powered transistors, electrical insulators and substrates, x-ray masks, magnetic disks and packaging assemblies. There are also exciting potential optical and optoelectronic applications as windows, lenses, mirrors, x-ray windows, heat sinks, ultraviolet detectors, wave guides, and so forth. In order for the most exciting of these applications to develop, there will have to be found methods of forming large area heteroepitaxial films, high conductivity n-type doping and high growth rates for large diamond single crystals. These are innovations that are still waiting to happen. When they do happen, their commercial introduction will require changes in devices and even systems which augur a long time frame for substantial growth to occur in the nascent low pressure diamond film industry.

The National Materials Advisory Board has recently published a comprehensive report: *Status and Applications of Diamond and Diamond-like Materials: An Emerging Technology,* NMAB-445, National Academy Press, 1990.

SILICON NITRIDE STRUCTURAL CERAMICS

Silicon nitride is a chemical compound, Si_3N_4, but we shall include in our discussion the whole family of compositions or alloys which consist in very large part of silicon nitride. Articles are produced by typical ceramic processing beginning with the synthesis of fine particle size controlled composition powders. Additives are mixed with these powders and shapes are formed by a variety of processes. Consolidation occurs at high temperatures with or without the application of high pressures. Production of silicon nitride parts may be regarded as an infant industry. There are a dozen or more companies manufacturing parts and the annual market is perhaps two hundred million dollars.

These materials have a combination of properties which include good thermal, chemical and mechanical stability at temperatures up to 1200-1300°C, a superior resistance to thermal shock which results from high toughness, high strength, low thermal expansion and moderately high conductivity. The relatively high toughness for a ceramic results from a fibrous internal microstructure which is developed by carefully controlled processing. In addition to high temperature properties, silicon nitride materials have excellent wear resistance due to their strength, toughness and hardness. Materials are in production for cutting tool inserts, for wear resistant parts such as sand blast nozzles, seals and die liners, ball bearings and similar applications, glow plugs and swirl chambers for diesel engines, rocker arm wear pads for automotive use and turbocharger rotors are in production in Japan. In addition to these commercial applications, silicon nitride materials are leading candidates for other advanced automotive and gas turbine engine components.

The synthesis of silicon nitride was first patented in 1895. About this same time, George Westinghouse had installed the first large polyphase hydroelectric generators at Niagara Falls which allowed the manufacture of synthetic silicon carbide by the Acheson process. Within a few years G. Eggley applied for a patent in which silicon metal was mixed with the silicon carbide and fired in nitrogen to form a silicon nitride-bonded material. However, it was not until the early 1950's that C.E. Nicholson of the Carborundum Company patented silicon nitride-bonded with a process that was capable of practical application (U.S. Patent 2,618,563, 1952; U.S. Patent 2,636,828, 1953). About the same time J.F. Collins and R.W. Gerby developed slip cast reaction-bonded silicon nitride shapes for thermocouple tubes, crucibles and boats for molten aluminum. Haynes Division of Union Carbide Company offered these materials for sale in 1958. They had relatively poor strength, but good thermal shock resistance and resistance to metal corrosion. The market for this material proved to be very limited.

The British Admiralty Materials Laboratory perceived that silicon nitride might be made in complex shapes appropriate for constructing higher temperature more efficient gas turbines. The material operating conditions within a gas turbine engine are extremely severe. Materials must have a high resistance to thermal shock, a resistance to steady stage and transient stresses, exposure to severe oxidation and corrosion environments, mechanical methods of attachment to other materials, the capability of forming complex shapes and high strengths with good reliability. Government funding supported work at the A.M.L., the British Ceramic Research Association, at universities, at the Plessy Company and the Birmingham Small Arms Group. Forming complex shapes from silicon and then reacting them with nitrogen gas at high temperatures was suitable for forming complex shapes with relatively poor properties. Researchers at the Plessy Company found that with magnesium oxide as an additive it was possible to form fully dense silicon nitride by hot pressing. By 1970 or so, significant advances had been made which gave complex shapes with poor properties (RBSN) or small simple shapes with good properties (HPSN). There was a substantial research innovation in 1972 when Y. Oyama at Toshiba and K.H. Jack at Newcastle independently reported on solid solutions in the Si-Al-O-N system, inventing **sialon** compositions which were subsequently developed during the 1970's. Beginning in the late 1970's, hot pressed sialon compositions and then hot pressed silicon nitride compositions came into use as silicon nitride based ceramic cutting tools and were manufactured by several companies in the United States and Japan. However, the major inventions and innovations in silicon nitride technology were mostly driven by the desire to develop a gas turbine and other heat engine components. Over the years dozens of innovative improvements have occurred. Some of these which illustrate the number and variety required are illustrated in Table 1.

TABLE 1

KEY TECHNOLOGICAL INVENTIONS AND INNOVATIONS

IN SILICON NITRIDE STRUCTURE CERAMICS

Compiled by D.R. Richerson

I. Materials Developments

Innovation	Contributors	Factors Influencing Innovation
A. Reaction-bonded Si_3N_4 invention 1955, major development 1955-1972	Early work in England: Collins (1955), Parr et al (1957), Deeley (1965), Thompson (1967). Later work in Germany, Japan and the U.S. Suzuki, Washburn.	Interest of the British Navy in improved gas turbine materials and corrosion resistance. Desire for a material which could be fabricated with minimum dimensional change during densification.
B. Hot pressed Si_3N_4, invented 1961, major development 1961-1975	Early work in England: Deeley (1961), Lumby (1971). Later work on new compositions and fabrication refinement in the U.S. and Japan at Norton, AMMRC, Toshiba, Toyota. Richerson, Gazza.	Perception that reduced porosity would result in higher strength and improved oxidation/corrosion resistance. New compositions developed to improve high temperature strength and creep resistance. Major driver was potential for heat engine applications, especially an automotive gas turbine engine. Substantial influence of Ford Motor Company in U.S., Rolls Royce in England, VW and Daimler Benz in Germany, and Toyota, Nissan and Toshiba in Japan.

C.	Injection molding and slip casting of complex shapes of RBSN 1970-1976.	Major innovations at Ford Motor Company, Mangels, Ezis, Baker et al.	Requirement for a low-cost mass-production fabrication process suitable for automotive components. Substantial government support. Concerns about potential fuel shortages and the need for improved fuel economy.
D.	Brittle material design validation 1970-1976.	Major innovations at Ford Motor Company.	Requirement for a design methodology compatible with the brittle fracture behavior of ceramics and the probabilistic distribution of strength controlling flaws.
E.	Sialon, invention 1972, major development 1972-1980.	Discovered by Oyama (1972) in Japan and Jack and Wilson (1972) in England. Early commercialization developments conducted by Lucas Cookson, Lumby et al.	Technology arose from efforts to reduce porosity and increase strength by additives.
F.	Grain boundary and microstructure engineering 1970-1974.	Norton Company, AMMRC; Richerson, Washburn, Gazza.	Norton efforts driven by desire to commercialize new materials using hot pressing capability and capacity initially established for Boron Carbide armor production. Grain boundary engineering conducted to achieve improved high temperature mechanical properties.
G.	Grain boundary crystallization	Tsuge et al (1975-1978) in Japan.	Driven by desire to avoid problem of high temperature property degradation due to grain boundary sliding of softened glassy phases.
H.	Understanding of the effects and mechanisms of achieving acicular grain growth.	Lange (1973), Knoch (1980) Tani (1985).	Resulted initially from characterization studies at Westinghouse and AMMRC in support of the DARPA gas turbine program.
I.	Pressureless sintered Si_3N_4.	Earliest efforts appear to be in Japan by Niwa et al (1972), Kamigaito and Oyama (1973), Komeya et al (1974) Nishida and Nakamura and in the U.S. by Terwilliger (1974).	Driven by desire to achieve complex shape capability and low cost to allow commercial viability.

J.	Sintered RBSN	Giachello and Popper (1979), Mangels (1981).	Desire to reduce porosity to increase oxidation resistance and strength and also to achieve a dense complex shape with reduced shrinkage compared to pressureless sintering; resulted in some of the earliest high Weibull modulus material (m > 20).
K.	Gas Pressure Sintering	Initial work at AMMRC (Priest et al 1977).	Resulted from efforts to reduce decomposition of Si_3N_4 during sintering to achieve higher density and improved properties. Was tried at AMMRC under encouragement of N.R. Katz because an over pressure furnace was available.
L.	Two-step gas pressure sintering	AMMRC - funded study at GE (Greskovich, 1981). Applied to sintered RBSN at Ford by Mangels in early 1980's. Later adopted broadly in Japan with publications appearing in 1987 and later.	Efforts to optimize gas pressure sintering and to go to higher temperatures to obtain improved density with compositions having higher temperature stability.
M.	Hot Isostatic Pressing (HIP)	Initial innovation by Larker, Adlerborn and Bohman (1977) of ASEA in Sweden. Included development of a glass encapsulation technique compatible with fabrication of complex shapes such as integral turbine rotors.	Resulted from ASEA's efforts to develop markets for sale of HIP equipment.

II.

	Innovation	Contributors	Factors Influencing Innovation
N.	Cutting tools late 1970's, early 1980's.	Sialon cutting tools developed at Lucas Research Center England by Lumby et al. Technology licensed to Sandvik and Kennametal. Later active work on Si_3N_4 compositions by GTE, Norton and Iscar in the U.S. and by NTK and Kyocera in Japan.	Interest of Rolls Royce and others in reducing cost of machining of metals.

O.	Bearings	Feasibility demonstrated by Norton in about 1972. Not commercial until late 1980's by a joint venture of Norton and Torrington.	Interest of US military in a bearing that could run under lubrication starvation. Commercial application delayed until low cost HIP technology under ASEA license available.
P.	Diesel Engine Glow Plugs	Introduced by Isuzu in 1981 and Nissan in 1985.	
Q.	Diesel Hot Plugs	Introduced by Isuzu for naturally aspirated diesel engines in 1982 and by Toyota for supercharged diesels in 1984.	10% increase in power output.
R.	Diesel Swirl Chamber	Introduced for export by Mazda in 1986 in a 2 liter indirect injection diesel.	Reduction of particulate emissions by 2/3.
S.	Rocker Arm Wear Pad	Introduced into production by Mitsubishi in 1984 for overhead cam automobile engines fueled with kerosene, alcohol or liquid propane.	Need for a material that could be cast into the aluminum metal and that would solve severe wear problems.
T.	Turbocharger Rotor	Introduced by Nissan in 1985 in the 300ZX.	36% faster response than a metal turbocharger rotor.
U.	Rotor, stators, shrouds of automotive size gas turbine engine.	Rigs and engine operated at up to 2500°F by Ford Motor Co. Mid to late 1970's.	Demonstration of feasibility to operate a gas turbine at increased temperature through the use of ceramic components.
V.	Demonstration of increased performance of a gas turbine engine through the use of ceramic components.	Garrett Turbine Engine Co., 1980.	Demonstrated 30% increase in power and 7% decrease in fuel consumption in a nominally 1000 horsepower turboprop engine.

In 1971 the U.S. Department of Defense Advanced Research Projects Agency (DARPA) funded a program with Ford and Westinghouse aimed at a high temperature gas turbine. Major innovations were developed at Ford Motor Company for brittle material design validation. Injection molding and slip casting methods were developed. From characterization of Ford-Westinghouse materials an understanding of airborne grain growth was obtained which gave greater toughness. Intensified silicon nitride research in both Japan and the United States was a result of the 1973 oil crisis. Grain boundary microstructure developments, grain boundary crystallization and pressureless sintering with additives of alumina and yttria were developed more or less in parallel in the United States and Japan during the 1970's. The Ford-Westinghouse program funded by DARPA and administered by the Army Materials and Mechanics Research Center demonstrated that an automotive ceramic gas turbine engine was feasible; Westinghouse demonstrated that large stator veins suitable for power generation turbine would withstand simulated conditions. The Cummins Engine Company demonstrated an adioabatic uncooled diesel engine using zirconia based ceramics. The Garrett Turbine Engine Company operated a turboprop engine containing reaction-bonded and hot pressed parts. Detroit Diesel Alison operated a diesel truck engine incorporating ceramics components. Kyocera Corporation operated a three cylinder silicon nitride diesel automotive engine.

These technological accomplishments did not address questions of cost, markets, long-term reliability or reproducibility. They had a significant effect in demonstrating that engine Components can survive thermal and mechanical stresses in realistic environments and encouraging continued materials development and characterization research. Engine development programs continue at Garrett and at the Alison Gas Turbine Division of General Motors.

The Ford Automotive Engine program administered by the Army Materials & Mechanics Research Center had a strong influence in developing the perception that advanced materials and brittle design could work together for the production of ceramic engine components. Dr. R.N. Katz was in charge of the AMMRC program. His perceptions follow:

THE ROLE OF THE ARPA/FORD/WESTINGHOUSE/AMMRC BRITTLE MATERIALS DESIGN: HIGH TEMPERATURE GAS TURBINE PROGRAM IN STIMULATING SILICON NITRIDE TECHNOLOGY - A RETROSPECTIVE

R. Nathan Katz, Worcester Polytechnic Institute

Introduction

This paper is a retrospective of the role of the ARPA/FORD/WESTINGHOUSE/ AMMRC PROGRAM[1] in the development of silicon nitride technology. As shown in Table I, silicon nitride, which does not occur in nature, was first synthesized and identified in about 1895. Since that time it has been the object of much scientific and engineering research, and starting around 1980 it has seen increasing commercial exploitation. The ARPA program provided a stimulus for much of the technology development that lead to eventual commercialization. The ARPA program also provides an interesting case study of the role of the vision of key individuals in several diverse organizations converging at a unique point in time; technologically, institutionally and economically. This paper will highlight some of the significant achievements in the technology of silicon nitride that were attributable to the ARPA Program. As is evident from an examination of Table I, the history of silicon nitride technology is also a fascinating case study of technology transfer on a global basis and thus, a fitting topic for this conference.

Background

The Advanced Research Projects Agency (ARPA) sponsored Brittle Materials Design:

---

[1] (Hereafter simply referred to as the ARPA Program, ARPA is currently called DARPA, the Defense Advanced Research Projects Agency).

TABLE I

SILICON NITRIDE - A BRIEF HISTORY

1895	MEHNER PATENT, GERMANY
1950's	RBSN PARR, GODFREY, MAY ET AL., MTL, UK
1960's	HPSN LUCAS CO., UK
EARLY 1970's	IMPROVED RBSN IMPROVED HPSN DARPA/FORD/WEST./NORTON/AMMRC, US
MID 1970's	SINTERED SN GAZZA, AMMRC; JAPAN
LATE 1970's	HIP Si_3N_4 LARKER ASEA, SWEDEN S-RBSN GIACHELLO AND POPPER, FIAT, ITALY 2-STEP GAS PRESS SINT. GRESKOVICH AND GAZZA, GE/AMMRC
1980's	APPLICATIONS AND PROCESS DEVELOPMENT COMMERCIALIZATION - MAINLY JAPAN
1989-1990	2nd GENERATION FULL DENSITY - SELF-REINFORCED Si_3N_4 $K_{Ic} \approx 10$, $S \approx 1$ GPa - DOW, ALLIED-SIGNAL, NKK, US, JAPAN

High Temperature Gas Turbine Program (hereafter referred to as the ARPA Program), initiated in 1971, was a major event in the history of the development of silicon nitride science and technology. It was clear from the very inception that this was to be both a very challenging and an unorthodox program. What was to become clear only with the perspective of hindsight was that this program was highly successful in attaining ARPA's original goal and that within the US Department of Defense's materials research environment, it was a unique transdisciplinary program.

The ARPA Program's goal was to encourage designers to utilize brittle materials. Brittle materials such as ceramics, carbon composites, and intermetallic compounds provide extraordinary high temperature strength, high specific strength, and a variety of other attractive properties. By 1970 it was clear to the Materials Science Division of ARPA that there were many advanced concepts for military systems which required the unique combinations of properties that many of these brittle materials could provide. Yet, designers were reluctant to use these materials. This was an entirely understandable position. Brittle materials, by their very nature, are prone to unpredictable catastrophic failure if subjected to tensile stresses. No designer dare risk the possibility of such failure, nor will society accept it. What was required was to simultaneously develop a methodology for reliable design with brittle materials and to learn how to improve the materials themselves, not to make them nonbrittle, but to make them provide consistent and predictable behavior.

With regard to the goal of utilizing brittle materials to enhance systems performance, the history of technology gives us considerable cause for optimism. The enabling material for the industrial revolution's steam engines and other machinery was cast iron - a brittle material. With the advent of steel and other high strength, ductile structural metals, designers no longer utilized brittle materials as tensile load bearing components. As a consequence design practices, "rules of thumb", and experience built up over centuries were no longer transmitted from senior to junior design engineers. However, we are now at a stage of technological development where the high strength, ductile metals are reaching their intrinsic limits with regard to temperature and specific strength properties. Thus, further development of systems such as gas turbines, industrial heat exchangers, metal cutting and forming tools, missile guidance domes, or bearings require the exploitation of materials that are brittle. Looking at the state of the art of brittle materials design in 1970 ARPA saw that there was considerable understanding of brittle materials behavior and the beginning of a coherent design philosophy [Dukes]. ARPA also noted the considerable progress that had been made in computer hardware, and in software for applying numerical methods of thermal and stress analysis (finite element and finite difference techniques). Thus, ARPA asked the question: "can emerging brittle materials design philosophy be married to computer based numerical analysis at this time in an effective manner and thereby encourage wider use of this class of materials?".

Concurrent with progress in computer aided numerical design techniques, improved ceramic materials such as silicon carbide and silicon nitride were transitioning from research laboratories into commercial production. Silicon nitride in particular had been the object of over a decade of intensive research supported by the British Admiralty Materials Laboratory. The fruits of this research were a variety of processes for fabricating reaction bonded silicon nitride (RBSN), the development of fully dense hot pressed silicon nitride utilizing a MgO densification aid, and a preliminary data base on the materials properties and behavior; all of which clearly justified further development of the material. Moreover, components of RBSN had been fabricated by several methods that seemed commercially viable. Dr. David J. Godfrey of the Admiralty Laboratory had even made a small, low performance diesel engine with a piston, rings, wrist pin, and cylinder all of RBSN [Godfrey]. This engine powered a lawn mower which was used to cut the Admiralty Laboratory's lawn ca 1970.

Dr. Maurice J. Sinnott, who in 1970 was the Director of the Materials Sciences Division at ARPA, believed that the goal of encouraging designers to use brittle materials could be achieved if a sufficiently dramatic demonstration of the effective use of such materials were made. The materials would have to demonstrate their ability to function and to provide systems advantages at temperatures far beyond the capabilities of conventional engineering metals. The result was that ARPA chose to demonstrate a "ceramic" gas turbine, operating without cooling, at a 2500°F (1375°C) turbine inlet temperature.

The program was unorthodox in many regards. Perhaps the most unorthodox aspect, was that a program to advance materials technology would depend to a very large extent on novel gas turbine engine design, the development of turbine component test rigs operating in temperature regimes beyond previous experience, and the use of component design and reliability prediction methodologies that were developing in parallel with the materials. We did not realize it at the time but we had implemented one of the earliest concurrent engineering programs. Some aspects of the interdisciplinary team approach that was implemented to facilitate this "concurrent engineering" approach are discussed in the paper by McLean in these proceedings [McLean]. The fact that a large portion of the ARPA materials science budget was going to engine hardware caused considerable consternation to many materials scientists who did not appreciate the fact that research funds that they viewed as potentially "theirs" should be directed toward other technical areas. However, it is clear, again with the benefit of hindsight, that forcing materials to be tested in real engine environments, from the start, focused the research effort so that the "show stopping" materials deficiencies were addressed early and with reasonable resources. This focusing of R&D activities was, I believe, one of the principal reasons that the ARPA program contributed so much to silicon nitride technology in such a relatively short time span.

To recapitulate, by the end of 1970 ARPA had: 1) identified the need to stimulate brittle materials utilization; 2) identified a strategy for addressing this need; 3) had determined that the requisite enabling technologies were available (even if not fully developed); 4) decided that the program would have to be interdisciplinary; and 5) programmed a reasonable amount of funds to start such a project (approximately 10 million 1970 dollars, roughly equivalent to 30 million 1990 dollars). ARPA's next task was to assemble a team of contractors and a government laboratory to act as program monitor, so that their vision could be translated into reality.

The Enabling Organizations

As the result of a lengthy and extensive process of soliciting competitive proposals, reviewing them utilizing panels of technical experts assembled from a variety of Department of Defense and NASA laboratories, and subsequent negotiations, ARPA selected a group of organizations that would carry out the program. The Ford Motor Co. was selected to be the prime contractor. Westinghouse Corp. was to be a subcontractor, and the Norton Co. was selected as the principal materials supplier. The Army Materials and Mechanics Research Center (now known as the Materials Technology Laboratory, MTL) was designated as the government's contracting agency and technical monitor. Each of these organizations brought the strengths shown in Tables II and III.

The Army Materials and Mechanics Research Center (AMMRC) in Watertown, MA, had the largest ceramics research laboratory in the US DoD and had research programs on silicon nitride ongoing as of 1969. In 1970, AMMRC arranged for Dr. D.R. Messier, the principal investigator of this program to spend four months as an exchange scientist at the Admiralty Materials Laboratory in the UK. This exchange conducted in 1971, helped to bring the AMMRC work on silicon nitride to a state of the art level very quickly. The exchange also initiated an ongoing relationship between these two laboratories and resulted in a very effective two way technology transfer which was to remain effective for well over a decade. Thus, by 1971 when the ARPA program was initiated AMMRC already possessed considerable technical capability and a demonstrated management commitment to this key technology for the project. AMMRC also possessed a strong design and mechanics group with experience in the use of numerical methods to solve complex materials systems design problems. AMMRC also had a Director, Dr. A. E. Gorum, who was a strong proponent of this approach to research as well as to the technology itself. Thus, AMMRC was ARPA's choice for monitor of the program. Dr. Gorum agreed to AMMRC assuming this role but only on the condition that AMMRC be a "participative monitor", that is not merely manage the contract and provide technical oversight (the traditional government laboratory's role in sponsored research) but that we should play a supporting role in technology development. This participative monitoring was another unusual aspect of the ARPA program.

The Ford Motor Co., the prime contractor, had a strong commitment to vehicular gas turbines in the late 1960's - early 1970's. Small gas turbines for automobiles must operate at turbine inlet temperatures well beyond the capabilities of superalloys if they are to have fuel economies equal or superior to reciprocating engines. Moreover, the costs of manufacturing an automotive gas turbine from superalloys which require intrinsically expensive metals, such as Ni, Nb, or Co, would be prohibitive. Furthermore, the manufacturing costs attendant to manufacturing small turbine blades and vanes with complex internal air cooling that the use of superalloys would require, were they to be utilized, would be unacceptable economically and aerodynamically. Thus, as early as 1967 Ford had determined that because of the limitations of superalloys a viable automotive gas turbine would have to be largely manufactured from ceramic components. Ford had an internally funded program headed by Mr. Arthur McLean aimed at developing a 250 hp, uncooled ceramic gas turbine suitable for a Lincoln sized automobile. Thus, the prime contractor had internal corporate goals and commitment in total harmony with those of ARPA. Ford wanted the program in order to accelerate what they were already doing and not just to develop technology for a fee as is often the case with government sponsored R&D. In part because of this commitment, and in part because of a corporate culture that was leery of taking government contracts (because of a belief the government would interfere with the way they did business) Ford elected to cost share the program approximately 50 percent. They purchased all materials and equipment, provided all test equipment, and absorbed

TABLE II

| **ARPA - BMD PROGRAM** |
| ENABLING ORGANIZATIONS: GOVERNMENTAL |

- ARPA:
    - VISION
    - DOLLARS
    - "THE WILL"

- AMMRC:
    - SHARED THE ARPA VISION
    - CERAMICS & MECHANICS BACKGROUND
    - MANAGEMENT COMMITMENT
      (THROUGHOUT THE ORGANIZATION)

(ca 1970)

TABLE III

| **ARPA - BMD PROGRAM** |
| ENABLING ORGANIZATIONS: INDUSTRIAL |

- FORD:
    TEAM IN PLACE WITH - TECHNOLOGY & MOTIVATION
    IN-HOUSE PROJECT UNDERWAY
    STRONG FINANCIAL COMMITMENT
    PERCEIVED NEED FOR TECHNOLOGY

- WESTINGHOUSE:
    IN-HOUSE PROJECT UNDERWAY
    PERCEIVED NEED FOR TECHNOLOGY

- NORTON:
    LICENSED HPSN TECHNOLOGY FROM LUCAS, UK
    ONLY U.S. PRODUCER OF MATERIAL

(ca 1970)

most of the overhead costs. The government only paid for direct labor. This was another unusual aspect of the program. Looking back, it was a very important aspect because it speeded up the pace of the research. Ford did not have to get a contracting officer's approval every time they wanted to get a major piece of equipment or evaluate a new material which may not have been foreseen six months or a year before when an annual operating plan was approved.

ARPA wanted to demonstrate that large as well as small components could be successfully demonstrated. Thus, Ford teamed with Westinghouse Corp., who had an interest in using ceramic turbine vanes in their 30 megawatt gas turbines for electrical power generation applications. Westinghouse Research Center, by 1970 had a small internally funded effort to utilize hot pressed silicon nitride (HPSN - a fully dense high strength form of the material) in this application. The Westinghouse subcontract did not have the same cost sharing provisions as did the Ford prime contract. In part this may have been because Westinghouse Research was accustomed to the role of government contractor and thus, did not share Ford's wariness of this role. It may also have resulted from the realization that the technology was risky and since they had a commercially viable product by using air cooling of vanes (a technology not economically viable for Ford) they had less time urgency with regard to this opportunity. Nevertheless, at the program start Westinghouse had a high level of commitment to making this technology viable.

Norton Co., while not a subcontractor, was the only domestic supplier of both RBSN and HPSN. Norton also had unique technology for producing high strength, fully dense silicon carbide. Their licensing of technology to make high strength HPSN, from Jos. Lucas Co., in the UK, provided both a domestic source for this key material and a technology base from which significant improvements to the material were to be made in response to needs identified in the ARPA program test results.

In this brief review of the roles of the various enabling organizations, I believe that a significant cultural difference between US and Japanese program management is evident. I have stressed the fact that we had three strong managers, or technology "champions", whose vision, personal management styles, and commitment were key to the ARPA program's uniqueness and success; namely Drs. Sinnott and Gorum and Mr. McLean. I am not sure that our Japanese colleagues would attribute so much importance to the personalities of the program managers for program success.

Before going on to review some of the major innovations to the science and technology base of silicon nitride, I'd like to briefly address the issue of why a program that at its outset had no preference between silicon nitride and silicon carbide, came to concentrate almost exclusively on silicon nitride by the end of the program.

Why Silicon Nitride Over Silicon Carbide ?

There were many reasons why silicon nitride gradually became the preferred material. The lack of sinterable silicon carbide powders ca 1970-1975 played a major negative role in retarding research on the material. The rapid rate of property improvement of silicon nitride vs. silicon carbide also played a role. (An excellent summary of the improvements in HPSN by the Norton Co., during the ARPA program has been provided by Torti.) However, I'd like to focus on one aspect of this issue which, I believe, has been unappreciated or overlooked. Namely, impressions of materials performance resulting from unplanned failures of components in test rigs. Early in the Westinghouse portion of the program a high temperature test of eight ceramic vanes, four of silicon carbide and four of silicon nitride, was carried out at 2300°F in a test rig. The rig consisted of a metal superalloy combustor and transition duct with the ceramic vanes downstream. During one run an overtemperature event caused the metal combustor to partially melt, break into pieces, be blown downstream and impact the ceramic vanes. Simultaneously the temperature instantaneously fell from about 2300°F to about 600°F (the compressor discharge temperature), an enormous thermal shock. All hot pressed silicon carbide vanes failed catastrophically, while the silicon nitride vanes suffered little, if any damage (see Figure 1). This unplanned failure and its result hada profound effect on the perceived viability of silicon carbide compared to silicon nitride for many associated with the program (including myself). After this event, and several other similar in kind (if not in scale) events in the Ford portion of the project, silicon nitride work was emphasized at the expense of silicon carbide.

Fig. 1. Comparison of damage to silicon nitride and silicon carbide turbine vanes resulting from failure of a superalloy combustor

Research Innovations Resulting from the ARPA Program

The ARPA program resulted in a great many advances in silicon nitride technology. I have chosen to highlight advances in RBSN, HPSN and processing technology which I believe to be of enduring significance. Table IV lists three major advances in reaction bonded silicon nitride technology. These advances involved fundamental understanding of the factors effecting the nitridation of silicon and how to control these factors to produce an improved microstructure with resultant property improvements. Table V lists highlights of work which elucidated the central role of the grain boundary phase in hot pressed silicon nitride. The ability to understand the relationship of the grain boundary chemistry and phase composition in HPSN in developing high temperature strength established a grain boundary engineering strategy for improving fully dense silicon nitride which is still utilized.

TABLE IV

| ARPA - BRITTLE MATERIALS DESIGN PROGRAM |

ADVANCES IN HPSN

- IDENTIFIED IMPORTANCE OF ALKALI
    IMPURTIES IN G.B. GLASS - RICHARSON (NORTON)
- IDENTIFIED IMPORTANCE OF
    SiO_2/MO ADDITIVE RATIO - LANGE (WESTINGHOUSE)
- IDENTIFIED Y_2O_3 ADDITIVE - GAZZA (AMMRC)
- ELUCIDATION OF GRAIN BOUNDRY
    ENGINEERING APPROACH - KATZ (AMMRC)

TABLE V

| ARPA - BRITTLE MATERIALS DESIGN PROGRAM |

ADVANCES IN RBSN

- EFFECTS OF ADDITIVES AND ATMOSPHERE DETERMINED
    - FISHER, MANGELS, EZIS (FORD)
- CONTROL OF NITRIDING EXOTHERM:
    NITROGEN DEMAND CONTROL SYSTEM
        - FISHER, MANGELS (FORD)
        - MESSIER, WONG (AMMRC)
- FUNDAMENTAL STUDY OF NITRIDATION KINETICS
        - MESSIER, WONG (AMMRC)

Table VI lists several advances in processing science which are the basis for much of today's commercial silicon nitride processing technology.

TABLE VI

ARPA - BRITTLE MATERIALS DESIGN PROGRAM

ADVANCES IN PROCESSING SCIENCE

- INITIAL WORK ON N_2 OVER PRESSURE GAS SINTERING OF Si_3N_4 - GAZZA (AMMRC)
- INITIAL WORK ON ADAPTIVE CONTROL TECHNOLOGY APPLIED TO MATERIALS PROCESSING - INJECTION MOLDED GREEN BODIES FOR RBSN - BAKER ET AL (FORD)
- N_2 DEMAND CYCLE PROCESS CONTROL
  - FISHER, MANGELS (FORD)
  - MESSIER, WONG (AMMRC)

Technology Transfer

Prior to the program's initiation ARPA and AMMRC jointly decided that, in order to fulfill the goal of stimulating the use of brittle materials, the maximum involvement of the wider technical community was desired. One mechanism to achieve this was to hold periodic open review meetings to which the technical community was invited, and at which the ARPA program results and the results of related non-ARPA programs were also presented. Starting in 1974 AMMRC began working with a group responsible for heat engine development for the Environmental Protection Agency (this group was later to be transferred to the Dept. of Energy, and became the DoE office of highway transportation) to transfer relevant technology to the civilian sector. Technology developed and demonstrated in the ARPA Brittle Materials Design Program was transferred to all subsequent ARPA and DoE engine programs, as suggested in figure 2.

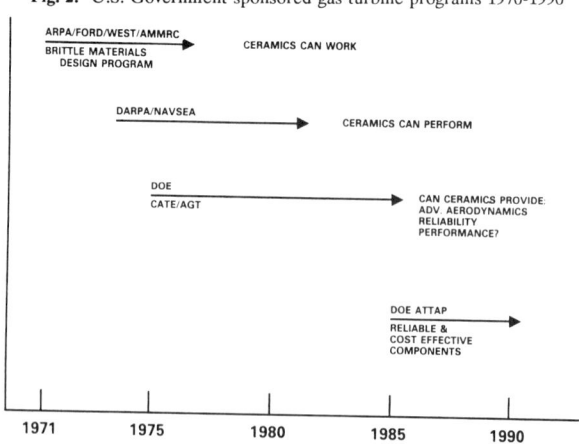

Fig. 2. U.S. Government sponsored gas turbine programs 1970-1990

Technology developed in the ARPA Program also played a major role in the development of silicon nitride cutting tools and bearings. The yttrium oxide additive for silicon nitride [Gazza] is used in most silicon nitride cutting tools manufactured in the US. Similarly the current ceramic bearings manufactured by the Cerbec Co. (a Torrington-Norton joint venture) are made of a material which is a further development of NC-132 silicon nitride which was developed for the ARPA Program.

Was the ARPA Program a Success?

There are those who maintain that because no one is driving a vehicle with a commercial ceramic gas turbine today, the program was unsuccessful. Such a view shows a fundamental misunderstanding of the ARPA program's goal. The goal was to encourage the wider use of brittle materials. In this regard the program was eminently successful. Had the program been unsuccessful there would not have been the follow-on ceramic gas turbine programs shown in figure 2. For the subject of this paper, the role of the ARPA program in the development of silicon nitride, it is clear that the existence of a major, focused project, largely dependent on this material significantly accelerated its development.

Conclusions

The ARPA program played a critical role in the development of silicon nitride. In hindsight one can attribute this to several factors some of which were unusual or non-traditional in US government supported programs. Some of these were:

* Provision of clear goals (i.e., operate in a turbine at 2500°F)
* Testing in engine test rigs provided rapid feed-back on show stopping
    materials deficiencies. This allowed research on the material to be focused.
* An unusual level of financial commitment by the prime contractor.
* An unusual level of technical participation by the government laboratory monitoring
    the program.
* An upfront commitment to technology transfer.
* The vision and commitment of the key people in each participating organization.

Each of these factors were important in the role the ARPA program played in the development of silicon nitride. I believe this role was important in transforming silicon nitride into a commercially useful material.

Acknowledgements

The author wishes to acknowledge the many fruitful discussions with his colleagues Dr. D.R. Messier and G.E. Gazza, MTL, Prof. E.M. Lenoe, USMA West Point, and A.F. McLean. These discussions not only helped to sweep the cobwebs from my brain, they also helped recall the excitement of shared creative endeavors.

References & Bibliography

Burke, J.J., Gorum, A.E., and Katz, R.N., *Ceramics for High Performance Applications,* Brook Hill Publishing Co., Chestnut Hill, MA, (1974).

Burke, J.J., Lenoe, E.M., and Katz, R.N., *Ceramics for High Performance Applications - II,* Performance Applications - II, Brook Hill Publishing Co., Chestnut Hill, MA, (1978) Dukes, W.H., *Handbook of Brittle Materials Design Technology,* AGARDograph #152, North Atlantic Treaty Organization, Feb. 1971.

Fisher, E. A., and McLean, A. F., Brittle Materials Design; High Temperature Gas Turbine" Final Report AMMRC TR81 - (1981).

Gazza, G.E., "Hot Pressed, High Strength Silicon Nitride", U.S. Patent No. 3,830,652, August 20,1974 (see also, *J. Amer. Ceram. Soc.,* **56**, 662 (1973), and *Bull. Amer. Ceram. Soc.,* **54**, 778-781 (975).

Giachello, A., et al, "Sintering and Properties of Silicon Nitride Containing Yttria and MgO", *Ceram. Bull.,* **59**, No.12, p. 1212-1215 (1980).

Godfrey, D.J., and May, E. W. R., "The Resistance of Silicon Nitride to Thermal Shock and Other Hostile Environments" in *Ceramics in Severe Environments,* eds. W.W. Kriegel and Hayne Palmour III, Plenum Press, NY, (1971).

Katz, R.N., and Gazza, G.E., "Grain Boundary Engineering and Control in Nitrogen Ceramics", *Nitrogen Ceramics,* ed. F.L. Riley, Noordhoff International Publishing, Leyden, The Netherlands (1977), pp. 417-431.

Lange, F.F.," Phase Relations in the System Si3N4-SiO2-MgO and Their Interrelation with Strength and Oxidation", *J. Amer. Ceram. Soc.*, **61**, 53-56 (1978).

Larker, H., Hot Isostatic Pressing of Silicon Nitride Parts", in *High Pressure Science and Technology,* eds, K.D. Timmerhaus and M. S. Barber, Plenum Publ., NY, (1979) pp. 651-655.

Lenoe, E.M., Katz, R.N., and Burke, J.J., *Ceramics for High Performance Applications - III Reliability,* Plenum Press, New York and London (1983).

Mangels, J.A., "The Effect of Silicon Particle Size on the Nitriding Behaviour of Reaction Bonded Silicon Nitride Compacts", *Progress in Nitrogen Ceramics,* ed. F.L. Riley, Martinus Nijhoff Publishers, Boston and the Hague, pp. 135-140, (1983).

McLean, A.F., see paper in these Proceedings.

Mehner, H., German Patent 88999, Sept. 30, 1896.

Messier, D.R., and Wong, P., "Kinetics of Nitridation of Si(Powder Compacts", *J. Amer. Ceram. Soc.,* **56**, 480-485 (1973).

Mitomo, M., "Pressure Sintering of Silicon Nitride", *J. Mat. Sci.,* **11**, 1103-1107 (1976).

Priest, H.F., Priest, G.L., and Gazza, G.E., "Sintering Silicon Nitride Under High Nitrogen Pressure", *J. Amer. Ceram. Soc.,* **56**, 395 (1977).

Richerson, D.W., "Effect of Impurities on the High Temperature Properties of Hot-Pressed Silicon Nitride", *Bull. Amer. Ceram. Soc.,* **52**, 560-562 (1973).

Wong, P., and Messier, D.R., "Procedure for Fabrication of Si3N4 by Rate Controlled Reaction Sintering", *Bull. Amer. Ceram. Soc.,* **57**, 525-526 (1978).

Torti, M.L., "Processing Hot Pressed Silicon Nitride for Improved Reliability: HS-110 to NC-132", in *Ceramics for High Performance Applications - III*, Plenum Press, NY and London, pp. 261-274, (1983).

The Norton Company in Worcester Massachusetts was one of the key American manufacturers providing prototype components and test samples for the AMMRC program. David Richerson was one of the key technical participants at Norton Company and subsequently worked with Garrett Turbine Engine Company development. His perceptions of the American corporate perspective from the(point of view of a technologist follow:

AN AMERICAN CORPORATE PERSPECTIVE

David W. Richerson, Ceramatec, Inc.

The listing of inventions and innovations in Table 1 above suggests that the U.S. and Japan have both been effective at invention and technical problem solving, but Japanese industry has been more effective at identifying and implementing applications. These trends have been influenced by many factors, some which are cultural and some which are structural. Key factors are identified in Table 2. Some are discussed in this section based upon experiences of the author at Norton Company (where some early U.S. material and manufacturing development occurred) and at Garrett Turbine Engine Company (where some early turbine development occurred). The initial factors to be discussed include (1) frame-of-mind of the company, (2) key core technologies/competencies, (3) personnel interactions, and (4) resources. Other factors and elaboration are included later in the Si_3N_4 conference summary.

TABLE 2

SOME FACTORS INFLUENCING INVENTION AND INNOVATION

* Frame-of-mind of organization
* Core technologies / competencies
* Personnel interactions
* Resources (source and level)
* Attitude toward use of technology
* Timing
* Commitment
* Continuity / duration
* Collaboration / partnerships
* Goals and milestones
* Market fit
* Comfort zone
* Innovation window
* Prior successes or failures
* Perception of acceptable market size
* Image
* Legal / regulatory issues
* Protection of proprietary position

Experiences at Norton Company 1969-1973

Table 3 summarizes some key factors at Norton Company during their early Si_3N_4 development. The environment was excellent for achieving rapid progress. Management provided strong support. They perceived that Norton had key core competencies and enhanced these by obtaining a key license on hot pressed Si_3N_4 from Lucas Co. (Solihul, England). Furthermore, they could envision that Si_3N_4 markets could offset projected reduction in boron carbide armor sales (as the Vietnam conflict subsided) and could be an important diversification beyond abrasives and grinding wheels.

The technology implementation approach at Norton in 1970 is identified in Table 4. This was started at low level with a couple of collaborating engineers and technicians and rapidly grew to an interdepartmental team with pilot scale responsibility. Communications were directly between individuals, cooperation was unrestricted, and substantial decision making was encouraged at the working level. This was a dynamic environment and led to rapid technology improvement and size scale-up. Key technology contributions of the team are listed in Table 5.

Norton Company had the best Si_3N_4 powder and hot pressed Si_3N_4 in the world in the early 1970's and a superior reaction-bonded Si_3N_4. Norton sold many test bars and prototype heat engine components, primarily as a vendor to government programs. In addition, Norton demonstrated that hot pressed Si_3N_4 had better rolling contact fatigue life than the best metal bearing. However, in spite of an apparent technical lead, Norton did not achieve

TABLE 3

NORTON COMPANY 1969-1973

FRAME-OF-MIND
* Top management interest in diversification beyond abrasives and grinding wheels
* Focus on advanced industrial ceramics
* Desire to find new products for hot pressing facility initially established for armor
* Willingness to develop or buy technology

KEY TECHNOLOGY
* Large-scale hot pressing
* Silicon oxynitride and SiC
* Si_3N_4 license from Lucas
* Diamond grinding

PERSONNEL INTERACTIONS
* Individuals and small teams had technology responsibility from R&D through pilot scale
* Marketing group pursued applications and prototype orders
* Characterization through service group

RESOURCES
* Company-funded
* 5-year commercialization goal

TABLE 4

1971

TECHNOLOGY IMPLEMENTATION APPROACH AT NORTON

* LICENSE KEY TECHNOLOGY (Lucas)

* CONDUCT DEVELOPMENT TO IMPROVE TECHNOLOGY AND ADAPT TO NORTON STRENGTHS
    * Identify deficiencies by strength, microstructural, and fracture surface evaluation
    * Conduct material and process improvement iterations guided by these evaluations
    * Establish both patent and proprietary technology positions

* VERTICALLY INTEGRATE
    * Synthesize improved powder, including improved design of synthesis apparatus
    * Establish low-contamination powder processing in clean-room facility
    * Develop improved machining tools and procedures and establish facility
    * Establish QC procedures including specifications and certifications

TABLE 5

KEY TECHNOLOGY CONTRIBUTIONS
OF THE NORTON TEAM 1970-1973

* Synthesized the best quality Si_3N_4 available in the early 1970's.
* Improved the high-temperature strength of hot pressed Si_3N_4 (MgO sintering aid) by greater than a factor of 2.
* Increased the stress rupture life of hot pressed Si_3N_4 by nearly a factor of 10.
* Achieved above by obtaining an understanding of the factors controlling these properties, i.e., specifically the influence of composition of the grain boundary phase.
* Identified through fracture analysis the fracture-initiating flaws and made major progress in eliminating these by process control.
* Optimized heat treatments to achieve maximum beta phase content and enhanced strength and fracture toughness.
* Established pilot production of a reproducible material (NC0132 Si_3N_4) and supplied the needs of many programs in the 1970's.
* Also established a source of high quality hot pressed SiC and reaction-bonded Si_3N_4.
* Demonstrated successful Si_3N_4 bearing performance.

production of Si_3N_4 products during the 1970's or early 1980's. There are a number of reasons for this, some of which are relevant to the factors listed in Table 2. The following are some speculations of the author as to possible reasons.

1. Emphasis in the U.S. on turbine engine components feasibility demonstration programs; the market was for a limited number of prototypes.

2. Hot pressing could not produce complex shapes economically competitive with other technologies until the early to mid 1980's.

3. Although Norton had a superior Si_3N_4 powder and some requests for sale of the powder, they chose not to sell the powder for two reasons: (1) to protect proprietary technology and (2) because they did not foresee a large market potential outside of Norton.

Experience at Garrett Turbine Engine Company 1973-1984

Table 6 summarizes some key factors at Garrett during the early 1970's which affected their decision to become active in Si_3N_4 technology. Key drivers were the large benefits that could be achieved in engine performance improvement and fuel reduction through the use of ceramics, concerns that competitors might achieve these benefits, and potential for substantial government funding to achieve the technology. Initial Si_3N_4 activities at Garrett were cooperative between an engineering sciences matrix organization and a project management organization. This resulted in a systems approach. Focus was on demonstration of feasibility rather than development of a commercial engine.

The technology implementation approach is illustrated by Table 7. This approach was utilized through 1984 when the author left Garrett and is still utilized. The matrix team and system approach are illustrated in Figures 2 and 3.

TABLE 6

GARRETT TURBINE ENGINE CO.
1973-1975

FRAME-OF-MIND
* Interest at top level of corporate management
* Defensive position
* Potential for Government funding
* Initial interest in short life engines

KEY TECHNOLOGY
* Broad engine design options
* Strong aerothermal and applied mechanics capabilities
* Extensive rig and engine test capabilities

PERSONAL INTERACTIONS
* Matrix organization
* Corporate ceramic review board

RESOURCES
* IR&D
* Plans to obtain Government funding

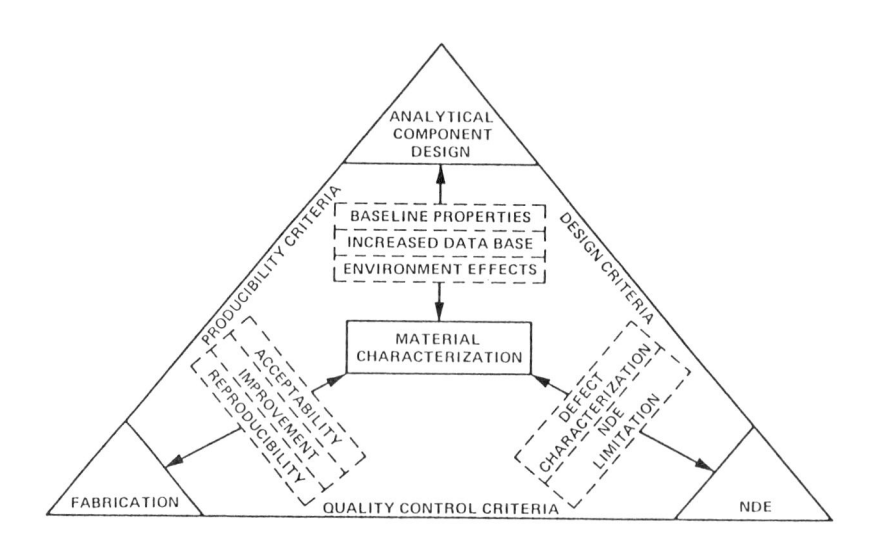

Fig. 2. Material and Process Development Approach at Garrett for DARPA/Navy Ceramic Turbine Engine Demonstration Program. 1976-1980.

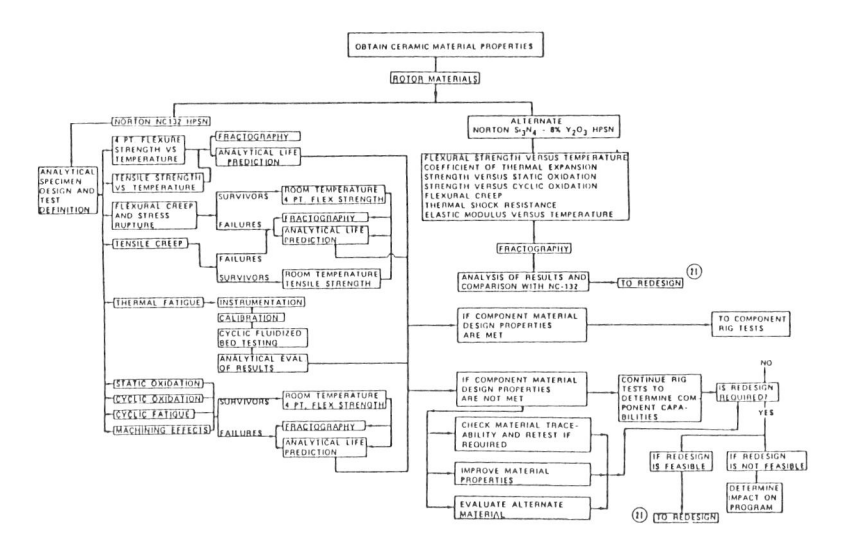

Fig. 3. Rotor Material Characterization Plan for Garrett DARPA/Navy Ceramic Turbine Engine Demonstration Program

TABLE 7

TECHNOLOGY IMPLEMENTATION APPROACH AT GARRETT

* Conduct all design, assembly, test and post-test inhouse
* Conduct characterization to evaluate materials options
* Select vendors for fabrication development of turbine components
* Establish inhouse component fabrication capability
* Development non-destructive evaluation criteria and capabilities
* Guide development of improved materials.

Initial efforts at Garrett were directed toward the use of Si_3N_4 components to demonstrate increased performance in an existing engine. The engine selected was a turboprop rated at 715 horsepower. Fig. 4 shows the original schedule and the initial ceramic design of the hot section of the engine. All static structure components were fabricated from reaction-bonded Si_3N_4 (sintered Si_3N_4 and HIP Si_3N_4 were not yet developed). Rotor blades were fabricated from hot pressed Si_3N_4 (by very expensive profile grinding). At the start of the program, the only organization who had successfully fabricated complex shaped turbine components was Ford Motor Company. Ford was not willing to be a commercial supplier, so other sources had to be developed. Norton was subcontracted as the key commercial source and slip casting and injection molding were initiated within the Garrett organization at the AiResearch Casting Company. Two ex-Ford employees were hired at Garrett who provided fabrication technology transfer and continuity from an earlier Ford/DARPA program (described later by Art McLean and Bob Katz). The fabrication

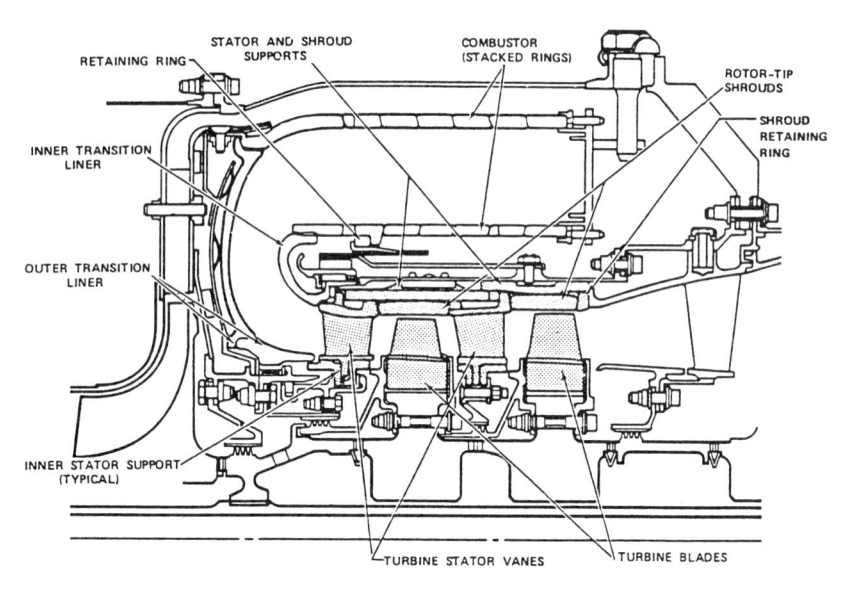

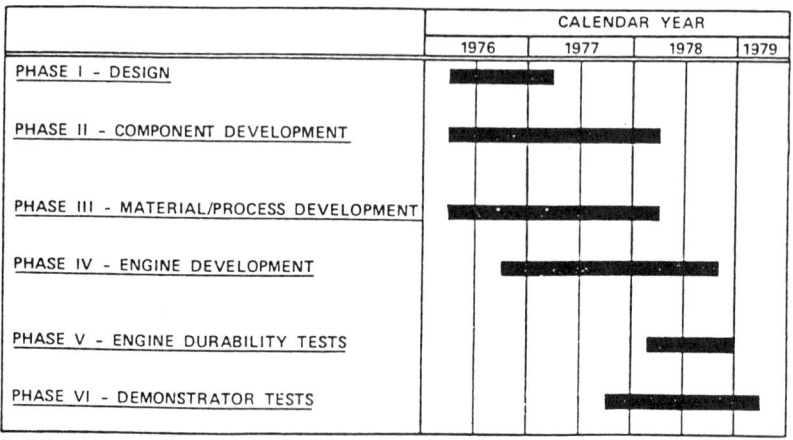

Fig. 4. Initial Ceramic Hot Section Design and Program Schedule for Garrett/DARPA/NAVSEA Ceramic Engine Demonstration Program

efforts were successful. Twenty seven engine builds were achieved, each containing 104 Si_3N_4 ceramic components. A 30% improvement in power and a 7% decrease in fuel consumption were demonstrated. However, long term cyclic operation was not achieved due to a contact stress problem at some ceramic-ceramic interfaces.

A version of the Garrett engine was redesigned and operated with one ceramic-blade rotor stage for 15 hours with no problems. Another redesign was conducted (under Air Force funding) to solve the contact stress problem. This engine design was also successfully operated.

The Garrett program of the late 1970's addressed many key technical issues of Si_3N_4 and use in gas turbines: net-shape fabrication, flow reduction, non-destructive evaluation, proof testing, brittle material design, property data base (flexural and tensile strength, stress rupture life, cyclic fatigue, contact stress resistance, oxidation/corrosion in burner rigs, impact tolerance, response to vibrational modes), rig and engine testing, and failure analysis.

The Garrett effort of the 1970's was a strong interdisciplinary team effort with close cooperation of Norton Company. Effort in the 1980's shifted to a cooperative program on an automotive gas turbine with Ford Motor Company (AGT Program funded by DOE). Sintered $Si3N_4$ and SiC were now available, so that effort shifted away from hot pressed and reaction bonded Si_3N_4. Primary fabrication efforts were conducted at AiResearch casting Company and Carborundum and later with NGK Insulators and Kyocera.

The AGT program stimulated major improvements in fabrication and properties of Si_3N_4 and SiC and resulted in successful engine operation. However, extensive development and testing are required to achieve a cost-affordable, high reliability engine. A follow-on program (ATTAP Program) is presently in progress to address these and other issues. A major objective of the ATTAP Program is to establish commercial sources of ceramic engine components in the U.S. Norton Company, Garrett Ceramic Components (outgrowth of AiResearch Casting) and GTE are all active in Si_3N_4 development for the ATTAP programs. Effort is focused on pressureless sintered and HIP Si_3N_4.

Conclusions

Major advances have occurred in the U.S. and Japan in Si_3N_4 technology. However, in spite of technical feasibility demonstration, Si_3N_4 has not yet reached application in a production gas turbine engine. Si_3N_4 has been spun off in other applications, though. Examples include cutting tools (Lucas in England; Iscar, GET Valeron, Kennametal, Norton and Industrial Ceramic Technology in the U.S.), bearings (Cerbec, joint venture of Norton and Torrington, Kyocera; NTK), glow plugs, Kyocera, NGK Spark Plug, swirl chambers (Kyocera, NGK Spark Plug), rocker arm wear pads (NGK Insulators) and turbocharger rotors (NGK spark plug), and wear parts (GTE WESGO, Norton). In addition, Norton and TRW have formed a joint venture to develop and market Si_3N_4 for heat engines, and GET and Eaton have announced a similar collaboration.

There remained a key problem of forming complex shapes to high density which would give rise to high toughness, high strength and good thermal shock resistance. Larker and others at ASEA (See Table 1) developed a method of encapsulating unfired shapes in impermeable glass for hot isostatic pressing at high temperature. In 1976, M. Mitomo of NIRIM reported on sintering at high temperatures under an atmosphere of nitrogen gas to prevent decomposition; Priest et al. developed the same process independently at AMMRC at about the same time. In the two step gas pressure sintering process, most porosity and all open porosity are eliminated in the first step, making the surface gas tight; then the gas pressure is raised to carry out hot isostatic pressing without the necessity of encapsulation. These processes led to the feasibility of silicon nitride bearings that could run under lubrication starvation, would have a long rolling-contact fatigue life. Norton Company in the United States demonstrated this in the early 1970's and by 1980 Kyocera Corporation and NGK Spark Plugs in Japan were offering these as commercial products. Norton Company had joined with Torrington Company in a joint venture, Cerbec, which produced these bearings commercially in the United States beginning in the late 1980's. Achieving high quality reliable silicon nitride parts requires the availability of high purity uniform powders. A number of processes had been developed including the imide preparation and decomposition process of Ube Industrials Ltd. in Japan. Several companies in Japan, but no one in the United States is producing silicon nitride powder. Also, the use of silicon nitride for devise applications required the development of silicon nitride/metal joining technology achieved by NGK Spark Plug Company, Ltd. and others.

Beginning with the joint development of diesel engine glow plugs by Kyocera and Isuzu in 1981, a number of automotive components are now in commercial production. Diesel hot plugs were introduced by Isuzu in 1982 and Toyota in 1984. Diesel swirl chambers were introduced by Mazda in 1986. Silicon nitride rocker arm wear pads were introduced by Mitsubishi in 1984. The most complex part, the turbocharger rotor was developed by Nissan and NGK Spark Plug Company and introduced in 1985. All of these components are now being commercially produced in Japan by Kyocera and by NGK Spark Plug Company.

During this period Dr. Kazuo Kobayashi was director of the MITI Regional Laboratory in Kyushu. He has prepared some comments on silicon nitride innovations and the influences of culture:

SILICON NITRIDE STRUCTURAL CERAMIC INNOVATIONS AND THE INFLUENCE OF CULTURE

Kazuo Kobayashi, Nagasaki University

Introduction
In Japan a "fever" of interest in structural ceramics developed during the middle 1970's. Many companies and government research institutes started R&D on silicon nitride and silicon carbide ceramics. This seems to have been motivated strongly by R&D achievements in the United States. The first national project in Japan in which high temperature structural ceramics researches were incorporated started in 1978. It was initiated by the Agency of Industrial Science and Technology (AIST), Ministry of International Trade and Industry (MITI) but lagged behind similar projects in the United States and West Germany.

The first national "High Efficiency Gas Turbine Project" aimed to establish a combined gas and steam turbine with an output power of 100 MW. However ceramic parts developed in this project were unsatisfactory because of poor strength and poor reliability exhibited in the test of a prototype plant in 1987. At present three national projects: "Fine Ceramics" (1981-1992), "Ceramic Gas Turbine" (1988-1996) and "Automotive CGT" (1988-1996) are carrying on as shown in Fig. 1. There are close relationships between private sectors and government research institutes.

```
1980          1985          1990          1995
  |             |             |             |
1978       1985  1987
```

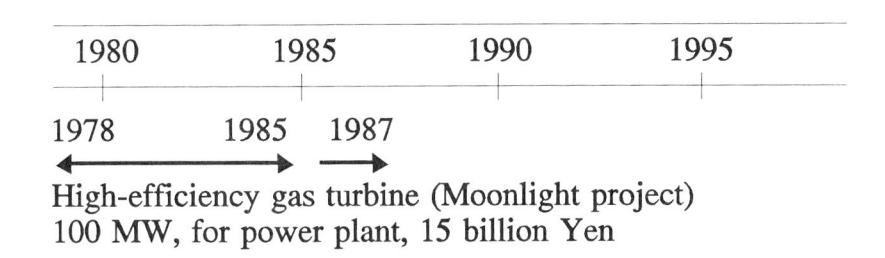

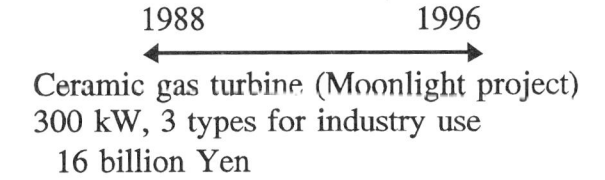

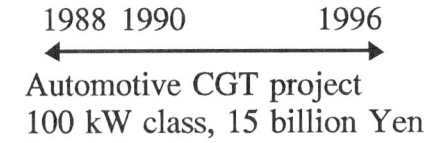

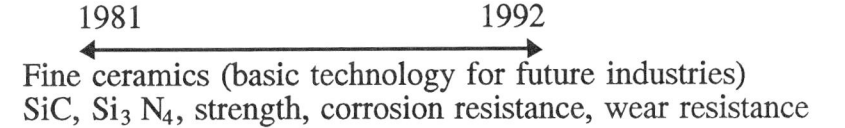

High-efficiency gas turbine (Moonlight project)
100 MW, for power plant, 15 billion Yen

```
              1988                    1996
```

Ceramic gas turbine (Moonlight project)
300 kW, 3 types for industry use
16 billion Yen

```
              1988 1990               1996
```

Automotive CGT project
100 kW class, 15 billion Yen

```
    1981                    1992
```

Fine ceramics (basic technology for future industries)
SiC, Si_3N_4, strength, corrosion resistance, wear resistance

Fig 1. National projects on high temperature ceramics in Japan

Trend of the Progress of R&D on high Temperature Structural Ceramics and Innovation
Progress seems to occur in stages from the past to the present and to the future. In each stage several technological innovations have been observed which had an influence on the rate of progress.

1) Stage of R&D on improvement of strength (middle of 1970's)
This was the stage of competition to show outstanding strength data for test pieces. In order to obtain higher strength it was considered important to fabricate the sample with higher density and it was also known that the use of fine submicron powder was favorable to obtain dense and strong specimens.

Various kinds of sintering aids such as MgO, Y_2O_3, Al_2O_3, etc. were examined for sintering of silicon nitride and silicon carbide and some of them were found to be effective for densification. As the hot-pressing sintering method was easy to operate to obtain dense samples, that method became popular on a laboratory scale and also large automatic apparatus was developed for mass production. Considering the problems of cost and complex shape for production of hot-pressing, pressureless sintering techniques were also developed. The importance of powder processing and microstructure were recognized.

2) Stage of R&D on improvement of high temperature strength (Late 1970's)
Degradation of strength was observed at high temperature for some samples, though they showed excellent properties at room temperature. Some sintering aids were known to have an adverse effect at high temperature. The properties of the grain boundary phase was known to be important and grain boundary technology was developed. On this point of view technology to diminish the amount of grain boundary phase was examined. Sialons which are solid solutions in the Si-Al-O-N system were extensively developed. Technology of crystallization of the grain boundary phase was also developed for improvement of strength. As impurities also cause degradation of high temperature strength, processing for pure and fine powder was developed. Gas pressure sintering was invented and HIP technology was gradually applied for production. In addition to strength, reliability (the scattering of the data) was recognized to be impor|ant and efforts to fabricate homogeneous fine microstructures were done for the purpose of commercialization.

3) Stage of start for commercialization, and improvement of toughness and other properties (Early to middle of 1980's)
Glow plug (1981) swirl chamber (1984) and turbocharger were commercialized. For commercialization it was known that problems of lower cost, products with high reliability and processes to manufacture parts with complex shape were important. It was recognized that improvement of toughness, oxidation and corrosion resistance and improvement of resistance to exposure in a long-time severe environment were also important for high temperature heat engine.

4) Stage of steady progress and R&D on composites, surface modification, gradient composites, nano composites, etc. (Latter half of 1980's to present). The fever for fine ceramics has settled down compared with previous days. There is recognition that it is not so easy to commercialize and to develop a market in high temperature ceramic materials. While some companies have cut down or given up their R&D projects, companies with a steady strategy are making continuous advances. A characteristic feature of this stage is that cooperation among different industrial fields and cooperation among industry, government and university have become more active.

Fig. 2[1] shows the change of the production value of fine ceramic parts since 1982. The total value is gradually increasing year by year and it is predicted to be 1176 billion Yen in

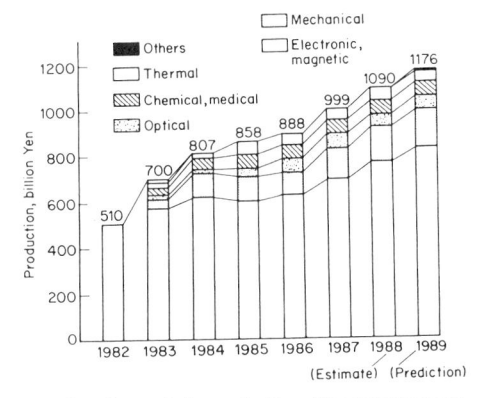

Fig. 2 *Change in the production of fine ceramics parts in Japan*

Fig. 3 *Estimation of the market size of the fine ceramic industry in Japan*

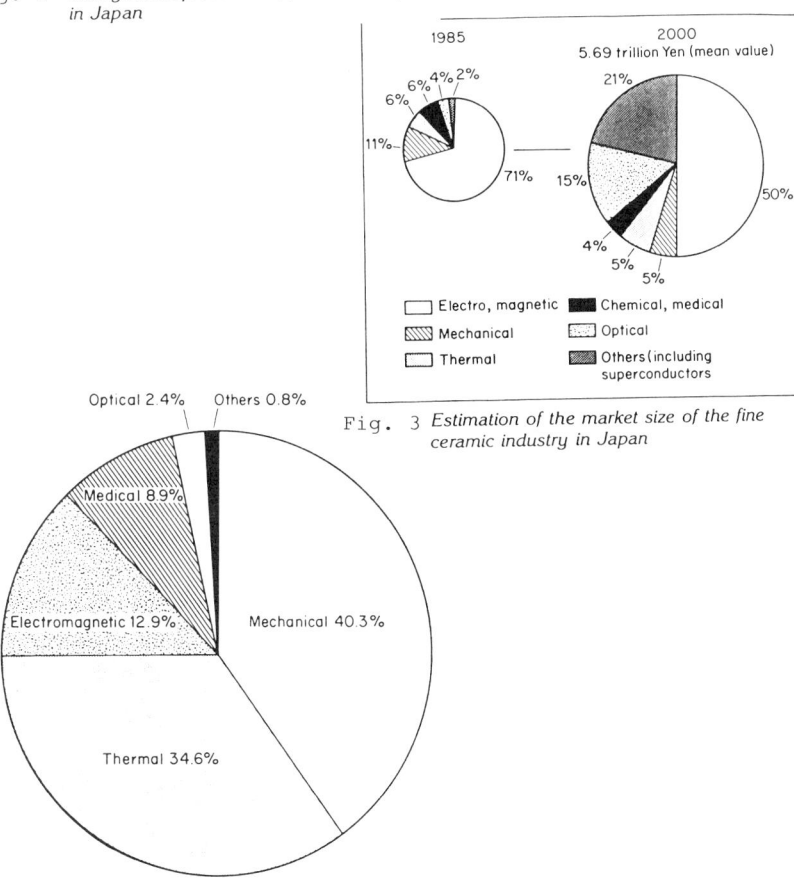

Fig. 4 *Fields expected in near future by users (March 1989 — MITI)*

1989. Electromagnetic ceramics occupy a 70% share, but the growth rate of structural ceramics is relatively higher than others. Fig 3[2] shows an estimation of the market size of the fine ceramic industry in the year 2000. Total production is estimated at about 6000 billion Yen in 2000 from about 1200 billion Yen in 1989. Growth of the market for optical and super conductor ceramics are expected to be comparatively great in the future. The value of mechanical and high temperature ceramics are estimated to be still about 10% of the total. However, as many users want mechanical and thermal applications, as shown in Fig. 4[2], it is thought that high temperature structural ceramic markets will become greater in the 21st century.

As for R&D on high temperature ceramics, one direction is a move from homogeneous fine microstructure to heterogeneous fine microstructure in monolithic ceramics. Grain growth of whisker or fiber-like crystals in a fine matrix improves toughness. Another approach is whisker or fiber reinforced ceramic composites. Those composites are expected to be a big innovation in the near future. Powder/powder composites with different ceramic phases still have many unexplored fields. Surface modification techniques, researches on nano composites with atomic molecular level control and gradient composites have also started. These new approaches are expected to have a big influence on future ceramic technological innovation.

With the progress of advanced technology, boundaries between fields are weakening and interdisciplinary researches become much more important. In the structural ceramic field we should consider not only structural properties but also incorporating other functional properties for developing intelligent ceramics with multi-functions. In this sense it is worthy to consider what we can learn from the mechanisms of living bodies for future ceramic technology.

3. Culture and technology in R&D on structural ceramics

Some differences of culture are observed between Japan and the U.S. affecting scientific and technological process. T. Motokawa[3] describes the differences between east and west in his paper "Sushi Science and Hamburger Science." Some of these considerations are described here.

1) Different ways of thinking about the scientific process.

Fig. 5 shows different ways of thinking about the scientific process between Japan and the West. Generally in Japan researchers consider that the experimental result is primary, though they also investigate theory and previous data. Particularly, ceramic science has not yet established mature laws and still has many unexplored fields. Previous theories may change and new ideas appear through experiment. Therefore, experimental facts are most important and learning theory is done through doing experiments. Interpretation comes after experiment. For this approach researchers need practical and precise observations and skillful experiments. Scientists and technicians work together as one body. New phenomena are often discovered by precise observation and new creative ideas with know-how comes as a result of this way of thinking.

On the other hand, western researchers seem to think more important interpretation and hypothesis aspects of research. If the hypothesis is proved, honor comes to the researcher's achievement. The theoretical approach is considered the most effective way to develop new ideas. By this way of thinking, new creative ideas come with theory. Technological innovation occurs in both of these different ways, but are a little different: one is with know-how and the other is with theory.

2) Different ways of R&D practice

Fig. 6 shows different ways of R&D practices between Japan and the West. In Japan there is a close relationship and almost no gap between researcher and technician, between

Japan

- Experiment result is first
- Fact is important / Avoid interpretation
- Practical Observation / Skillful experiments
- New creative idea with know-how

West

- Interpretation is first
- Hypothesis is important
- Theory
- New creative idea with theory

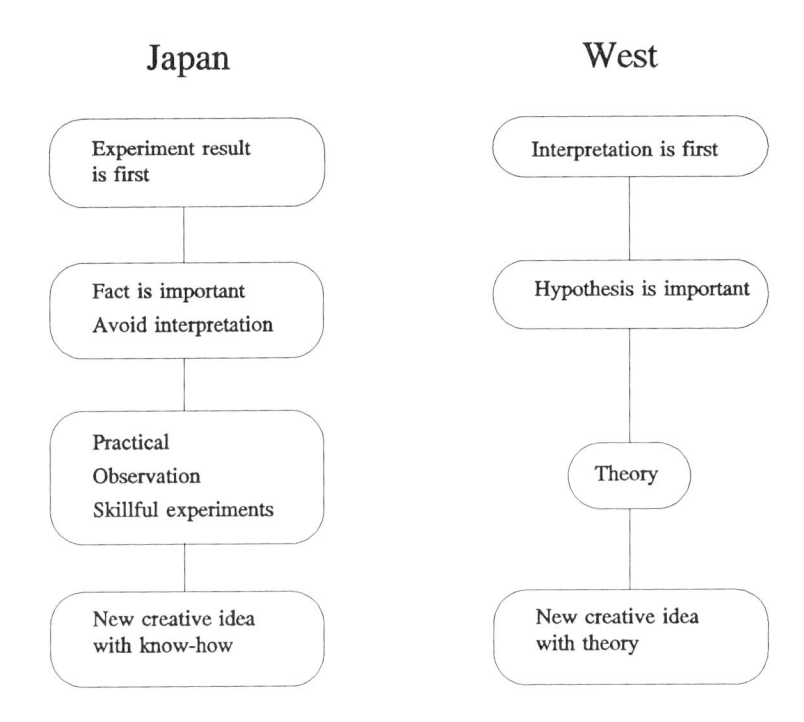

Fig. 5 Different ways of thinking about scientific process.

scientist and laborer. There is not so big a difference in salary between them, as the salary is usually decided by age and working period. Joint work of researcher and technician is often carried out together from planning through experimental works. Technicians frequently join planning discussions and they understand the essential experimental purpose. Efficiency of this way of work is not so good. However, there are many possibilities to find new phenomena in the course of experiments. Technicians and laborers do willingly the work for improvement of technology and product quality.

In western countries there seems to be a gap between researchers and technicians; they seem to work separately. The researcher makes a plan on his desk and the technician or laborer does the work according to the researcher's directions. The researcher learns of a result after the experiment and the technician has no concern about it. Efficiency of this method of working is better than the Japanese method, but there are fewer possibilities to find unexpected new phenomena.

3) Differences in the sense of values in research organizations.

Fig. 7 shows differences in the sense of value between Japan and Western organizations. In Japan most private companies and government organizations adopt life-time employment and promotion is done mostly by seniority. They think that achievement is done by a group

```
            Japan                              West

   ┌─────────────────────┐          ┌─────────────────────┐
   │ Researcher ~ Technician │      │ Researcher / Technician │
   │ Scientist ~ Laborer │          │ Scientist / Laborer │
   │       no gap        │          │         gap         │
   └─────────────────────┘          └─────────────────────┘
              │                                │
   ┌─────────────────────┐          ┌─────────────────────┐
   │ Planning ~ Experiments │       │ Planning / Experiments │
   │     joint work      │          │   separated work    │
   └─────────────────────┘          └─────────────────────┘
              │                                │
   ┌─────────────────────┐          ┌─────────────────────┐
   │  Efficiency is worse │         │  Efficiency is better │
   └─────────────────────┘          └─────────────────────┘
              │                                │
   ┌─────────────────────┐          ┌─────────────────────┐
   │ Many possibilities to find │   │ Fewer possibilities to find │
   │    New phenomena    │          │    New phenomena    │
   │ Improvement of quality │       │                     │
   └─────────────────────┘          └─────────────────────┘
```

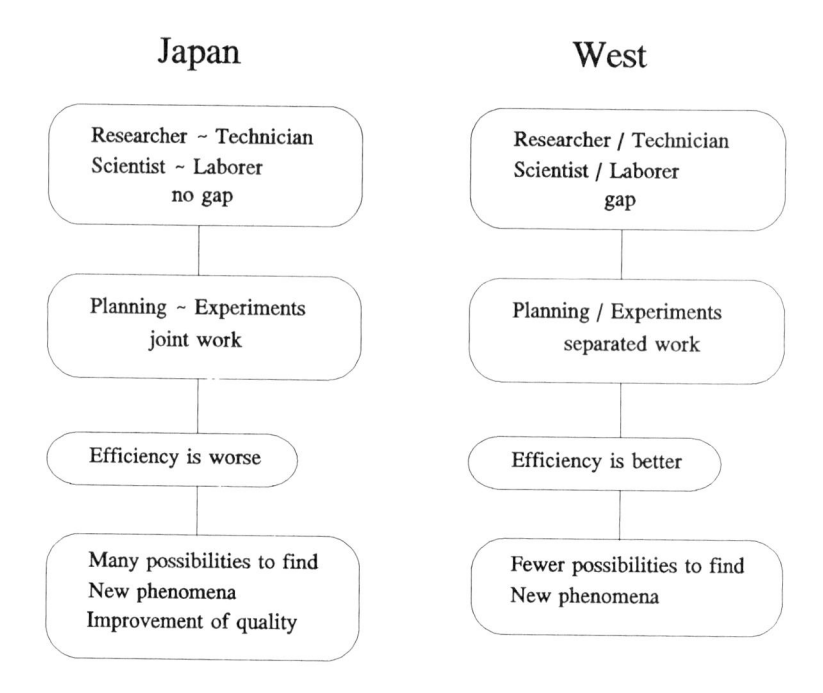

Fig. 6 Different ways of R&D practice.

effort and that success by any person is only achieved by the help of his associates. It is thought to be a cooperative result. Honor for achievement is usually given to the company, group or group head. The company is like a family, so the company must take care of employees. People think that company and leaders must always think about their welfare. Therefore, employees work hard. In addition, there is a mind of confucianism or mind of Do (such as in Ken-Do, Ju-Do, etc.) in which people aim to approach a condition of perfect personality, to obtain other people's admiration and to be in a state of harmony. These ideas affect people's sense of values.

In western countries employment is usually done by contract and promotion is done in accordance with a person's ability and accomplishments. Individuality is important and there is a recognized difference among individuals. When an achievement is made, honor comes to the individual. Therefore they work hard. Both Japanese and Western results of "working hard" are the same, but the process leading to that result is different. A distribution of the number of the people working hard is narrow and sharp in Japan, while in western countries, it is wide and broad. In western countries the number of researchers working extremely hard may be larger than in Japan.

Japan

- Life time employment
  Mostly promotion by seniority

- Achievement by group
  man
  ╱╲
  man ══ man
  Cooperation results in equality

- Fame ──→ company
       ↘ group
       ↘ group head

- Family = Company
  Company should take care of employee
  merit ──→ individual
  work hard!

West

- Contract employment
  Promotion by achievement

- man | man | man
  individual
  Differences among individuals

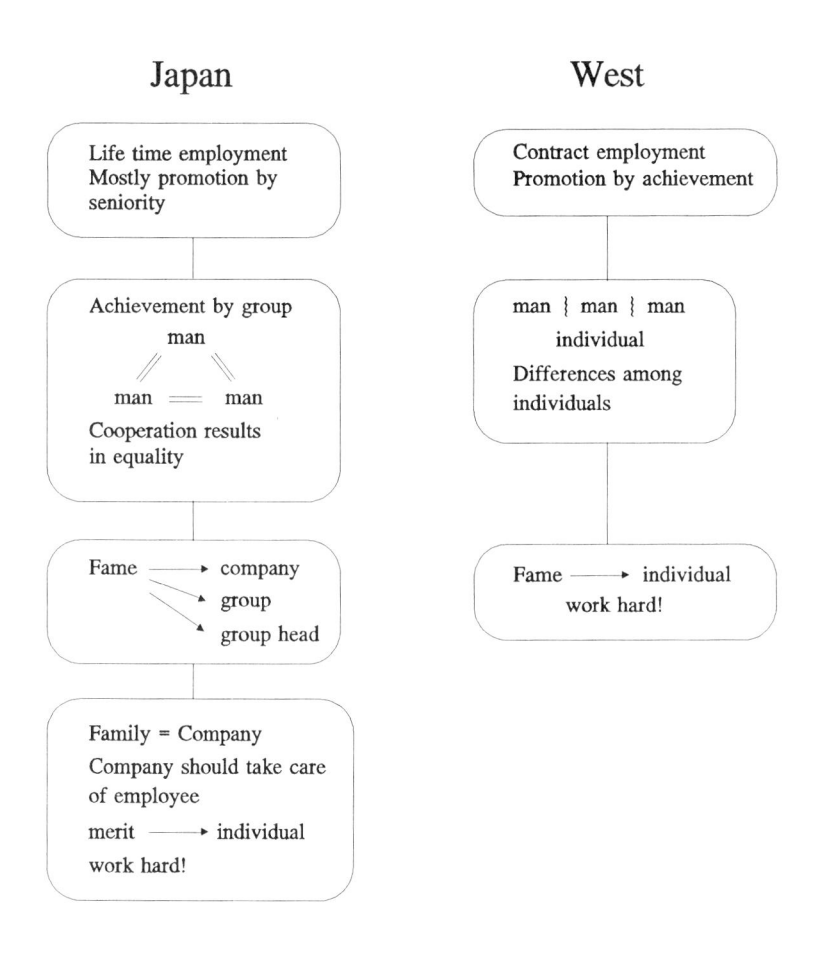

- Fame ──→ individual
  work hard!

Fig. 7 Differences in the sense of values in the organization.

References
1) Report of the Fine Ceramics Basic Problems Council, MITI, (1989).
2) K. Kobayashi, *Materials & Design,* Vol. 11, No. 2. April (1990).
3) T. Motokawa, *Perspectives in Biology and Medicine*, 32, 4 Summer (1989).

Dr. Kobayashi elaborates on the "Fever" for ceramics in Japan:

"FEVER" FOR STRUCTURAL CERAMICS IN JAPAN

Kazuo Kobayashi, Nagasaki University

Fever for structural ceramic industry

In Japan "fever" for structural ceramics occurred in the middle of the 1970's. While the electroceramic industry has about 40 years history, large scale R&D on structural ceramics started after the first oil crisis and it was triggered by researchers in the United States and European countries. Therefore its history is very short, 10-15 years. However, after the middle of the 1970's structural ceramics were strongly anticipated to become a main new industry along with electronics and bioindustries. News media propagated the idea that this new technological field would have a brilliant future and many private companies entered the field and started R&D on structural fine ceramics.

As a background of the fever, first there is a big influence of Japanese government policy along with the understanding and consensus of Japanese people. That is, the Japanese government made a policy to strengthen science and technology and most Japanese believe that science and technology, particularly advanced technology, should be promoted as an important strategy for present and future growth of Japan. We, most Japanese, believe that fine ceramic technology is precisely one of the suitable fields we should promote. Second, as Japan is a unitary nation, the understandings and consensus have spread widely throughout the public. Even common housewives and children have learned the words "Fine Ceramics" and have gotten interested in fine ceramics. When a fair of fine ceramics opened, visitors with family were often seen at the fair as well as professionals. Third, MITI started many projects to promote structural ceramic industry and many private companies wanted to be part of the trend.

Table 1 shows motivations of decision for private companies as to why they have

Table 1. Motivations for entering structural ceramic field; (from 168 private companies in MITI survey, 1989).

1) To deal with a slump of existing materials and products 18%
2) To improve performance of existing materials and products 27%
3) To keep an R&D potential for competitive companies 7%
4) Need to use new materials within the company 9%
5) By request of users 10%
6) Expectation for high profits 26
7) Others 3%

entered the structural ceramic field. This data shows that many companies aim to be high technological industry and to gain high profit. Table 2 shows the time when the company

Table 2. Time that companies entered structural ceramic field (From 168 private companies in MITI survey, 1989).

Time	Number of Companies
1) Before 1975	36
2) 1975-1980	14
3) 1981-1985	30
4) 1986-1988	10
5) Since 1989	11

entered the structural ceramic field. From this data it is known that about half started their activities on structural ceramics after the year of 1975. This tendency is also known from the number of patents on silicon carbide, silicon nitride and zirconia as shown in Fig. 1. The data shows that the fever started and many companies supplied manpower and assumed R&D expenses in structural ceramics after 1975.

Fig. 1. Change of number of patents granted with relation to structural ceramics (The number in 1973 is set as 100)

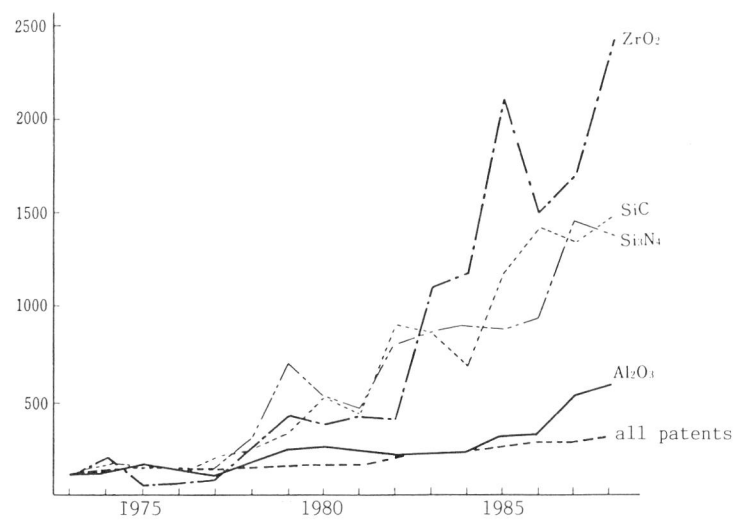

The users also have eagerness to use structural ceramics and this prompted the fever. Table 3 shows the reasons why the users want to use structural ceramics. From the data it is known that many user companies want to apply high technology by adapting fine structural ceramic. Table 4 shows the technological areas and number of items of fine ceramic parts used at present and those to be used in the future. About 75% of the total area is occupied by mechanical and thermal structural parts.

Many private local societies on fine ceramics have started since the middle 1970's with members of companies, government research institutes and universities. For example, Kyushu Fine Ceramic Technoforum started with about 170 private companies, researchers of government institutes and some professors of universities in the Kyushu area. The forum is now sponsoring activities such as seminars, symposia, exchange of information, etc. every year. The number of such societies or groups on fine ceramics is now about 50 throughout Japan from Hokkaido to Kyushu. The activities in local areas have given good stimulation to local industries, particularly to small and medium companies in local areas. Recently a council to connect those local societies has been established to promote their activities more efficiently.

The Japan Fine Ceramics Center Foundation established in 1985 has given support to the fever and the activity at all levels. About 250 companies are members of the center and the works are testing, examinations, standarization, consultant, collection of information, education, international exchange, etc. The center also holds a fine ceramic fair every March in Nagoya and the fair is very effective for business negotiations, exchange of information and also to prevail the real image of fine ceramics to the public.

As an academic society The Ceramic Society of Japan established in 1927 has a long history of contribution to all fields of ceramics. The members of the society are 7000 individual researchers. The society publishes academic journals and holds annual meetings. Presentations on fine ceramics are increasing in number every year at the meetings.

Japan Fine Ceramics Association was also established in 1986. Membership is limited to private companies and its number is about 220. The works of the association are collection and supply of information, survey of market, training courses and symposia, cooperation with related domestic and overseas group.

NGK Spark Plug Company of Nagoya began research on silicon nitride materials in the late 1960's. Dr. Yo Tajima joined the company in 1975 and has been associated with that research and development program since then. He reports:

SILICON NITRIDE AUTOMOBILE ENGINE COMPONENTS

Yo Tajima, NGK Spark Plug Company, Nagoya, Japan

As has been discussed in this meeting, silicon nitride has a number of good characteristics, such as high strength and high toughness, excellent thermal shock resistance, low density and good wear resistant. This makes it applicable as automobile engine components as well as cutting tools, ball bearings and it remains a prime candidate material for ceramic gas turbine engines which are off in the future. The most challenging application of engine components is turbocharger rotors which were first introduced in 1985 in the Nissan Z cars in Japan. Turbochargers are now produced by three ceramic manufacturers and one auto manufacturer in Japan and are commercially used by three auto manufacturers. The primarily advantage of ceramics for turbochargers is the low density of silicon nitride as compared to metals which gives a better response. Because of the low density the moment

Table 3. The reasons given by users as to why they want to use structural ceramics

The Reason	Number of answers
1) For the purpose of energy saving in production process	15
2) For the purpose of improvement in reliability of products and processing	48
3) For the purpose of improvement of life of products and process	86
4) To make products and process of lighter weight and smaller scale	37
5) To improve performance of manufacturing process	77
6) To reduce cost of manufacture of product	23
7) To improve on image of product	20
8) No other material except fine ceramics able to satisfy a required property	43

Table 4. Technological area and the number of fine ceramic items which are used at present and those to be used in near future

Field	items used at present	Items to be used in near future
Electro and magnetic parts	56 (22.4%)	16 (12.9%)
Mechanical parts	109 (43.6%)	50 (40.3%)
Thermal parts	58 (23.2%)	43 (34.6%)
Optical parts	4 (1.6%)	3 (2.4%)
Chemical and medical parts	16 (6.4%)	11 (8.9%)
Others	7 (2.8%)	1 (0.8%)
Total	250	124

As fine ceramics have a variety of applications, not only the traditional ceramic industry but also all material-manufacturing industries such as steel, non-metal, chemical engineering, textile, and the users industries of electronics, machinery, transportation and other related industries joined into this field. This brought severe competition among them and products with better quality came to be produced.

of inertia is small. Other engine applications are rocker arm pads which provide maintenance-free operations. This is particularly applicable for taxis which have a long idling time; during idling there is a lower oil supply and poor lubrication. Silicon nitride provides substantial performance improvements. A third application is glow plugs for diesel engines which were first introduced in 1981 by Isuzu. These consists of a tungsten heating element integral with a silicon nitride container. Turbocharger rotors grew to be about 20 million U.S. dollars in 1988 and are perhaps tripled that in 1991. The total market for silicon nitride automotive components is perhaps 100 million dollars.

There have been many barriers to the development of new ceramic components for automotive application. These are summarized in a Delphi Survey of Larson and Vyas shown in Table 1. The ceramic materials had unproven durability. There was no reliability

---

TABLE 1. BARRIERS TO UTILIZATION OF CERAMIC COMPONENTS

(INTRODUCTION)
(PENETRATION)

1. UNPROVEN DURABILITY
2. INADEQUATE RELIABILITY RECORD
3. INADEQAUTE NDE TECHNOLOGY
4. UNFAVORABLE ECONOMICS
5. ENGINE MANUFACTURERS' RESISTANCE TO CHANGE
6. LOW YIELD PROCESSING TECHNOLOGY
7. LACK OF DESIGN DATA BASE
8. INADEQUATE COATING

Source: R.P. Larsen and D. Vyas, "The Outlook for Ceramics in Heat Engines 1990-2010: Results of a Worldwide Delphi Survey," SAE Paper No. 880514 (1988)

---

record, inadequate nondestructive evaluation and particularly unfavorable economics. In addition, engine manufacturers were generally resistant to change. There was not very much processing technology, an inadequate design data base and an inadequate quality of ceramic powders. The innovations which have lead to the utilization of these materials have been discussed already and they consist of the development of suitable sintering aids, particular gas pressure sintering techniques, and the development of high quality powders, joining with metal parts and shaping methods for complex parts. Relative to interactions with national laboratories, we should mention that Dr. Yotomo of NIRIM filed a patent for the gas sintering technique and NGK Spark Plug obtained development funds from JLDC for the commercialization of gas pressured sintering techniques. We were successful in this manufacturing effort and have reimbursed the development fund.

A number of automobile manufacturers and ceramics manufacturers have combined to introduce silicon nitride parts as shown in Table 2. One of the main factors affecting market acceptance of silicon nitride parts for engine components are the customer requirements. The tax system in Japan is higher for larger engines which gives rise to the desirability of small engines. The cost of diesel fuel is lower which puts a premium on diesel engines. The roads in Japan and the parking conditions also put a premium on smaller cars. As a result, there is a strong demand for diesel engines with high performance.

In addition, corporate influences are important. At the NGK Spark Plug Company, silicon nitride and sialons research started in the 1960's while the first commercial adaption

TABLE 2. DEVELOPMENT TIME OF CERAMIC PARTS

PARTS	PURPOSE	CAR MAKER	CERAMIC MAKER	YEAR
Turbocharger	Response Power	Nissan Nissan Isuzu Toyota	NGK Spark Plug NGK Insulator Kyocera Toyota	1985 1987 1989
Rocker Arm	Maintenance Free	Mitubishi Nissan	NGK Insulator NGK Spark Plug	1984 1987
Glow Plug	Quick Start Clean Exhaust Gas	Isuzu Mitubishi Mazda Nissan	Kyocera Kyocera Kyocera NGK Spark Plug	1981 1983 1985 1985
Hot Plug	Clean Exhaust Gas Power	Isuzu Toyota Mazda	Kyocera Toyota NGK Insulator	1983 1984 1986

to supercharger parts by Nissan did not occur until 1985. There was a remarkable patience of top management and continuing support of the concept that silicon nitride and ceramic materials would be important to the future of automobile components which are the principle business of NGK Spark Plug. The corporation had a strong desire to be a leading manufacture in the ceramic engine components business and to maintain a significant market share. A second advantage of NGK's Spark Plug was its history and experience as an engine component's manufacture with engine test capability. This made it possible to work effectively in cooperation with Nissan in the development of automotive parts. A second necessary component is the engine manufacturers' willingness to adopt ceramic components. As the ultimate user of these materials, they were the final decision maker and were willing to expend time and effort to develop a proven liability and anticipate that the cost performance ratio would come to be justified. The engine manufacturers had to work in coordination with the ceramic manufacturer in a close relationship and spirit of trust. This interaction between ceramic manufacturer and auto manufacturer was an essential element in the development of silicon nitride engine components.

At the conference a group of technologists and managers who had close association with silicon nitride developments discussed the organizational factors which seem to have affected the course of events. They concluded that the frame of mind of the organization, the competence in core technologies, personal interactions, facilities, attitudes toward the use of technology, timing of technological developments, the commitment to continuing innovation, the nature of collaborative partnerships, the goals and milestones that were developed, the market fit of manufacturing and customer need, the history of previous successes, images and perceptions of innovation procedures, and legal regulatory issues all had a strong influence on the course of events. A summary of these discussions follows:

Si$_3$N$_4$ - STRUCTURAL CERAMIC INNOVATIONS

Richard C. Bradt, University of Nevada, Reno

The perspective from which I will address this topic is that of a university faculty member over the time period of interest, the mid-1960's to the present. During that time frame, I have continually conducted research on structural ceramics with both American and Japanese graduate students, as well as in cooperation with faculty colleagues from both countries. I spent a sabbatical in Japan (1978) and have visited Japan on more than thirty occasions. I believe that I have good to excellent familiarity with the structural ceramics programs and related activities in both the US and Japan. Perhaps I know all of the participating engineers and scientists in both countries on a first name and last name basis, respectively. For the purpose of this presentation, I'll adopt the definition of innovation that has been advanced by Professor Stephen Kline of Stanford, namely that the innovation process requires three features: (i) design, (ii) manufacturing and (iii) extensive use. I'll attempt to flavor this summary with my own personal impressions, influenced by my more than twenty years in the technical field and my fondness for sushi.

The silicon nitride influence on structural ceramics began with enthusiasm in the early 1960's in Great Britain with the Lumby and Lucas activities creating excitement for turbine blades, but achieving applications in the cutting tool field. Today, a couple of decades later, US ceramic manufacturers are making some automotive valves, some bearings and quite a few cutting tools, while Japanese producers are in the midst of major production runs of numerous artifacts, including glow plugs and rocker arm wear pads but most importantly and impressively, ceramic turbocharger rotors. This latter item is particularly significant in that it is the first truly complex shaped structural ceramic product that has achieved Professor Kline's three criteria. Ceramic turbocharger rotor production in Japan is now estimated to be about 4000 per month. Relative to Professor Kline's criteria, the Japanese have clearly innovated! Has the US? Hardly? Why this difference? I'll attempt to address the reasons as I perceive them to have evolved over the last two decades.

At the beginning of the 1970's, the US was clearly the structural ceramics leader in the world in my opinion. The seed for structural ceramics in heat engines was planted in the US and began with the Ford/Westinghouse ARPA (DARPA) programs. These were followed by the AGT efforts, each of which ground to a virtual halt with the end of US government funding. These programs left the US ceramic community and automotive producers with what can be best characterized as a mild enthusiasm for structural ceramics in heat engine-like structural applications. To be sure, significant progress was made in these US government funded programs, but they did not succeed in the fullest sense of innovation. Products were not being manufactured and used by consumers.

In Japan, there were government programs, too, most notably the early MITI "Moonlight" and "Sunshine" programs which paired up various ceramic producing companies and potential user concerns. However, once the government funding ran out, the Japanese seemed to have established long term relationships, not only between the companies, but the individual scientists as well. The consequence was that a "ceramic fever" developed in Japan and the general public had a consensus that the future industrial utilization of structural ceramics would bring about a "Second Stone Age". The contrast between Japan and the US was practically like day and night.

The consequence of the initial period, in the eyes of this university bystander, was that the US had clearly planted the initial seeds and began much of the technical effort;

however, Japan saw the potential in the US activities, believed in the concepts and proceeded forward. The US industry never really seemed serious, just spending government funds but not following through. For some reason, Japanese industry became committed and planned long term programs that have ultimately proven to be innovative and successful.

Let me next address each of Professor Kline's three criteria: (i) design, (ii) manufacturing and (iii) use. I'll attempt to separate them, but will apologize in advance if I fail. The design criteria can be divided into the material design, its chemistry, microstructure and processing and the final product (object) structural design. Each relates to the other in a complex intertwined fashion. While the initial Si_3N_4 structural studies originated with Professor Jack in England, both US and Japanese scientists quickly picked up on that research and rapidly extended it in nearly parallel advances. These types of technical items are readily interchanged at numerous international meetings. The use of additives to Si_3N_4 and processing fundamentals, including the complex phase equilibria rapidly advanced. In the US, in the 1980's, there was a strong emphasis on the fundamental science of ceramic processing which clearly established the US as the world leader in ceramic processing from a basic science point of view. On the other hand, the ceramic processing activities in Japan were more manufacturing technology oriented as opposed to the US basic processing science. A decade later, the consequence has been a resounding Japanese success in manufacturing of ceramics.

As structural ceramics are brittle materials, they could not be successfully applied as one-for-one substitutes for metals in actual applications. This necessitated the evolution of a whole new brittle materials design methodology. It required the cooperation of ceramic scientists and mechanical engineers to develop new design concepts and numerous iterations, as the methodology advanced. The US established the early directions, applying concepts of fracture statistics and probabilistic design. NASA with an in-house program at Lewis-Cleveland and a NASA sponsored program at the University of Washington (materials and mechanical engineers) was instrumental. However, these programs never really made significant inroads to the industrial ceramics laboratories and structural ceramics never made a big impact in the field. It was in many ways similar to the basic ceramic processing science in that the US actually seemed to initiate the fundamental effort, but somehow Japan carried through with the concepts to the final technological applications.

Both the US and Japan have made significant advances in the manufacturing (processing) of ceramics in the past several decades. However, the two efforts may be contrasted as primarily first class university fundamental research (US) versus strong industrial technology applications oriented development (Japan). This is not to say that the US is without its industrial successes, for certainly the catalytic converter supports have been a huge success, but like spark plugs, they are not a true structural ceramic application. Japan successfully created industrial involvement and applied the technology to manufacturing very early in the developmental process. While anything as complex as a turbocharger rotor was initially out-of-the-question, the Japanese ceramic manufacturers produced ceramic fish-line eyelets for fishing rods and numerous ceramic scissors (Who doesn't have several?). These seemingly mundane objects established the industrial experience base which lead to the development of the ceramic processing of high-tech structural ceramics. The consequence is that today most knowledgeable ceramic scientists and engineers believe that Japan leads the technological world in the slip casting and the injection molding of ceramics, the two manufacturing processes which most readily lend themselves to the production of complex shapes of structural ceramics for heat engines and other structural applications. So much for manufacturing!

The use feature is the last of Professor Kline's criteria for innovation. There is certainly some use of structural ceramics in the US in automotive applications. There are glow plugs, bearings, rockerarm pads, valves, valve seats, etc. However, it is the Japanese who are producing thousands of ceramic turbochargers per month in the NGK/Nissan and KYOCERA/Isuzu cooperative efforts. Ceramic turbochargers are available in cars in Japan at local auto dealers. It is clear that the Japanese have succeeded and closed the loop on Professor Kline's definition of innovation for structural ceramics.

Although figures are not publicly available, it hardly seems possible that ceramic turbochargers are economical, yet the Japanese are installing them and have them available for purchase by the general public. Perhaps the only current advantage over metal turbochargers is one of weight, as they achieve maximum RPM faster than their metallic counterparts. Why then are the Japanese so committed to turbochargers of structural ceramics? It seems that the Japanese must envision the same long term advantages that the US and everyone else does for ceramics. However, they are willing to make the investments to improve the technology and they appear to be willing to accept the profits later. This is in contrast to a US attitude that we won't use it unless it's cheaper or better (or both) and in the US we must make a profit now. It's clear which strategy has proven to be the best for structural ceramics over the past two decades.

In summary, perhaps it's appropriate to highlight those features which I believe have allowed the Japanese to successfully innovate in structural ceramics over the past two decades. Both US and Japan had early government funding and involvement. In the US, it practically stopped when the government funding stopped, but somehow in Japan, a synergism developed and the industrial associations which were initiated by the government continued and prospered. Was it simply financial? Somehow, I doubt it! In terms of Professor Kline's three criteria for innovation, the teamwork cooperative approach to *design* has been critical to the NGK/Nissan and the KYOCERA/Isuzu turbocharger successes. Materials design and product design are intimately connected. Somehow the Japanese government (MITI) made the industrial connections and made them work. The *manufacturing* difference between the US and Japan has been clearly a fundamental basic science of processing approach versus an industrial technology applications one. Many industries are able to prosper without a truly fundamental scientific understanding of all of the principles, the disciplined Japanese approach to the technology of manufacturing has certainly been adequate in this instance. Of course it probably didn't hinder the Japanese progress that the US was simultaneously and vigorously pursuing a basic science of ceramic processing theme, the results of which were available to all who attended the international meetings and read the scientific journals. As for actual *uses* of structural ceramics, the Japanese had a longer term viewpoint, but did not completely abandon immediate, more simple applications such as the ceramic scissors, etc. These intermediate term applications probably had critical effects on the development of high yield manufacturing capabilities; the full extent and benefits of which are unknown. However, there is no doubt in my mind that the longer term approach will ultimately lead to an improved, higher level of technology. Not only will that technology prove innovative in terms of the products that are originally envisioned, but it will lead to a captive situation for successive generations of new products. The future tragedy is that one cannot just jump on the structural ceramics bandwagon to produce these successive generations of new products, for it requires the long term investment of resources to be competitive. In my opinion, few, if any in the US, have been willing to make that long term investment to achieve innovation and the current state of structural ceramics is one result.

FACTORS AFFECTING SILICON NITRIDE INNOVATIONS

David Richerson, Ceramatec Corporation

The objective of this section is to use the silicon nitride development history to compare the cultural and structural factors influencing invention and innovation in the U.S. and Japan. For the purposes of this discussion, innovation is equated to a goal to achieve a saleable product. Based on this definition, three conditions must coincide to achieve innovation: (1) a customer need, (2) technology maturity adequate to meet the customer need, and (3) a business decision to commit the required resources to achieve the development and production implementation. The following paragraphs address these conditions based on the factors listed in Table 1.

Frame-of-Mind of the Organization
The frame-of-mind of the organization determines strategic planning and thus has a major influence on achieving innovation. The effect is moderate regarding invention. Invention can occur on an individual basis and does not necessarily require a strong commitment of the organization. However, a strong commitment from the organization can provide the resources to accelerate R&D and enhance the likelihood of invention and innovation. Companies and government organizations have demonstrated a favorable frame-of-mind in both the U.S. and Japan regarding development of Si_3N_4 technology.

Table 1

Some Factors Influencing Invention and Innovation

* Frame-of-mind of organization
* Core technologies/competencies
* Personnel interactions
* Facilities
* Attitude towards use of technology; Comfort zone
* Timing
* Commitment
* Continuity/duration
* Collaboration/partnerships
* Goals and milestones
* Market fit/Innovation window
* Prior successes or failures
* Image
* Legal/regulatory issues

Core Technologies/Competencies
An existing technology base can provide the right environment for recognizing the potential for invention or innovation. The broader the technology base, the easier that a critical team can be assembled to address system requirements and achieve innovation. Norton Company progressed rapidly in Si_3N_4 technology in the early 1970's because they had all the key core technologies (and equipment) in place relevant to reaction bonding, hot pressing, powder synthesis, diamond grinding, and property characterization.

Personnel Interactions

Invention can occur with an individual working independently. Innovation typically requires interdisciplinary input and a close interaction between individuals and often organizations. The Japanese society has a strong tradition of team effort both within an organization and between partner organizations. Efforts are cooperative and non-competitive. An example is the cooperation of NGK Spark Plug and Nissan to develop and commercialize turbochargers containing a Si_3N_4 rotor. Cooperative efforts do occur in the U.S. also, but there tends to be more competition and emphasis on individual contribution.

Facilities

Invention can usually be accomplished with relatively modest equipment and facilities. Innovation generally requires substantial facilities and equipment, especially if large scale production is involved. Experience in the U.S. with Si_3N_4 in the 1970's was to use or modify existing equipment, even if results indicated that development of a new piece of equipment would be beneficial. No government support was available to conduct process equipment development. Also, since Si_3N_4 development was directed towards heat engines and was considered high risk, industry was reluctant to make substantial equipment investments.

The philosophy in Japan was different. High priority was placed on developing and implementing improved processing equipment. For example, as soon as NGK Spark Plug individuals heard about the overpressure sintering studies at NIRIM in Japan and at AMMRC and GE in the U.S., they applied for and received funds from MITI to work with a Japanese furnace manufacturer to build an overpressure sintering furnace. As a result of this funding, they had a working furnace earlier than U.S. Si_3N_4 manufacturers and accelerated achieving commercial sintering capability. This is only one example. In general, Japanese companies were much more aggressive than U.S. companies in developing and procuring advanced processing equipment such as injection molders, sintering furnaces and HIP furnaces suitable for advanced Si_3N_4.

Attitude Towards Use of Technology; Comfort Zone

Examples during Si_3N_4 development indicated a strong difference in philosophy between U.S. and Japanese companies in the use of technology. U.S. companies appeared to have a preference to develop technology internally and only consider externally-developed technology if it fit into a narrow comfort zone. For example, Norton was comfortable with hot pressing and licensed the Lucas Si_3N_4 hot pressing technology and much later the ASEA hot isostatic pressing technology. However, Norton did not license the Lucas pressureless sintered Sialon technology. Furthermore, some of the literature indicates no significant effort in pressureless sintering at Norton, even though they were very close geographically and technically to AMMRC where early significant sintering advances were accomplished. "Not invented here" and "comfort zone" may have been factors.

The philosophy in Japan was different. Japanese individuals systematically searched the literature and attended international meetings looking for alternatives with potential for commercialization. They by choice expanded their comfort zone by duplicating reported studies and using this new capability as a baseline for expanded efforts. Japanese companies were encouraged to do this by MITI. For example, MITI provided support to establish powder synthesis capabilities at Denka and Toshiba and sintered RBSN at NGK Spark Plug.

Timing

Timing is extremely important in achieving innovation. If customer need, technology maturity, and a business decision do not coincide, innovation of a marketable product may not occur. A good example is ceramic bearings. Technology feasibility and a customer need were demonstrated in the early 1970's, but the technology was not mature enough to allow cost effective fabrication. This did not become viable until the 1980's when near-net-shape processing became mature through overpressure sintering and HIP techniques.

Commitment/Continuity/Duration

Commitment, continuity, and duration are important to achieving innovation. This is illustrated in Figure 1 above. To progress from an idea or an initial invention to an established product historically (on the average) has taken 10-20 years. The commitment in resources is relatively small during the early stages of development, but increases rapidly as product development, marketing and processing scale-up are required. Invention can be achieved with a short term commitment, but innovation requires a long term commitment.

A number of factors affect commitment and continuity. Key factors are the nature of business planning, the continuity of personnel and the duration of the funding source (industrial or government). In the U.S., concern in most companies is the near term return on investment and having a favorable bottom line at the end of each year. Planning cycles are generally 3-5 years. Where government funds are involved, the financial commitment is even shorter, i.e., generally a maximum of three years with incremental funding yearly. In Japan, government programs are generally 5-10 years. Industry programs also appear to be longer term than in the U.S.

In addition, Japanese companies in the Si_3N_4 area have pursued "soft markets" to gain production experience and "learn by doing." A soft market is defined as one that either (1) does not have large growth potential, (2) will run at a loss for a number of years, but may become a large market, or (3) based on current technology, does not have prospects for high margin. Examples of some of the soft markets that have been pursued by Japanese companies for Si_3N_4 have included knives, scissors, rocker arm wear pads and turbocharger rotors.

Another factor is continuity of personnel. Continuity of personnel can affect technical progress as well as continuity of business decisions. The Japanese approach is for an individual to remain at a company throughout his career. This leads to long term technical and business continuity and generally to no interruption in the commitment to achieve an innovation. The tradition in the U.S. is personnel mobility. Companies continually change at the management level. Technical personnel move from company to company and to universities or government laboratories. Management changes can interrupt the duration and continuity of new product planning and implementation. Technical personnel changes can either delay progress or accelerate progress. For example, when the principal investigator for Si_3N_4 at Norton left, planned efforts on material reliability optimization, composite microstructures, and HIP were delayed or not pursued. Ultimately, Norton hired experienced engineers with interdisciplinary industrial and university experience, and during the 1980's has had a strong effort. When individuals from Ford moved to Garrett, the Garrett effort was immediately accelerated. Thus, mobility in the U.S. has had short term negative and positive affects on advancement of Si_3N_4 technology, but the long range affect on innovation and market penetration is not yet known.

Collaboration/Partnerships

Invention generally results from an individual or small team effort. Innovation generally requires substantial collaboration of an interdisciplinary team or of a manufacturer and an end user. Relationships during early Si_3N_4 development in the U.S. were primarily vendor-customer or contractor-subcontractor and were directed towards government-funded engine demonstration programs. Many of the Japanese Si_3N_4 efforts have been collaborations or partnerships directed towards a commercial product. Examples are Mitsubishi-NGK Insulators for the Si_3N_4 rocker-arm pad, NGK Spark Plug-Nissan for the Si_3N_4 turbocharger rotor and Kyocera-Isuzu for a prototype diesel engine. Japanese have a strong philosophy of networking and are comfortable with partnerships. Some of these partnerships are an extension of personal ties; justification for collaboration can be either business or personal ties.

Collaboration has increased in the U.S. within the past five years. This has partially been a result of changes in antitrust regulations which now permit greater cooperation between competitors.

Goals and Milestones

Japanese government and industry cooperate. Both have a primary goal of commercialization of products and economic development. U.S. industry also has a product goal, but the government can only support industry if the product is in selected military, energy or medical categories. This is changing. There are presently government initiatives which are attempting to establish support for technologies which will stimulate or accelerate commercialization.

Market Fit/Innovation Window

Innovation has enhanced potential for success if it is responding to a market pull. This can be thought of as an innovation window. If the company has the right technology needed for the technology window, then there is a market fit. Having a technology window and a market fit significantly reduces the risk of expending resources for innovation.

Early applications of Si_3N_4 in Japan resulted from an innovation window that did not exist in the U.S., i.e., specifically the light duty diesel market and liquid propane fueled taxis. The needs of the Japanese light duty diesel market stimulated development of Si_3N_4 glow plugs and swirl chambers, which became the earliest commercial Si_3N_4 heat engine components. The liquid propane fueled taxis were a second innovation window. These were broadly used in Japan, but not in the U.S. The combination of the fuel-burning characteristics and the frequent starts and stops resulted in inadequate lubrication of the rocker arm wear pads. Metal pads failed rapidly. Si_3N_4 provided greatly improved wear life and was the only ceramic material which could be successfully cast into the metal rocker arm.

Prior Success or Failure

Innovation requires a business decision and commitment. Such a decision and commitment is affected by prior experiences. An example is cutting tools. NTK (NGK Spark Plug) is a major manufacture of Al_2O_3-TiC and other cutting tools. Once sialon cutting tools had been successfully demonstrated in England, it was a logical and comfortable decision for NTK to explore Si_3N_4 materials for cutting tools. In contrast, Norton Company had all the capabilities and capacity to pursue Si_3N_4 cutting tools as early as 1973, but chose not to until many years later. Perhaps their decision was based on prior experiences with cutting tools. Norton had led early developments in hot pressed Al_2O_3 cutting tools, but had withdrawn from the market because of the inherent marginal performance of Al_2O_3 and the lack of the necessary cutting tool distribution network within Norton. Thus, the prior negative experience of Norton with cutting tools resulted in the decision not to pursue Si_3N_4 in the mid 1970's, whereas the prior positive experience at NTK resulted in the decision to pursue Si_3N_4 cutting tools.

Image

Because of the close, lifelong bond between an employee and company in Japan, image of the company is very important to each employee. As a result, each employee has a strong vested interest in helping the company to be successful and in projecting a good image. Company image is less important to U.S. workers. Priority is more on the image of the individual, i.e., the motivation is for individual advancement rather than company image.

The issue of image affects business decisions in Japan. For example, it is good for the image of a company to participate in a MITI program. Therefore, companies will strive for participation even if it involves a substantial cost share.

Legal/Regulatory Issues

Legal and regulatory issues are significant factors in business decisions in the U.S. A company will often delay a decision to pursue an innovation until legal and regulatory issues have been addressed and resolved. For example, Ford Motor Company fabricated and tested ceramic turbocharger rotors in the late 1970's, but did not proceed towards commercialization partly due to concerns regarding potential liability.

Legal and regulatory issues in the U.S. typically add substantial cost to the development effort. This adds to the challenge of competing with companies in countries where regulations are not as stringent and where litigation is less prevalent

---

The high cost of suitable silicon nitride powders and particularly the high cost of reliably processing shaped parts must be substantially lowered for large scale cost competitive replacement of metals by silicon nitride structural ceramics. Silicon nitride turbo charger rotors cost about twice as much as metal rotors. Some twelve thousand per month are now being produced in Japan. The manufacture of automotive parts in Japan is not profitable but is said to have reached a break-even point. A long period of processed development and incremental innovations is still required to develop substantial markets for silicon nitride structural ceramics. After nearly fifty years of research innovations, it remains an infant industry largely centered in Japan.

An excellent account of the Japanese program is given in J.B. Wachtman, Jr., R.C. Bradt, R.F. Davis, R. Raj, D.W. Richerson and N.J. Tighe, *Japanese Structural Ceramics Research and Development,* Foreign Applied Sciences Assessment Center Report, Science Applications International Corporation, McClain, Virginia, July 1989.

MANAGEMENT OF INNOVATION

During the 1970's, Japanese industry demonstrated a capability of improving productivity through technology development while at the same time enhancing the quality of production. As a result, there has been enormous Japanese success in export markets for steel, ships, watches, cameras, semiconductors, automobiles and consumer electrical appliances. Management of technology is not a neglected subject. A recent bibliography lists more than 1500 references on technology strategy[1] and there have been dozens of studies on differences between corporate cultures and management systems in Japan vis-à-vis the United States.[2-7] Conference presentations on the general question appear in contributions of Poncelet, Kline, Cutler and Kii which appear earlier in this volume. A number of analysts have suggested that one of the key factors accounting for Japanese success is "superior management". Others have pointed out that there are a number of well-managed successful, effective American firms as well.

One specific management area suggested as explaining the cost differential between Japanese and Western automobile industries is the use of human resources. Japanese companies are able to obtain better contributions from their employees through a system of management in which all employees have responsibilities often considered as managerial, and there is a bottom-up participative management style in contrast to the usual western top-down authoritarian approach. It has even been proposed that the superior-subordinate relationship is cooperative in Japan and tends to be more adversarial in the United States. A second specific Japanese achievement is in-process inventory control, exemplified by Toyoto's *kanban* system of just-in-time inventory management. A third area of the Japanese success is in the use of quality circles and of production worker responsibility to achieve outstanding quality control. The fourth area of Japanese success has been in technology management where the Japanese are purported to have spared no efforts to maintain their industrial plants at the highest technological level. This is seen as resulting from Japanese concern with building market share to achieve results in the long-term as compared with the short-term profit orientation of American industry.

In addition to specific management practices, national and corporate strategies of Japan have been directed toward developing export markets. The Ministry of International Trade and Industry (MITI) has been influential in guiding Japanese firms with regard to exports, foreign plant establishments and industrial development. MITI has stimulated companies to catch up with foreign technology by encouraging cooperative ventures and collaboration even though fierce competition for internal markets has not been restricted. In contrast to the American Federal Trade Commission, which is essentially a regulatory agency, MITI functions as a supporter of industry and industrial firms. MITI has been supportive of what it considers key industries of the future but also has adjusted to parameters of the macroeconomy, encouraging low cost labor intensive imports for example. Amongst other things MITI has encouraged and supported the development of advanced materials technology.

There have been a number of analysts of Japanese-style management, which has its roots in social and cultural characteristics of the Japanese people and Japanese society. A primary distinction is the Japanese emphasis on the group as compared with the American emphasis on the individual. In the corporate world, the Japanese enterprise is a social as well as an economic organization and follows the pattern of other social organizations in Japanese culture.[7,8] Japanese society tends to be organized on "vertical" family relationships in which a family consists of parents, grandparents, children and grandchildren who make family decisions by consensus as opposed to an "extended horizontal" family consisting of siblings, in-laws, cousins and so on. This same organizational structure is seen in companies where relationships between corporations and their suppliers are strong and long-term. There are close ties between superiors and subordinates in a given unit and consensus decision-making is preferred. In contrast American culture puts strong emphasis on individual achievement. While a Japanese might describe his occupation as a "Hitachi man", his American counterpart would almost certainly reply with his professional category. If the ideal Japanese organizational structure is the vertical family, the American ideal is probably a football team in which each member has his individual responsibilities; the game plan and coordination are left to the coach.

The emphasis on group versus individual, the emphasis on human relationships as opposed to functional relationships and the expectation that managers are generalists rather than specialists all lead to behaviors that characterize a particular management style. The emphasis on group leads to a corporate strategy that favors a long-term outlook for marketing and product development, a preference for growth, and an emphasis on well established market share building on the permanence of each worker's relationship with the company. This is radically different from Japan of the 1930's, where primary emphasis was placed on rapid return on investments. Matsuchita Electric Company is given the credit for explicitly emphasizing the importance of building market share by reducing prices and deferring profits both to increase market share and to establish barriers for competitors anxious to enter the market.[9] Further building of market share by superior quality, availability of patient capital and technology/manufacturing leadership followed.

These different management styles have been subjected to a comparative analysis by Murayama[10] who has shown how particular approaches of Japanese/American management are aimed at achieving norms, that is common expectations, values, standardized views as local business systems become integrated with an international business system.

Melcher and Aroggaswamy[11] have contrasted the decision and compensation systems in the United States and Japan, feeling that these are the key to understanding Japanese and American managements. These authors feel that the management systems that have been developed function effectively within a particular environment and are only likely to work within that culture. In the United States, for example, vague authority and responsibilities and reward systems not based on individual performance tend to promote inefficiency and low productivity. In order to understand the productivity level of quality and competitiveness of Japanese firms, one needs to look at public policies that impact the costs structure, operating constraints and general support of private enterprise.

While there seems to be an increasing convergence in corporate research and development strategies in Japan and the U.S., significant differences remain that affect technological innovation. There are many exceptions on both sides, but in general there are the following contrasts:

R & D structure: Japan has more decentralized R&D with a greater part at manufacturing sites while the U.S. has more autonomous and centralized corporate laboratories and many small high-tech venture firms in which development is more closely allied with marketing needs.

Employee specialization: Japan has rather little specialization with company training in a variety of functions while the U.S. has a high level of specialization with training by specialized institutions outside the company.

Employee orientation: Japanese employees function in groups with group and company recognition for achievements while U.S. employees generally have an individual career orientation with professional norms dominating.

Technology transfer: Japan has an extensive system of technology transfer actively encouraged by government and occurring between families of suppliers, manufacturers and customers. In the U.S. cooperation between suppliers, producers and users has been rare; so have close interactions of small businesses and large corporations.

Dr. R. A. Swalin, former vice president for research at Allied Signal Corporation focused discussion on an incisive definition of innovation given by P. F. Drucker[12] "Innovative organizations first know what innovation means. They know that innovation is *not science or technology*, but value. They know that it is not something that takes place within an organization, but a change outside. The measure of innovation is the impact on the environment. Innovation in a business enterprise must therefore always be *market* focused. Innovation that is product focused is likely to produce miracles of technology but disappointing rewards." With regard to achieving innovation corporate culture is more important than corporate structure. Two kinds of companies exist as end members of a continuous spectrum. These may be termed entrepreneurial versus professionally managed.[13] as illustrated in Table 1. From the 1920's through the middle 1980's, there was a change in American corporations from an entrepreneurial style to one of professional management. In the late 1980's and continuing, the trend seems to be in the opposite direction, in part, because corporations are flatter in structure because of the information technology revolution which has made middle management redundant and the need to innovate more rapidly an environment of global manufacturing, global science and increasingly, global technology. In this environment the role of corporate technology is changing. More and more, it is necessary for corporate technology to serve an information function with regard to developments occurring outside of the corporation, even outside the United States, as opposed to internally generated new technology.

Table 1. There is a spectrum of different kinds of companies

ENTREPRENEURIAL	PROFESSIONALLY MANAGED
Internal Controls	External Controls
Creativity	Conformity
Individual Autonomy	Central Control
Intuitive	Rational/Logical
Right Brain	Left Brain
Gut Feeling	Scientific Management
Decentralized	Centralized
Networks	Hierarchies
Adult-Adult	Adult-Child
Product Differentiation	Low Cost Product
Person Centered	Organization Centered

Source: DeLisi, *Sloan Management Review,* Fall 1990.

A number of discussants suggested that there was not such an enormous difference in management *per se* between Japanese and American companies, but there was a range of behavior. In a comparative study of technological innovation in Japan and the United States, Edward Mansfield[14] analyzed successful projects based on questionnaires submitted by Japanese and American companies with 125 corporations replying: 75 in the United States and 50 in Japan. He found that the Japanese innovate more rapidly than American corporations and also they do it more cheaply. The difference, however, is not enormous: about 23% faster in Japan than in the United States. In the electronic industry the Japanese innovation was about 50% more rapid but the difference was small in the chemical industry. The most surprising difference is related to the innovations with externally acquired technology as compared with internally developed technologies. While internally generated innovations in Japan and the United States seem to go along at about the same rate, the rate at which externally acquired innovations proceeded is much faster in Japan. He reported that Japanese and American companies spend about the same amount on research and development, but the Japanese spend about twice as much as American companies on process development while American companies spend twice as much on product development and have much larger manufacturing start-up costs. Differences seem to be focused more on research objectives than on major differences in management effectiveness.

References
1. Paul S. Adler, "Technology strategy: A guide to the literatures," pp. 25-152 in R.S. Rosenbloom and Robert A. Bungelman, *Research on Technological Innovation, Management and Policy,* Volume 4, JAI Press Inc. Greenwich Conn. 1989.
2. Carl Pegels, (1984), *Japan vs. the West,* Kluwer-Nigholl Publ. Boston.
3. Sang M. Lee and Gary Schwendiman, eds. (1982), *Management by Japanese Systems,* Praeger Publ. New York.
4. William D. Wray, ed. (1989), M*anaging Industrial Enterprise,* Harvard Univ. Press, Cambridge, MA.
5. P.F. Drucker, (1981), "Behind Japan's Success" *Harvard Business Review,* Jan/Feb, pp. 83-90.
6. W. Ouchi, (1981), *Theory Z,* Addison-Wesley Publ. Reading, MA.
7. R.T. Pascole and A.G. Athos, (1981), *The Art of Japanese Management,* Simon and Schuster, Publ. New York.
8. Dick Kazuyuki Nanto, (1982), "Management, Japanese Style" in Ref. 2, pp. 3-24.
9. C. Nakane, (1970), *Japanese Society,* Univ. Calif. press, Berkeley,
10. Motofusa Murayama, (1982), "A Comparative Analysis of U.S. and Japanese Management Systems," in ref. 2, 219-238.
11. Arlyn J. Melcher and Bernard Aroggaswamy, (1982), "Decision and Compensation Systems in the United States and Japan: Contrasting Approaches to Management", in ref. 2, pp. 302-319.
12. P.F. Drucker (1973)
13. DeLisi, *Sloan Management Review, Fall, 1990.*
14. Edward Mansfield (1988).

One of the successful technical innovations discussed in the section on silicon nitride structural ceramics was the DARPA Program on ceramics for gas turbines. The manager of that activity for Ford Motor Company, Arthur McLean, commented on the management influence on technical innovation as follows:

MANAGEMENT INFLUENCE ON TECHNICAL INNOVATION

Arthur F. McLean, Consultant, Tucson, AZ

Introduction

As a conference participant, I was asked to discuss, from a corporate perspective, examples of technology innovation in the field of advanced materials and the influence of management style and cultural differences, especially between Japan and the U.S.A. This paper, which is a follow up from my earlier paper presented at the Session: Silicon Nitride Structural Ceramic Innovation, focusses on the aspects of management which establish a framework for involvement in technology innovation. I believe it is this creative aspect of management that can have a favorable impact on individuals and groups involved which can really make the difference in terms of, not only nurturing R&D process, but achieving success in the overall process of technology innovation, from research through production to sales.

Goal Setting and Involvement

Fig. 1 shows the interrelationships of many disciplines toward the goals of the DARPA/Ford Brittle Materials Design Program (1). Even the generation of such a chart at the onset of the program involved the interfacing of many different disciplines, thus providing an awareness in the group of how one's own efforts might help (or hinder) the whole; the presence of the chart and its updating served as a continual reminder of the responsibility

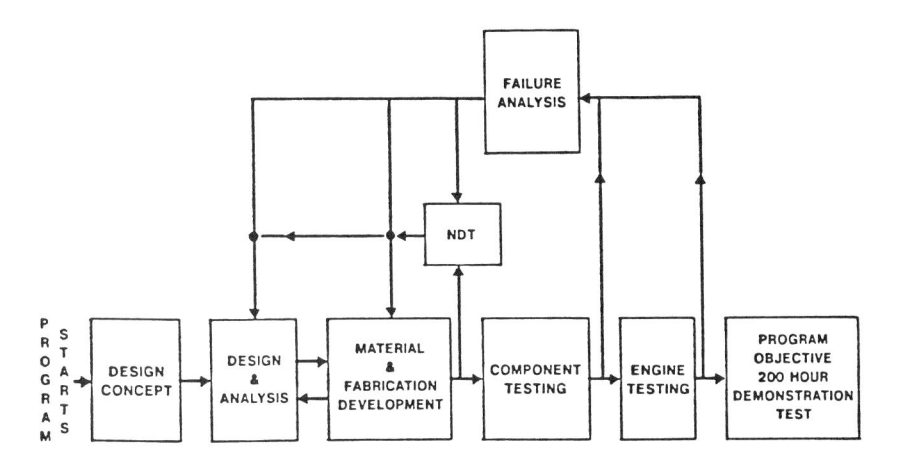

Fig. 1. Interrelationships involved in DARPA/Ford/Westinghouse Program

of each member of the group. Furthermore, it provided a framework to establish and change priorities during the program.

In any major program, there are usually mini-objectives for each step; these also frequently involve different disciplines, though perhaps with finer differentiation. For example the interrelationships between different scientists on a project -- one may be involved in compositional experimentation, another is an electron microscopist and yet another specializes in materials characterization. Success is more likely if they all feel committed to the objective and work in an involved manner.

Establishment and working toward the goal will determine the various disciplines needed to be involved. In general, the bigger the goal, the broader the multidiscipline involvement. The overall goal could be as big as a company goal or even a national goal. Clearly, the Japanese goal to be first to commercialize structural ceramics for heat engines (2) was a big goal and must have involved more disciplines than the DARPA/Ford Program or the many other U.S. R&D programs underway. In particular, their commercialization efforts must have included marketing, production, costing and an appreciation of their competitor's (national and international) development/potential strategies.

Perhaps a most important question to ask with respect to selecting a goal is "will the goal, if achieved, lead to a successful program?" This then prompts the question "what is a successful program?" In this field of endeavor, part of the answer is to research, develop and demonstrate the application of advanced materials to one or more components. Another part is to develop production techniques and to produce and sell such components (i.e., commercialize). Real success, however, is the measure of the sustained benefit of the particular undertaking to the customer. In our free market society, this translates to "acceptable profit" or more specifically "acceptable return on investment." The problem is that in a new material development, it takes many years (often >20) from the scientific research stage to significant production and acceptable return on investment. Furthermore, societies (e.g., Japan vis-a-vis U.S.A.) may have different views on what return on investment is acceptable and these views may vary with time! So, establishment of the goal or of contributory mini-goals is a difficult management task, needs interdisciplinary involvement and almost certainly will require, at least to some degree, a leap of faith.

Operational Involvement

Perhaps an important area where Japanese and U.S. methods have differed is the degree to which day-to-day involvement is achieved in carrying out the tasks at hand after the goal has been selected. Many of us have had the privilege of witnessing or knowing of examples of this; without referring to specific names, I would like to relate two such examples.

The first is very simple. Three representatives of a Japanese company were visiting me and others of the Ford Research Staff when the question of time dependent material properties came up. To explain their views, they decided to draw a graph on the blackboard. One of them drew the horizontal and vertical axes, each with an appropriate scale. Simultaneously, the second of them crossed in the points of a curve and third circled in the points and drew another curve. It seemed as if each finished his part of the task at the same time. Viola! -- a very neat plot was produced rapidly by three people in unison.

The second example is more complicated and deals with the time it takes to build a newly-designed experimental automotive piston engine. It was hard to believe that a Japanese company was able to do this and have the engine ready for test in three months! It would normally take us over a month just to design and fabricate the patterns for the castings, let alone pour the casting, make machining fixtures and complete finish machining. And then there are all the other parts that make up the engine. It turns out they were able to do this by working in an interdisciplinary and simultaneous manner, rather than a compartmented, sequential manner. At the same time as the original engine was being designed, a patternmaker was making his pattern drawings and a machining expert was working out his fixture methods. Not only that, but all other parts of the engine were being worked on in this interdisciplinary and simultaneous manner. The advantages of this approach are not only a much faster time to completion, but also an improvement in quality or cost of the finished work. For example, if a problem to do with machining is spotted, it can be addressed quickly and by all concerned and corrected at the source, even if it means changing the design.

To achieve this type of multidisciplinary involvement, it is management's responsibility not to pick and check at the many details of a program, but rather to foster the involvement and trust of the people on the program, their commitment to the goal and their feelings of pride and responsibility to achieve it. This means that operational decision making gets pushed down to lower levels in an organization's hierarchy, thereby facilitating quicker and more accurate actions and, furthermore, a smaller management focussing on bigger strategic issues.

There is still the question of the cultural differences between Japanese and U.S. employees. It may well be that their smaller country with limited raw materials, their more dense population, their more entrenched ethnic background and their respect for authority enhances the Japanese ability to work in a close team. Considering the U.S. on the other hand, it may well be that their individualistic innovative capabilities, their entrepreneurial style and their multi-ethnic background all combined with good, flexible management can make a well-managed U.S. team successfully compete with the best, be it in research or elsewhere.

Organization Involvement

The Ford Motor Company was perhaps the first U.S. automotive company to formally focus on the importance of company goals and employee involvement. Let's review Ford's "Company Mission, Values and Guiding Principles" shown in Fig. 2. Some points are worthy of special comment. First, Ford is a worldwide company and, therefore, so are its goals. People are listed first in the three Values: People, Products and Profits. Quality and the importance of "consumer and not producer orientation" have been stressed in all disciplines, including research. An interesting manifestation of the overriding importance of quality that has been introduced at Ford is that anyone working on the assembly line can push a button to stop the line in the event of a quality problem. This procedure also speaks to the greater involvement of employees and the increased management trust of employees. The process of employee involvement at Ford is combined with management participation (PM/EI) and entails personnel from within and between groups regularly getting together to interchange views and recommend actions. PM/EI at Ford is very active and has been credited with many successes. Similar methods have been active in Japan and are now being used in many large U.S. companies.

With respect to national R&D programs, it would seem that Japan's government, industry and academia have a common and rather singular goal to stimulate technical innovation and increase international trade. In fact from time to time, MITI and various Japanese corporations announce specific goals of where they should be by a specified time in a particular field of technology. In the U.S., on the other hand, our materials R&D goals are not so focussed and include such diverse aims as strengthening national defense, increasing U.S. productivity, conserving energy and easing environmental problems, with defense, historically, having the strongest effort. An important question is to what degree is U.S. organizational involvement used in determining, prioritizing and executing U.S. materials research and development programs? In West Germany, many national R&D programs were structured around significant corporate cost sharing (as was the DARPA/Ford/Westinghouse program) which, of itself, tends to foster continued corporate involvement. In Japan, there is an intricate web of involvement (3) between government (including national laboratories), industry (including trade associations) and academia which, though seemingly overly complex, may well serve to establish common goals and organizational feelings of pride and responsibility.

MISSION

Ford Motor Company is a worldwide leader in automotive and automotive-related products and services as well as in newer industries such as aerospace, communications, and financial services. Our mission is to improve continually our products and services to meet our customers' needs, allowing us to prosper as a business and to provide a reasonable return for our stockholders, the owners of our business.

VALUES

How we accomplish our mission is as important as the mission itself. Fundamental to success for the Company are these basic values:

People — Our people are the source of our strength. They provide our corporate intelligence and determine our reputation and vitality. Involvement and teamwork are our core human values.

Products — Our products are the end result of our efforts, and they should be the best in serving customers worldwide. As our products are viewed, so are we viewed.

Profits — Profits are the ultimate measure of how efficiently we provide customers with the best products for their needs. Profits are required to survive and grow.

GUIDING PRINCIPLES

Quality comes first — To achieve customer satisfaction, the quality of our products and services must be our number one priority.

Customers are the focus of everything we do — Our work must be done with our customers in mind, providing better products and services than our competition.

Continuous improvement is essential to our success — We must strive for excellence in everything we do: in our products, in their safety and value — and in our services, our human relations, our competitiveness, and our profitability.

Employee involvement is our way of life — We are a team. We must treat each other with trust and respect.

Dealers and suppliers are our partners — The company must maintain mutually beneficial relationships with dealers, suppliers, and our other business associates.

Integrity is never compromised — The conduct of our Company worldwide must be pursued in a manner that is socially responsible and commands respect for its integrity and for its positive contributions to society. Our doors are open to men and women alike without discrimination and without regard to ethnic origin or personal beliefs.

Fig. 2. Ford Motor Company Mission, Values and Guiding Principles

Arden Bement (4) points out the winds of change are upon us, including the powerful upcoming European market, continued strengthening of the Japan/Pacific Rim markets, major reforms in the USSR, and a reduction in U.S. defense expenditures. He contends that a new partnership is needed between government and industry.

The writer agrees that it is time for a new relationship between U.S. government, industry and academia to foster technical innovation, including, for example, improving the way materials R&D programs are selected, prioritized and executed. Many issues need to be addressed, such as: How can corporate management be involved in and committed to national goals? Do national R&D programs always have to be considered as government programs? Why not corporate programs with government support -- or just national programs with governmmnt and corporate support! How should national programs involving different countries and multinational companies be weaved together into the international arena? How can small companies be committed and their entrepreneurial qualities be best utilized? Should the military-emphasis of U.S. R&D be changed? How can college graduates be educated to better appreciate the business aspects of R&D and technical innovation? Industry, government and academia must plan to handle change. They must cease acting in advocatory roles and learn to work effectively together in the field of technical innovation, thereby enhancing the prospects of sustained future growth and improved living standards.

References
1. A.F. McLean and E.A. Fisher, "Brittle Materials Design - High Temperature Gas Turbine." Final Report July 19 to August 1979. AMMRC TR 81-14, 1979.
2. J.B. Wachtman, Jr., R.C. Bradt, R.F. David, R. Raj, D.W. Richerson and N.J. Tighe, "Japanese Structural Ceramics Research and Development," FASAC Technical Assessment Report, July 1989.
3. Japan Fine Ceramics Association, "Annual Report for Overseas Readers: Fine Ceramics for Future Creation," 1987.
4. A.L. Bement, Jr., "Winning the Technological Race in the 1990's," Ceramic Industry, September, 1990.

Dr. Seiichi Watanabe, Director of the Sony Corporation Research Center described Sony policies in developing a Center of Excellence. There is an ancient Chinese story of two armies, one smaller than the other, facing each other across a river. The smaller army voluntarily crossed and positioned itself with its back to the water. This provided a do-or-die situation and enabled the smaller army to emerge victorious. Sony attempts to have its research groups voluntarily put themselves in the same position where, after extensive consultation and discussion, an objective is selected for which hard work and dedication is necessary. As an illustration of the continued support, he cited the twenty year research program to develop a high definition CCD video device. For a period of ten years, this dedicated research program had a research and development cost of about 0.15 to 0.3% of total company sales, which finally resulted in technological success. In his view this long term commitment is necessary to achieve difficult technological accomplishments. An important part of the Sony approach is to actively encourage multicultural experience of the research and development managers. In addition to research activity, employees are expected to work at production jobs, at sales jobs, and encouraged to have foreign experience. Outside influences are developed through collaborative research with universities and participation of foreign researchers and trainees within the Sony corporate research center.

A third factor Sony considers extremely important is the relationship with customers. "Our researchers are encouraged to maintain constant cooperation and advice from customers. We look forward to customer complaints as a valuable source of innovations. Nothing is more important for our management than maintaining effective customer feedback." Some time ago, Sony was engaged in a joint venture with an American company, sharing production sales and American research and development. With regard to product performance, the American partner took the viewpoint that existing performance and policies should be satisfactory for consumer applications and they did not see needs for continual improvements and it was on this basis that the joint venture fell apart. "I lived in the United States many years ago and it is my observation that American companies have little or no regard for customer opinion." Daniel Button, Director of Electronics at the DuPont Japan Technical Center, commented that his company's ceramic materials sales are a greater fraction of total business in Japan than in the United States and that parallels the intensity of the Japanese DuPont interaction with customers. "In Japan, we live with our customers, our technological people are constantly with customers visiting them and likewise we are visited by customers." He indicated that far more customer complaints are received in Japan which is a value because it provides a basis for understanding critical requirements. In Japan it is necessary to develop a structure that enhances transferred information, technical service person and researcher being one and the same. In the United States, the technical service group is completely independent of the research group. Research people tend to look down upon technical service people. In Japan the cycle times are "extraordinarily shorter" and Button attributes a lot of this to the direct researcher-customer interactions.

Dr. Watanabe proposes the effective R&D and product development affects corporate culture:

CULTURAL REVOLUTION AND R&D

Seiichi Watanabe, Sony Corporation Research Center

As I review the R&D efforts in Sony I have found that there is another, hidden mission of R&D other than the well-known mission of making something new and developing prototypes or samples; that is, to keep the company always young, flexible and challenging by constantly shaking up the whole company culturally through the introduction of new concepts associated with outputs of R&D. I will review some cases of research and development and see how they revolutionized and broadened the corporate culture. Sony was established in 1946 right after the war and is now a company of 20 billion dollars sales in 1989. The business covers a broad area. The growth of the company has been rapid, and has resulted from the introduction of new products creating new markets: tape recorders, transistor radios, televisions, videotape recorders and so.

The R&D spirit at Sony is summarized in the prospectus of foundation written by Mr. Ibuka as shown in Fig. 1. The Sony style of R&D has been to concentrate on selected targets, often at the risk of the whole company or at least a whole business group.

The first case is magnetic recording, and the first product was a tape recorder. It is a well known story that Mr. Morita who had been a researcher in physics and lectured at Tokyo Institute of Technology prior to joining Sony threw himself into marketing to sell tape recorders and transistor radios. He himself experienced a number of tremendous culture shocks during the course of establishing a global sales market network. The technology has evolved to walkman, and also to video tape recorders, and with the 8 millimeter videocam

Fig. 1. The Sony Spirit as written by Mr. Ibuka

- The establishment of an ideal factory—free, dynamic, and pleasant—where technical personnel of sincere motivation can exercise their technological skills to the highest level.

- We shall eliminate any untoward profit-seeking, shall constantly emphasize activities of real substance, and shall not seek expansion of size for the sake of size.

- We shall be selective as possible in our products and will even welcome technological difficulties. We shall focus on highly sophisticated technical products that have great usefulness in society, regardless of the quantity involved. Moreover, we shall avoid the formal demarcation between electricity and mechanics, and shall create our own unique products coordinating the two fields, with a determination that other companies cannot overtake.

products, the metal powder tape was introduced. The progress of recording density has been linear in logarithmic scale against the year as the semiconductor memory, and has experienced a number of generation changes (Fig. 2). The cultural changes affecting Sony caused by the introduction of the first phase of magnetic recording technology can be summarized as in Fig. 3.

Another example is the charge-coupled-device image sensor. The device was invented at Bell Telephone Laboratories and exploratory research was begun at Sony in 1970, in a so-called under-the-table manner. This kind of exploratory research under the table typifies the free and dynamic atmosphere of research at this company. Then in 1973, Mr. Iwama, the president at that time, picked it up as a corporate project. In 1978 the project was transferred to the business unit, Semiconduct Group, where a new clean room facility was built for this project. In 1980 the device was commercially introduced for ANA airplanes and a mass-production line for consumer products was built in 1982. In that year Mr. Iwama died before any applications to consumer products were successful.

It was only toward the end of the 1980's that the device business began to pay back the accumulated R&D and production expenditure, after nearly 20 years' intensive activities pushing up the company to be the largest share holder of the device. At the 1990 international solid-state circuit conference, we could successfully report on the fabrication of a CCD image sensor for HDTVs. Cultural revolutions initiated by CCD in Sony included the first ultra-clean technology which was revolutionary to the plant where they knew only discrete devices. The next was a corporate strategy that includes the huge investment which is characteristic to any semiconductor industry of very large scale integration.

The next case is the compact disc. Research of digital audio technology was begun by Dr. Heitaro Nakajima around 1965 at NHK, and he continued the research after he joined Sony. Various new technologies had to be researched and developed associated with compact discs. He had to negotiate with Philips about technical issues for standardization and also with Japanese companies. The laser diode was not available when Sony decided to employ it as the optical source for reading the signal on the disc. The whole project was successful

Fig. 2. The Passage of Magnetic Recording

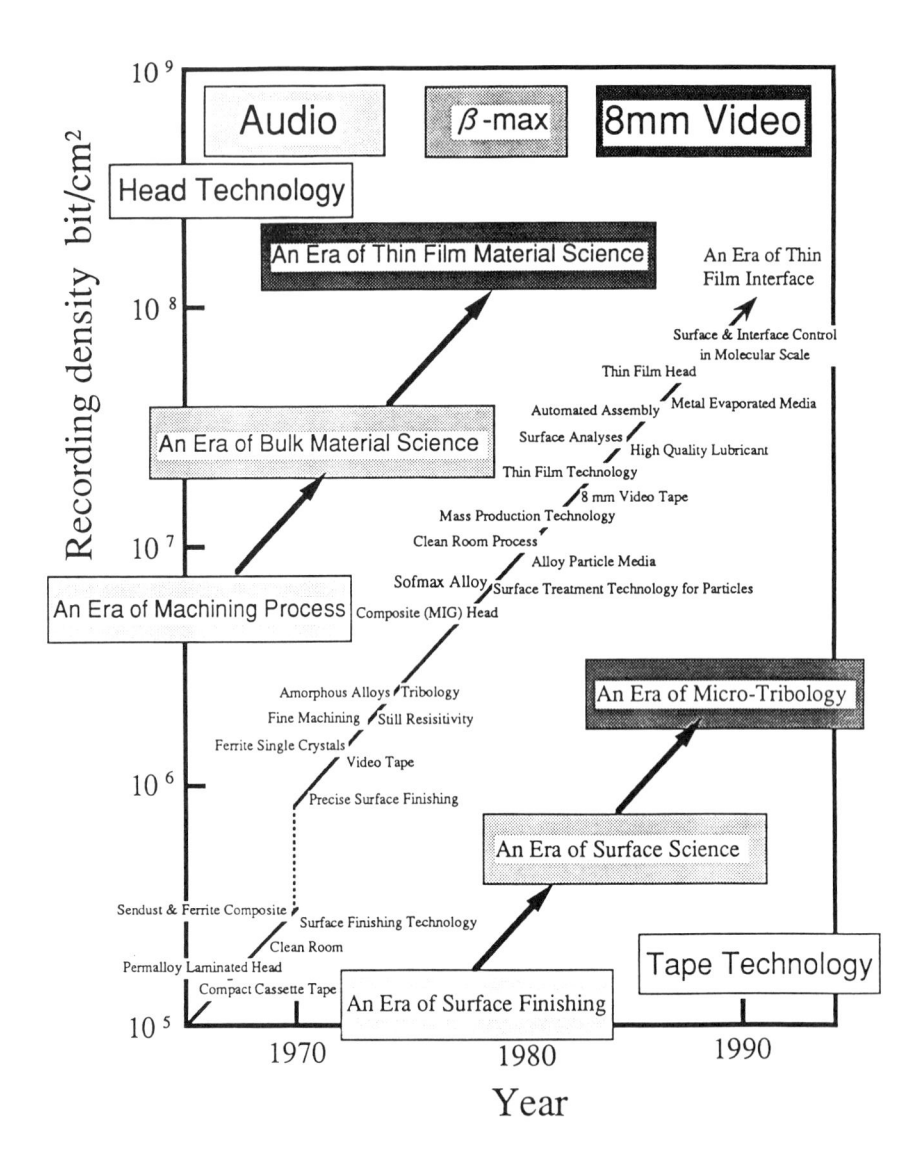

Fig. 3. Cultural Revolution Experienced Affecting Sony through the Evolution of Magnetic Recording

1. The Third Creativity is Creativity of Marketing
   (1) Scientific Creativity
   (2) Creativity through Technology and Products
   (3) Creativity through Marketing
2. Production of Materials and Devices
   Establishment of the Plant for Magnetic Material Devices (in Sendai)
3. Cooperation and Competition with Academic Circles and other Companies

in the end, replacing the older technology invented by Edison. The cultural revolutions initiated by this technology are summarized in Fig. 4.

Fig. 4. Cultural Revolutions Affecting Sony through Evolution of the Compact Disk

1. LSIs for Massive Consumer Application
   16 kbit Memory
   ASIP
2. Major Merchandise in Component Business
   (Optical Pickup, LSI, Semiconductor Laser, Servo)
   Cost
   Reliability
3. Leadership in Audio Business
   Price
   Product Planning
4. Infrastructure for Computer Peripheral Business

Through exciting and challenging R&D activities, a number of aphorisms have appeared which continue to encourage researchers and engineers. Some of them are as follows:
* Assign jobs to the busiest one.
  (because he is capable and being busy is not a reason to delay the project.)
* Place the troops with a river behind their backs.
  (so that there is no way back but to fight and win.)
* Don't let the manager look at interesting seeds.
  (until they take a form good for judgement.)
* Bury failed seeds in the darkness
  (so that risk-taking exploration may not be discouraged.)

It is true that R&D is not the only engine for cultural revolution. There are other motivating forces, such as so called customers' voice, voice from the floor, etc. Yet, nobody would deny that R&D is a very important driving force.

So, how can management enhance smooth embedding of R&D outputs into a stiff existing body and keep the company young, flexible and challenging? It is effective if managers take risks and encourage R&D, and the resulting cultural revolution as shown in Fig. 5. Second is to create an environment where researchers and engineers get exposed to

Fig. 5. R&D and Cultural Revolution

- To Accomplish the other Mission of a Research Center -

1. Positive revolution of the System and Management through Accepting R&D Results
2. Multicultural Experience and Understanding of the Persons in Charge of R&D
3. Construction of a free, generous and happy working atmosphere

various experiences so that they may have multi-cultural understanding, and may enrich their own views, enabling them to evaluate themselves from various aspects. Such experiences as production, sales, or working in other organizations outside the company should be encouraged. A third is, as stated in the prospectus of Sony foundation, creating a free and dynamic working environment where each member can fully demonstrate his or her capability, based on free, unrestricted ways of thinking.

To reach these objectives I have proposed to create a center of excellence in industry of our Research Center as shown in Fig. 6. A center of excellence in industry, in my definition, should have not only a physically excellent environment, including facilities and adequate funding, but also a positive integrating interaction among researchers gathering from all over the world. The environment should be basically open, although some consideration must be given to company secrecy. I think that trying to achieve this target which is not yet well established, is itself effective in keeping the company active.

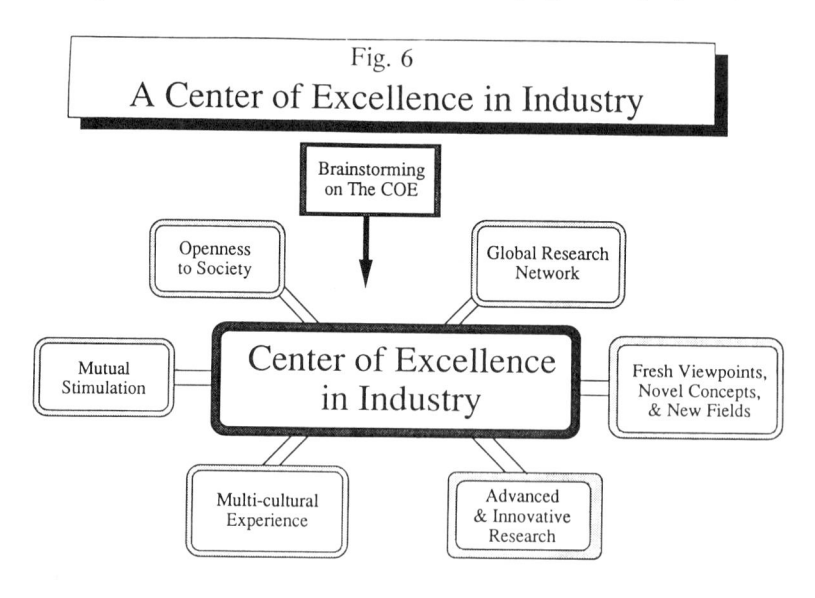

Fig. 6
A Center of Excellence in Industry

- Brainstorming on The COE
- Openness to Society
- Global Research Network
- Mutual Stimulation
- Center of Excellence in Industry
- Fresh Viewpoints, Novel Concepts, & New Fields
- Multi-cultural Experience
- Advanced & Innovative Research

A quite different management problem is faced by a government organization whose aim is to encourage the development of new high-value-added industry. Kazuo Kobayashi, formerly director of a MITI laboratory, gives his views of the MITI strategy.

MITI STRATEGY

Kazuo Kobayashi

The Agency of Industrial Science and Technology (AIST) is a part of the Ministry of International Trade and Industry (Fig. 1) MITI. AIST has 16 government research institutes

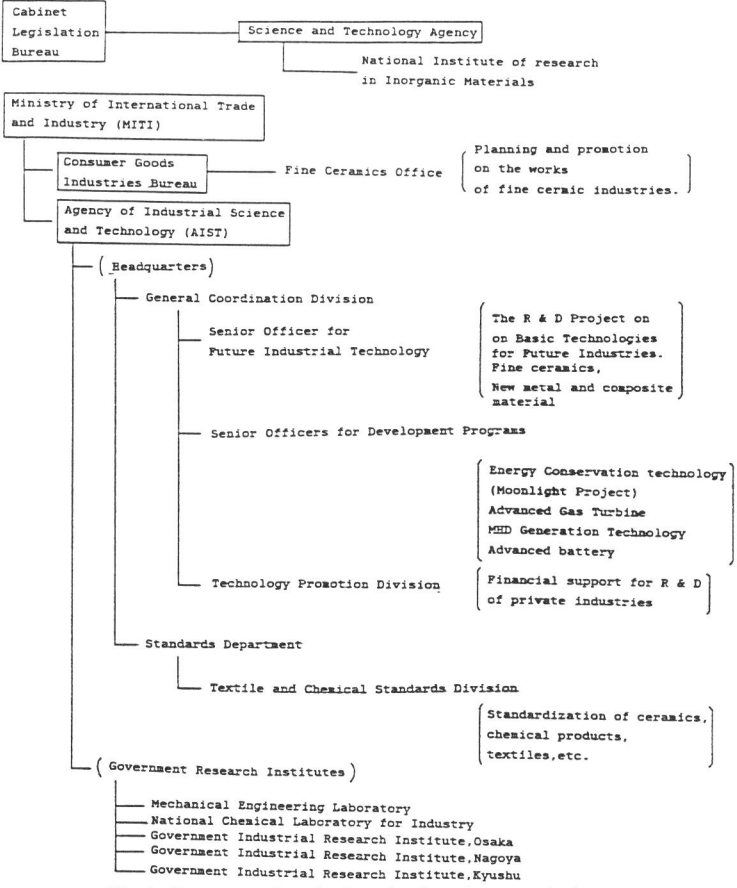

Fig. 1. Government Organization related to Fine Ceramics in Japan

with a total of about 2500 researchers, and plays a big role for technology in fine ceramics. In MITI the Fine Ceramic Office works on planning and promotion for fine ceramic industries. In the General Coordination Division in AIST senior officers work as coordinators for big national projects such as the Advanced Gas Turbine Project which is a cooperative project of industries and government research institutes. Nine of 16 institutes which were located in the Tokyo metropolitan area moved to Tsukuba area to form the research center of AIST and another nine institutes are scattered in local areas from Hokkaido to Kyushu to promote local industry as well as advanced technology.

Current national R&D projects related to fine ceramics promoted by MITI and the budget in FY1991 are as follows.

High-Performance Ceramics: (1981-1992, Budget for FY1991 is 1,222 million Yen). This project is one of the projects on Basic Technologies for Future Industries which aims at the development of innovative basic technologies essential for establishing new industries. The High-Performance Ceramic Project aims at development of high-strength ceramics for use at extremely high temperatures to be used as materials for gas turbine components. Silicon nitride and silicon carbide ceramics with high reliability which can stand high temperatures up to 1200°C were developed.

Ceramic Gas Turbine Project: (1988-1996, Budget for FY 1991 is 1,773 millon Yen). This project belongs to the Moonlight Project which is a comprehensive program of R&D for energy conservation. The project aims at development of a ceramic gas turbine applicable to co-generation and electric power generation systems. These turbines, which use non-petroleum fuels such as natural gas and methanol, offer thermal efficiency which may be increased to 42% by raising the turbine inlet temperature to 1350°C.

Superconducting Materials and Devices: (1988-1997, Budget for FY 1991 is 2,778 million Yen). This is one of the projects on Basic Technologies for future Industries. The project aims at development of new superconducting materials, processing technologies for applying superconducting materials to electric power equipment, e.g. magnets and wire, and technologies for fabricating superconducting electric devices. One R&D result is an automated fabrication apparatus developed for searching out new superconducting materials. Development of strongly pinned superconductors and direct observation of the magnetic field distribution on the superconductor were also accomplished.

High Performance Materials for Severe Environments: (1989-1996, Budget for FY 1991 is 1,699 million Yen). This is one of the Projects on Basic Technologies for Future Industries. This project aims at the development of carbon/carbon composites, intermetallic compounds and fiber reinforced intermetallic compounds which can be used to develop a space plane and SST/HST. At present, SiC fiber modified by the electron beam method was developed to stand high temperatures of 1500°C.

Advanced Materials Processing and Machining System: (1986-1993, Budget for FY1991 is 3,159 million Yen). This is one of the Large Scale Projects. This project is aimed at advanced surface processing using laser beams and ion beams and an ultra precision mechanical processing for advanced industries such as energy, precision machining and electronics. At present, elementary techniques for high power excimer laser, high current density ion beam and ultraprecision machining have been developed.

Fuel Cell Power Generation Technology: (1981-1995, Budget for FY1991 is 3,738 million Yen). This is one of the Moonlight Projects. This project aims at development of design concepts for systems adaptable to both dispersed and centralized power stations, using fuel cell power generating devices whose potential efficiency can reach as much as 40 to 60%. Natural gas, methanol and coal-derived gas are used as fuels.

Advanced Chemical Processing Technology: (1990-1996, Budget for FY1991 is 1,161 million Yen). This is one of the Large Scale Projects. This project is R&D on advanced chemical processing technology for producing new functional materials such as gradient materials, pure metals and polymers with fine alignment of molecules.

Superconducting Technology for Electric Power Apparatus: (1988-1995, Budget for FY1991 is 3,140 million Yen). This is one of the Moonlight Projects. This project aims at development of a more efficient and stable electric power system using superconducting power apparatus, among which generators are the closest target. The system will assist in overcoming problems such as power loss and lack of suitable sites for transmission lines which occur as power stations become bigger and more remotely situated.

Advanced Surface Modification in Material Processing: (1989-1993, Budget for FY1991 is 41 million Yen). This is one of the Regional Technology Development Systems. The system was started in 1982 for the purpose of promoting advanced technology for regional industries. Each of 9 government research institutes conduct one suitable project with cooperation of regional industries. The Government Industrial Research Institute, Osaka is the coordinator. The aim is to develop technology on advanced surface modification for materials such as metals, plastics and ceramics in order to yield mechanical, electrical, magnetical and optical surface functions. Equipment was developed for studying the effect of ion beam pre-bombardment on the material surface and high energy ion implantation was carried out to provide an intermixed layer at film/substrate surfaces.

Those mentioned above are big national projects with cooperation of industry, government research institutes and universities. In addition to those projects, all government research institutes under AIST are carrying on their own special research projects, about 23 themes in total, related to fine ceramics. Those special projects can be proposed by each institute.

It is important for Japan to promote creative technical development and to contribute to the world community through technical development. To consolidate research and development related to industrial technology, MITI established the New Energy and Industrial Technology Consolidated Development Organization (NEDO). This organization is given responsibility for undertaking industrial technology. One of the works of NEDO is the Research Facility Development Program. The program is to establish large and high-level facilities which are indispensable for promotion of creative R&D in advanced fields and have these facilities widely available for domestic and foreign researchers.

Improvement and operation of these facilities are performed by the third sector established for each facility. The third sector will finance one half of the initial capital investment and also half of the operating cost. Two thirds of the capital is provided by NEDO while the remaining one-third is provided by private and local government agencies.

Some facilities at present related to fine ceramic technologies are as follows.[2]

Ion Engineering Center located in Osaka. This facility opened partially in 1990 to study the technology of ion beams.

Japan Microgravity Center located in Hokkaido. This is a vertical drop facility which enables various non-gravity tests for about 10 seconds using existing vertical pits of old mines and opens in July, 1991.

Applied Laser Engineering Center located in Niigata. This is a facility to study the technology of applying laser for industry and opened in April, 1991.

Advanced Material Research Center located at Yamaguchi and Gifu. This is a facility to study and evaluate material physical properties and functions in super-high temperature environment and is scheduled to open partially in 1991.

References

1. Fine Ceramics Vision, The Fine Ceramics Basic Problems Council, The Fine Ceramics Office, MITI, 1990.
2. Handbook of AIST, 1991.

In the United States the 3M company is an exception in which new venture businesses are regularly developed within a large corporate culture. Other large corporations have had extreme difficulty with new ventures that are generally thought to be done more effectively in a small business entrepreneurial environment. Dr. Tait Elder, formerly a venture manager at 3M, describes what he has learned from his experience.

MANAGING SMALL INNOVATIVE VENTURE GROUPS

Tait Elder, Two Harbors, MN 55616

Small venture groups form new sub-cultural units which achieve long term viability through overcoming obstacles created in a wide variety of cultural interfaces. The following account illustrates this assertion from two personal perspectives: first, in the development of one particular business venture; and second, in a discussion of characteristic features of venture group management. The development of these opinions mirrored my own transition from being a physicist-scientist-supervisor into being a marketer-manager-executive. The common factor in all of these changes was productively dealing with cultural interfaces. Success in this effort depended on repeated assistance both from other people and from luck.

Transition from Research and Development

I joined 3M after seven years of industrial research elsewhere in semiconductor materials and geophysical measurements. In the Central Research Laboratory I first did research in magnetic materials and in organic dielectrics; following this I was made supervisor of a research group involved in several areas of physics and materials. This unit was measured by professional publications, patents/inventions and occasional internal consulting. By the time the group consisted of about thirty people, I was made a manager, and I developed a growing personal interest in identifying research directions. Broadening, which occurred during this period, was a natural consequences of my job responsibility. The people I spent time with introduced me to more physics topics, to materials new to me, to legal and marketing aspects of applied research and to a wider span of personnel or administrative situations.

My boss at that time was Dr. James R. Johnson, a president of the American Ceramic Society and a longtime contributor to technological innovation at 3M. Over the years, in many late afternoon discussions we considered the commercialization of technology and 3M ways of addressing this. In the mid 1960's, he initiated a very large cooperative project involving several of the groups in his laboratory, including nearly a third of my group, in a technically successful development of an electron beam stimulated laser with tunable emission throughout the visible region. It was a **laboratory** effort, set up to be a proprietary element in a concurrent development of a computer output printing system under design in an operating division. Unfortunately, our laser was not commercially operable in time to be included in that design, and by the time the laser prototype was completed there were competing technologies serving the same functions more economically. In analyzing this disappointment, we concluded that we should have had a committed product champion whose **business** was to find a selectively appropriate and early application before competitive technologies emerged.

This experience led me to value first hand knowledge of a whole array of what are called above cultural interfaces: with financial people, whose quantification of our dreams controlled funding; with marketing experts who passed judgment on the dreams themselves; with manufacturing professionals whose estimates were often regarded more highly than ours; and with actual customers, whose opinions, however formed, were more important than all the other factors combined. The transition from a research project (which is **not** a business) into a profitable new commercial venture (which **is** a business) involves a succession of new cultural encounters, frictions, clashes and adaptations. Generally it is the seller of a product or a concept who must learn and adapt, not the buyer.

The fact that I simply decided to leave a laboratory manager job and try to start a business tells a great deal about the pervasive and favorable climate for commercialization at 3M. A phone call to the secretary of the Vice President for the New Products Division, whom I did not know, arranged the first of what turned out to be a sequence of conversations devoted to what to do and how to get started. Initially I looked at ideas with proprietary prospects in my Central Research group to exploit; but I found they looked better to me when I had been selling them than they did when I was going to bet my career on one of them. Many of these derived from new materials or phenomena for which there were still serious problems with cost or reproducibility. Their patents mentioned many applications, but it was not clear that any had real customers at present price and performance. The simple linear model of innovation, with a concept being handed off from research to development to commercialization groups, was just not available to one with my status and expectations. Finding a current business use for a "cure for which there is no known disease" had a long and indefinite time scale. Fortunately, the vice president then suggested that there had been an unusually exhaustive market research report on the security industry recently prepared in his division.

I read the report, and what caught my attention was the idea of preventing article theft by use of an affixed target or tag which was detectable—actually, distinguishable—at a distance. There were already four such detection systems being sold, each depending on a different technology. Interviews with vendors and with actual users were reported which identified limitations, market potentials and even the range of prices paid. Unlike the linear model, I had here a selection of tangible objectives; I saw my assignment as being the broker between customer needs and a proprietary adaptation of any technology which could fit them. The result would have to be both cheaper and better than what was available by the time we could get to market. The convoluted sales arguments needed to prove that a more expensive new product was somehow a lot better did not seem to me to be a practical way to success.

At first I tried to read commercial, patent and technical literature pertaining to this concept of article protection. In a matter of weeks it became clear that this was no part time job. With nothing more than a conversational agreement, I relinquished my position in Central Research in early 1969. My group was divided into two, each headed by a capable new supervisor willing to do my former job permanently. In turn, I assumed the entry level title of Technical Market Analyst in the New Products Division at my same salary; I asked for and received a desk, a phone, a small shared office and occasional secretarial aid. I also asked for, and received later as appropriate, money when needed for tangible requirements of my venture as these emerged. Over the next year this informal arrangement worked very well for my travel expenses to visit vendors, customer prospects, the U.S. Patent Office, and industry/trade meetings. It also enabled me to fund targeted research and evaluation studies to quantify technical alternatives in both cost and performance. Eventually this incremental funding method provided me the protracted loan of people dedicated to pursuing practical implementation of both detection systems and target materials. Apart from this explicit funding there was available to me simply for the asking consulting aid from accountants, manufacturing engineers, attorneys, marketing managers, business planning and promotion experts, scientists, company librarians and outside consultants already retained by the company. The efficiency was phenomenal; I only took their time when they were helpful, and they appreciated having a listener who valued their suggestions.

There were some paradoxical features in this situation. By having no one report to me I became exceptionally free. By walking away from my expertise as a scientist I became a limited kind of generalist, focussed on a cluster of needs and a cluster of potential technological solutions. By being willing to use any technology to fit tangible customer needs I was able to evaluate even those technologies which I already knew in terms of meeting these needs better than ones new to me. And by giving up my limited power as a manager I had become broadly empowered to follow wherever my business aims would lead. I have seen this power of commitment since in many other people, even in organizations not so hospitable to innovation as 3M.

In summary, I had left a position in which craftsmanship was the key measure. I had entered an area where results measured by someone else's criteria—like sales growth and profit margins—were all that mattered. A workmanlike failure was no better than any other failure, and a lucky win was as good as any other win of the same magnitude. This value system change lay at the heart of the cultural changes I encountered; and I emphasize this to show some of the things which those who follow materials research into full commercialization must face. Perhaps someone present here may undertake such as assignment to improve the probability of commercial success of his/her practical ideas in the topic focus of this conference.

Development of the Business

The happy stage of learning and freedom gradually came to an end. I began to understand better the details of the technologies: their costs, their safety concerns, their efficacy difficulties when used in practical detection geometries, and their patent problems. This left fewer alternatives from which to select. I was originally prepared to shift market targets significantly to suit technical limitations or strengths. If satisfactory targets had turned out to be a little too costly but capable of several distinguishable signals, there could be non-security markets in automatic sortation or vehicle monitoring. These choices were not supported in our technology studies then, but this kind of expansive thinking laid the basis for product line expansion after initial commercialization.

Of all the security market segments for this kind of product, the value to customers was clearly greatest in library book detection. Librarians valued the usage of books by patrons highly; and about ten percent of the collection accounted for most of the usage. A three percent annual loss to theft—usually most desired books—could thus reduce the utility of the library by nearly a third. Furthermore, unlike merchants, librarians routinely would spend a great deal of time and money with each new acquired book to identify it and establish records for its future handling. Target cost and processing time were not excessive by this measure, and circulation of each book extended the value of the targets through repeated use. Most important in narrowing down our technology choice was the limiting characteristic of one effective and economic technology which required a thin seven inch long target strip. This would be difficult to affix on varied retail merchandise; but with existing 3M adhesive technology it could readily be concealed in both books and magazines. This singular feature in a proprietary technology led us to select the library market and the hysteresis loop technology for our venture; and the various patents to which we owned rights could be defined to cover what we would sell while not infringing patents of others.

Progress toward this defined objective tended to reduce the freedom to look broadly mentioned above. Before my first year ended, it became clear that electronic engineering was a critical skill for actually designing and manufacturing the detection and signal analysis circuitry required. I obtained the loan of a capable person with these skills, Don Wright; within a few weeks he was clearly my partner and, when it came to patent application filing, co-inventor. By then I had realized how much of the progress had been due to others. With Don I found that even the management of our success was as much his doing as mine. New people added in full time pursuit of our development tended to demonstrate two assets which Don first showed me: an independent commitment to the business goal and occasions in which they were individually indispensable to project success. No more was I the sole focus. Like Don, these people were selectively skilled; and adding them increased the breadth of capability as well as the number of people to handle the rising work load. Although a great deal of the work was still contracted to others, the small nucleus of dedicated people of necessity had to grow beyond my personal span of control or credit.

After the first patent application had been filed, we were free to be more specific in market research calls. The cultural interfaces with library personnel were varied and important. The nature of this product was to affect the key operation of the library with its patrons: the checking out process. Furthermore, both the geometry of the circulation counter and the building exit control paths usually were changed. The internal hierarchical relations between circulation and reference librarians, building engineers and library directors forced us to learn their culture and essentially to gain acceptance at every level.

Before our prototype detection system was completed we had developed the information for a business plan; for we had detailed statistics on the library market subdivided by library type and size. We even had customer prospect lists, based on our interviews, which included libraries waiting only for a capital authorization to be approved to be able to purchase a system. Our funding thereafter depended on a budget authorized on the basis of our projected sales and costs. Our first year unit forecast of sales was slightly under actual performance, and the anticipated product mix between targets and hardware also turned out to be conservatively stated. Sales and profits continued to increase faster than projected, a very fortunate situation when presenting the budget request for following years.

The business grew quickly because we had the technology of choice with an advantageous price in a time when our customers had money. In 1972 our sales exceeded one million dollars and the business became profitable. In the following spring I looked better than I ever did before or since, at an instant when the General Manager's position became vacant in the New Business Ventures Division. Luck leads to luck and I got the job.

Managing the New Business Ventures Division

My promotion was almost a complete discontinuity. As a venture manager I had evolved a team of subordinates to whom almost every recurring task could be automatically delegated. Furthermore, I had learned slowly but thoroughly all about our market, our capabilities, the pitfalls and the opportunities to consider in decisions. On a Friday I learned that my new job would start on Monday; and no successor was named for me until several weeks later. For the first few months in the new job, however, I was so consciously over my head that my mouth was constantly dry. I started with a full in-basket and a temporary secretary, with a stockpile of new materials for which decisions were pending.

After the first trauma, I began to evolve explicit guidelines for managing other ventures and their managers as well as to learn how to comport myself in the executive stratum. At the outset, there was little I could do with the existing ventures. My new subordinates had previously been my peers or seniors; they understood their businesses, and I was ignorant of them. Their current operations were set and approved in budgets; so my first priority was to learn enough to evaluate their directions and the direction-setting process for the division.

Directions

The objective of each corporate venture at the outset had been to build a new, proprietary business with large margins and potential to reach tens of millions of dollars in sales. The enabling freedom which I had enjoyed appeared to be a good model, wherein the ultimate direction had been allowed to evolve as understanding of the circumstances grew. The process seemed to be a sequence;
—Identification of ideas, opportunities, constraints
—Ranking of candidate ventures and their prospects
—Validation of product, customer, feasibility, timeliness
—Focus and commitment: development and implementation
—Identification of derivative businesses; exit strategy.

In fact, this was an iterative process rather than a sequential one. After proceeding through all the steps at one level of commitment, new insights led to reexamining the whole sequences before a higher level of dollar commitment could be made. It was similarly preferable that the venture manager delay full commitment to an early version of his business plan.

Venture Processes

There are other characteristics which distinguish venture unit management from established operating unit management. These differences tended to diminish as a venture became profitable and hence became qualified to leave the venture division.

—Learning was the paramount intermediate goal, for at the outset of a new business ignorance is honorable if recognized.

—Operating procedures are tailored to the particular business and to the stage of its development.

—Commitments are made or changed while uncertainty is high, and budget needs rise or fall suddenly.

—Structure and governance are correspondingly flexible, and freedom of managers varies with time and circumstance.

Controls

Direct management, through informing and discussions, helps keep visions current and obstacles visible. In addition, some built-in controls are especially needed because of the corporate risks involved in making mistakes with new areas of business:

Financial controls (such as budgets, selective signature authority) can be used to bring problems to light through prompt information reports on performance against forecast, cumulative expenditures and invoiced sales.

Management support for regulatory compliance must be strong and unambiguous, and coupled with staff assistance to acquaint all venture division people with elements of safety and legality as well as government agency rules (OSHA, EEO, TSCA, etc.).

Delegation

The constant change associated with increased understanding, varying outside influences and addition of personnel makes conventional delegation of authority very difficult. In fact, for the earliest stages of venture identification the best kind of delegation is relatively undefined and limited to specific tasks or inconsequential dollar exposure. One might designate these "training delegations", which are illustrated below:

Type	Order	Communication
Simulated	"Check with me first"	Testing
Erratic	"I didn't want that done"	Correction
Conflicting	"I asked Jack to finish up"	Dissatisfaction
Implied	"Please handle"	Trust

These techniques are used unconsciously by many venture managers on their subordinates; their continued use of these is a possible danger signal.

In the first search or market research more trust-based delegation can be practiced, so long as there is some follow-up to determine if responsibility is well handled. Performance here is a measure of both the subordinate's capability and the clarity of the objective as stated.

Conditional delegation is preferable to blanket authorizations as a business takes form. A classical example is in the military "Rules of Engagement", where a delegated authority to act quickly is automatic if and only if a certain set of conditions are met. Examples are authorizations to sign purchase requisitions to implement an approved expenditure for materials or work, or to extend credit to customers.

These Rules of Engagement in aggregate evolve into Policy. If a venture is well managed, its policy will be founded on the Key Factors of Success in the particular industry in which the venture operates, the style and culture of the particular venture management, and the general regulations of the corporation. Such a policy will allow coherent, broad and clear delegation, a vital need in a growing organization. Furthermore, basic policy statements in simple language can aid many aspects of the business besides delegation: hiring and new employee training, advertising, communications with top management, etc.

Corporate Motivation for Ventures

A frequently stated motivation for starting a venture group is to create new, large proprietary businesses. It is true that one big winner makes up for all the costs of many disappointing ideas, and the yield from such units is relatively small in terms of number of hits per try.

The ultimate benefit to a corporation from having had a venture unit, however, may well be the creation, if only briefly, of an environment in which proprietary capability can be tailored to form a product which meets the needs of real customers. These customers, if they can be identified, represent many potential markets for variants of the base capability. A venture unit mechanism is particularly appropriate for technology-based capability, for the proprietary coverage of a patent lasts long enough to cover large undiscovered market opportunities which may only appear later. Although operating companies may also serve this purpose, the venture unit has the advantage of being a relatively minor element in the corporate power structure—and consequently easily disposed of when no longer needed.

Entrepreneurial skill has been called the last holistic business specialty. Venture units tend to build people with wide, entrepreneurial experience who may thereafter be used elsewhere in the corporation. Because this broad experience of venture participants occurs in the environment of their employer, it is selectively appropriate to that corporation's operation.

The group discussing management and the conference participants did not come to any agreed conclusions. As seen in the discussions above, there was general agreement on good practices and a recognition that innovation management was done well in both American and Japanese contexts. Rather than management *per se*, it seemed to be other cultural variables that affected results. Amongst others, there is the differing view of the science-engineering nexus discussed elsewhere in this report. There is the attention to the literature and stronger technological interactions within a specialty and less interactions outside a specialty on the American scene. There is a much greater networking between disparate elements in the socio-technical system in Japan with a resulting increase in feedback loops leading to the recognition of needs and opportunities. Perhaps it is inevitable that the Japanese are more effective at incremental innovations while Americans at all levels are more entranced with breakthroughs.

COMMENTARIES

Several participants prepared and presented commentaries based on discussions at the conference. Those are collected in this section along with a brief collation of conference discussions based on notes taken at the conference and transcriptions of the formal conference interactions. A very concise presentation is thought appropriate since most of the content appears in the general presentation, focussed discussions and individual commentaries.

SMITHSONIAN HORIZONS

Robert McC. Adams, Secretary, Smithsonian Institution

Anyone who shops for items like electronics, cameras or toys is aware of how dependent on foreign manufacturers we have become for new consumer ideas. American prowess in basic scientific discoveries notwithstanding, economic studies have provided convincing evidence that our declining competitiveness in international markets is at least partly attributable to a lagging rate of innovation with new products. Why is this so? What can be done about it?

The usual academic answers to these questions marshal statistics to contrast American with, say, Japanese industrial performance. But lacking the vividness of immediate experience, they are for the most part not very helpful as guides for particular firms or even industries.

In December, at the University of Arizona, I had the pleasure of participating in a conference that took a different, more illuminating approach. It examined innovation processes in American and Japanese industries by focusing on three types of advanced materials that are still in an early stage of commercial development. Made most familiar to us by the media are high-temperature oxide superconductors, but silicon nitride structural ceramics (used in some car engines) and synthetic diamonds produced at low temperatures are also finally on the threshold of significant commercial exploitation--with, in each case, the Japanese well in the lead.

All are materials of high commercial potential. All are the results of scientific research extending over generations, although it may have periodically shifted in direction or lain dormant for a while. Also over many years, all have required massive, risky investment to identify marketable uses and to develop reliable, less costly production processes. It is in these latter stages that American firms have generally abandoned the chase.

Participants at the conference, convened by materials scientist W. David Kingery, included mostly Japanese and American scientists and engineers who have played leading roles in basic or applied research on one or more of these technologically demanding substances. Most knew one another's work well, and there had been a surprising (to me) degree of direct contact in one another's laboratories. Further enriching the mix was a sprinkling of generalists in science policy or organization theory, and of anthropologists like myself with a presumed sensitivity to the social and cultural context of innovation.

These contextual factors are in many ways the most elusive, dynamic and difficult to deal with. Our higher cost of capital and inability to sustain long-term development without firm assurance of an existing market have been widely noted. Less often taken into account is the way in which U.S. Government needs (particularly those of the defense establishment) have turned corporate effort and the stream of talent recruitment toward developing high-performance materials rather than reducing costs. Further contributing to Japanese commercial success have been much larger investments in industrial laboratory instrumentation and in process and production engineering. More uncertain is the effect of a closer integration of the Japanese scientific and engineering communities, and even of their priorities, although this may be disputed.

The problem is that contextual factors are, precisely, those most deeply embedded in the larger social order and hence the most difficult to alter. This applies to the Japanese no less than to ourselves. One can only speculate whether the traditional stability of their corporate careers can be retained as rising incomes give new salience to quality-of-life issues. Or again, can the heavy reliance on the Japanese family as the guiding metaphor for corporate organization retain its force when the virtual exclusion of women from scientific and managerial careers leads to increasing tension?

Prominent among the factors accounting for superior Japanese competitive performance is an organizational approach strikingly different from its counterpart here. The basic units in Japanese laboratories, and even on production lines, are cooperative work groups. Ranging widely across related disciplines, these groups typically include craftsmen and technicians as well as senior scientists and engineers. There is a consistent emphasis on consensus-seeking; on the making of decisions by those as close as possible to the point at which they must be implemented; and on the opening of channels of communication at all levels. Individual workers are given every opportunity to attain their full potential by being cross-trained in a variety of skills, rather than being forced into routine assignments in which they are essentially viewed as expendable parts of a machine.

The result, clearly, is a deeply felt attitude of mutual respect and loyalty that results in consistently superior responsiveness, accountability and performance. The lesson is, as one conference participant observed, that "people make a difference." That lesson is as disarmingly elementary as it is universal.

[Reprinted from the *Smithsonian* Magazine, February, 1991]

CULTURAL INFLUENCES ON TECHNOLOGICAL DEVELOPMENT IN THE UNITED STATES AND JAPAN: AN ARCHAEOLOGIST'S MUSINGS

Michael Brian Schiffer, University of Arizona

This brief paper brings an archaeological perspective to discussions about the cultural contexts of technology in the U.S. and Japan. An archaeological framework suggests that emphasis be placed on (1) concrete activities, (2) long-term processes involving structural relationships, and (3) behavioral and cultural variation within societies. Certain issues are identified that perhaps can be addressed more fully in future multi-cultural meetings.

The Concept of Culture

Anthropology's major theoretical contribution to the social and behavioral sciences (and beyond) is the concept of culture. The Englishman, E. B. Tylor, regarded in some quarters as the father of anthropology, furnished the first modern definition of culture in the late nineteenth century. For Tylor and most other early anthropologists, culture referred to the lifeway of a particular society--its characteristic ways, for example, of obtaining food, building houses, reckoning kin, settling conflicts, and relating to the supernatural. This concept of culture, which explicitly embraces technology, was useful as long as anthropologists studied mainly small-scale, reasonably well bounded societies such as the Andaman Islanders, Greenland Eskimo, and Hopi. However, when anthropologists began to probe more complex, differentiated societies, including the modern U.S., additional definitions were needed. Such definitions usually confine culture to the ideational realm: a group's worldview, values, and attitudes, or rules for appropriate behavior. Moreover, culture can refer to any analytically distinguishable aggregate of people. Thus, anthropologists now discuss the culture of Afro-Americans, of yuppies, of IBM, or of high-energy physicists.

An appreciation for the diverse meanings of the culture concept can help investigators to frame questions about "cultural" influences on technological development. Obviously, traditional definitions still have an important role to play. For example, one can query, How have particular circumstances of Japan's history and natural resources shaped its technological policy and priorities, especially in recent decades? Ideational definitions of culture can also orient investigations of culture-technology relationships. A typical question might be, Do different corporate cultures in the U.S. or Japan have characteristic styles of technological research and development? Similarly, Are contrasts in Japanese and American worldviews reflected in the way the two nations prioritize technological research? Although diverse concepts of culture can generate productive questions, it should be kept in mind that different questions imply different methods of investigation.

An Archaeological Perspective

As anthropologists, American archaeologists also employ various concepts of culture. In addition, for many archaeologists another important concept is that of *activity;* the latter consists of patterned interactions between energy sources (often people) and artifacts (for elaboration of the activity concept, see Schiffer n.d.). In addition to their material constituents (people and things), activities also have social organizations (specifiable relationships among the participants) and ideologies (appropriate attitudes, beliefs, and rationales), and tend to be carried out repeatedly in the same places. The goals of specific activities are diverse, ranging from creating a diamond transistor to propitiating the spirits of ancestors. Regardless of whether their goals are mainly technological, social, or religious, activities do not occur in isolation but are linked to other activities by movements of people and artifacts. In addition, activities are nested within--and crosscut--organizational units such as households, clans, villages, palaces, churches, government laboratories, and corporations. Clearly, different research problems require that a society's activities be partitioned analytically in different ways. For example, one can study how a society's patterns of settlement across a landscape are influenced by subsistence, religious, and technological activities. Similarly, activities such as technological research and development (R & D) can be examined across corporations, industries, or even nations.

Artifacts play many roles in activities, from the mundane storing and converting of matter and energy to symbolizing the social roles of participants; artifacts even embody a group's most sacred ideas. Indeed, the majority of artifacts in all societies have both utilitarian *and* symbolic functions; and the very same artifact may have different functions in different activities (on the functions of artifacts, see Rathje and Schiffer 1982; Schiffer n.d.).

The activity perspective of archaeology calls attention to the complex ways in which technology is embedded in human societies. An appreciation for these multidimensional relationships may help us to formulate fruitful questions about the cultural context of technology.

In addition to a concern with concrete activities, an archaeological perspective entails two additional components: (1) examination of long-term processes and (2) consideration of cultural and behavioral variability--in the present case, within nations. Archaeologists reconstruct and study the behavior of societies over spans of centuries, even millennia; thus, archaeologists take a long-term view of change processes (Plog 1974). We are especially apt to look for long-term effects of demographic and natural-resource variables on activity patterns, organization, etc. These kinds of effects may be pervasive, thoroughly affecting even worldviews, yet be far from obvious if investigators confine their observations and analyses to just one point in time.

Archaeologists also pay considerable attention to variability--differences in activities, artifacts, organization, and ideational phenomena (Rathje and Schiffer 1982; Thomas 1989). In large and heterogeneous nation-states like the U.S. and Japan, there are likely to be wide and perhaps substantially overlapping ranges in the specific behaviors of interest. For example, in both nations corporations range from tiny start-up firms to enormous multi-national entities. In the U.S. and Japan one also finds diversity in corporate cultures, even among organizations of similar size. In addition, both nations engage in research activities that run the gamut from the most esoteric "pure science" to the most mundane projects in applied technology. Clearly, though ranges of behavior may be similar in the two nations, modes may differ substantially. For example, in research activities, Japan does far less and the U.S. far more "pure science." Such descriptions of modes or norms, which may be characterized as cultural differences, require explanation. We must seek to identify and understand the long-term processes that may have brought about the differences in modal behavior.

One final point about the archaeological perspective. Though archaeologists have traditionally studied long-dead societies, more recent periods--including the present-day U.S.--have come within the discipline's purview (Rathje 1979; Reid, Schiffer, and Rathje 1975). Today, archaeologists even study recent garbage (Rathje 1989, 1990) and portable radios (Schiffer 1991)! In effect, archaeologists focus on (and seek to explain) behavioral variability over time and space; we recognize no arbitrary boundaries on our subject matter.

Cultural Differences and Technological Innovation

Eric Poncelet's fine synthesis alerts us to many differences in U.S. and Japanese values, attitudes, and worldviews (i.e., culture as ideational phenomena). Although these differences are substantial, one must avoid turning such cultural descriptions into stereotypes or even caricatures and invoking them mechanically to account for purported differences in R & D activities in the two nations. As Bryan Pfaffenberger pointed out, those seeking to explore the cultural contexts of technology must be wary of falling into the trap of a simplistic cultural determinism. Though such cultural explanations are intellectually fashionable today, they can nonetheless easily become a stultifying reductionism.

An archaeological perspective forces us to look at cultural processes over time, to recognize that values and attitudes are not constant but change along with changes in activity patterns. Let us examine some long-term processes in the U.S. and Japan that might be relevant for studying the cultural contexts of technological invention and innovation.

From about 1900 to about 1930, U.S. society underwent a dramatic transformation under the impact of factories powered by electric motors and new mass media (general circulation magazines, movies, and radio). America became the world's first industrial consumer society where even the humblest factory worker could aspire to possess the same kinds of products as the factory's owner (for overviews of this transformation, see F. Allen 1952; Fox and Lears 1983; McElvaine 1984). Growth of this consumer society was facilitated by abundant natural resources as well as a well developed infrastructure of rail transport and electricity distribution systems.

Although a consumer society was consistent with America's long-standing commitment to democratic values, the spend-and-consume ethic was at odds with other traditional values, deeply held among middle and working classes, that had arisen in the context of agrarian, non-urban lifeways. However, under the tutelage of social commentators and the mass media, Americans began to move away from (or reinterpret) values of thrift, frugality, deferred gratification, and sacrifice for family and community (Horowitz 1985). By the end of the twenties, Americans were growing up believing that mass consumption was good for them and good for the country, and was as much their right as the franchise.

During the first half of the twentieth century, consumers developed as a potent political force in U.S. society. In effect, consumers gradually formed the largest and most powerful interest group (that seemingly transcended conflicts between labor and big business). Legislation for the "public good" was in reality legislation in support of mass consumption, which ostensibly benefitted both producers and consumers. Gradually, however, Congress and a succession of administrations fashioned a framework of constraints within which companies engaged in product development and manufacturing had to operate. For example, beginning at the turn of the century, the ability of U.S. business to form huge horizontally and vertically integrated corporate entities was gradually weakened, because these organizations did not seem to operate in the consumer's interest. Similarly, after the Second World War, free-trade policies opened up U.S. borders to imports from around the globe (cf. Office of Technology Assessment 1983; Vatter 1963); unrestricted trade did work to the consumer's advantage, though it obviously did not benefit U.S. firms whose products were displaced in the marketplace by (usually cheaper) foreign substitutes.

In the mid-fifties, the U.S. consumer society acquired a permanent Pentagon booster thanks to the Cold War. Military spending not only provided a "safe haven" for companies tiring of fickle consumers, but it also set R & D agendas throughout academia, government labs, and industry (cf. Melman 1965, 1974).

Japanese industry and society underwent a different developmental trajectory under different circumstances (on Japan's industrialization, see, e.g., G. Allen 1958; De Mente and Perry 1968; Reischauer and Craig 1978; Tsurumi 1976). Japan is a densely populated island nation with scant natural resources. When in the nineteenth century Japan decisively embarked on a course toward industrialization, it was appreciated that to pay for imports of raw materials and machinery, manufactured goods would have to be exported. Thus, concomitant with industrialization came an export imperative. To this day, every Japanese child learns in school that the nation must export manufactured goods to survive (Bolling and Bowles 1982:68).

Following the Second World War, while the U.S. was refining a consumerism buttressed by military spending, Japan was creating an advanced producer society. Although ostensibly a constitutional democracy, Japan in fact has become a technocracy; significant economic and technological policy is forged and implemented by the Ministry of Finance and the Ministry of International Trade and Industry (Johnson 1982). The latter agency in particular, which maintains close ties to industry through the movement of senior technical people, sets agendas for technological development, establishes the parameters of competition among Japanese firms, regulates the import of technology, and protects Japanese industry from foreign imports. In Japan, technological policy is, in fact, the highest form of national policy.

With this organizational structure in place, Japan has created a framework in which manufacturing firms, many conglomerated into huge and diversified multinational corporations (lineal descendants of *Zaibatsu*), have thrived making products for export. As has become obvious in recent decades, Japan's multinational conglomerates are especially well adapted to conditions of international competition (cf. Davidson 1984; Kotler et al. 1985; Ozawa 1974). For example, such a firm can easily parcel out production among its foreign subsidiaries and collaborators to take advantage of low labor costs in virtually any corner of the globe, and can also sustain losses for long periods while slowly capturing a targeted market. Similarly, these companies can use the internal Japanese market--where goods are sold at high prices--to secure some cash flow while the same products (like color TVs) are dumped abroad to put pressure on competitors (Office of Technology Assessment 1983). Clearly, Japan's "economic miracle" owes much to these institutional and organizational forms.

In Japan's producer society, domestic consumers clearly are not a powerful political force. But to Japanese companies the consumer does matter--especially the foreign consumer. Products are made with meticulous concern for the user; that is why Japanese firms employ so many researchers (what Americans would call scientists and engineers). This research is carried out to create new products and to improve old ones, seldom to generate abstract science. Japanese firms also make many marketing innovations, often successfully defying established conventions in the countries where they do business. In the U.S., for example, early Japanese transistor radios succeeded so well in part because they were sold in myriad retail outlets, such as hardware stores and drug stores, that had never before carried an electronic product (Baranson 1981; Schiffer 1981). (These radios, for the most part not technological marvels, were also assembled with cheap labor using components built with equally cheap labor, and so could be priced well below comparable American radios.)

A Simple Model

In discussions of contemporary differences in the cultural contexts of American and Japanese technological development, one must not lose sight of those historically and environmentally based cultural factors. Labeling Japan a producer society and the U.S. a consumer-military society does not itself explain specific technological research activities, but these significant cultural differences may help us to understand varying modalities in technological invention and innovation in the two nations. The following section presents a model for tracing influences of these large-scale cultural differences on R & D activities.

In seeking cultural influences on technological invention and innovation in the U.S. and Japan, it is convenient to divide up the development process into three stages: (1) choice of problem, (2) basic research, and (3) search for applications. In the following section, hypotheses about cultural influences are framed with respect to these fundamental stages.

Choice of problem

Most organized research begins with explicit formulation of a problem. In the choice of problem, national priorities and imperatives as well as institutional and organizational structures can have a significant influence. In the U.S., a scientific bias in favor of abstract knowledge and publication may influence problem choices even in the most down-to-earth company. In contrast, one sees in Japan that problem choice is nearly always made with an eye toward practical applications. New technologies are pursued if there is reason to believe that they can improve existing products or lead to new ones. A few specific hypotheses follow.

I begin with the obvious: in the U.S., much technological development is pushed by military spending and priorities. Thus, not only do military needs provide ready-made technological problems, but much other research--perhaps not even funded by the military--is undertaken with an eye toward military applications.

In both nations, accepted scientific theory doubtless plays a role in identifying potentially promising problems. However, it also appears that theory exerts a more constraining influence in the U.S. than in Japan. Predictions from theory, both correctly and incorrectly drawn, have served in the U.S. to close off prematurely certain research avenues. The BCS theory, for example, may have discouraged researchers from pursuing high - T_c superconductors, just as thermodynamic theory (or incorrect implications drawn from it) dissuaded most investigators from seeking low-temperature, low-pressure regimes for making synthetic diamonds. Nonetheless, in the former case IBM and in the latter John Angus--both of the U.S.--undertook pioneering research in the new technologies, theory notwithstanding. These examples seem to suggest that because of the great size and diversity of research efforts in the U.S., there will always be some firms and individuals willing to choose problems seemingly beyond the fringe of feasibility.

In the United States, the choice of problem is also sometimes influenced by the desire to create a result that serves mostly for technological display: it is valued more for its symbolic function than its practical utility. U.S. firms seemingly are more apt to create a product or process that has obvious newsworthiness, which can serve to attract investors and polish the company's image before consumers. The result is flash-in-the-pan projects that, finally, lead nowhere--like gas turbine engines for automobiles. In Japan, faith is put in developing a promising technology, often beginning with something very modest or mundane, like a pocket transistor radio or a silicon nitride glow plug. The goal is to accumulate technological know-how that can be applied gradually to more complex products. Such modest products are sometimes at the beginning of important technological trajectories, such as consumer microelectronics, that build up considerable momentum.

In offering hypotheses about cultural effects on the choice of problem, it is important to keep in mind that the enormous diversity in the institutional base for conducting technological research in both the U.S. and Japan (e.g., small and large firms; start-up and established firms; poor and wealthy firms; government and university laboratories; independent inventors) perhaps guarantees a comparable diversity in selecting problems. Similarly, technologies in various stages of development may also give rise to different kinds of research problem, and in both the U.S. and Japan, research is carried out on technologies spanning the entire range of maturity. Future research should strive to sort out the influence of these factors on problem choice.

For example, in Japan as compared to the United States, a greater percentage of technological research is carried out in corporate contexts. Could this factor account for dominance of a practical orientation in so much of Japan's research? Of course, that still reflects a higher-order influence of culture, in that in Japan--the consummate producer society--technological research efforts are disproportionately concentrated in organizations strongly focused by the export imperative. The need to compete in the world marketplace forces research choices in the direction of more practical technologies.

Basic Research

After problems are chosen, often by managers, they are taken up by researchers--scientists and engineers in the U.S. My impression is that cultural differences at the national level have little effect on concrete research activities. This is so because the cultures of science and engineering cross-cut nations. If anything, cultural differences between disciplines or subdisciplines are likely to be much more profound than international differences (controlling for the institutional basis of the research, stage of the technology, and level of support for the research). The easy movement of individuals from country to country, international journals and conferences, comparable laboratory equipment, the universality of team organization, shared worldviews and work habits, and general adherence to scientific methods, lead to an unusual degree of uniformity in research activities, wherever they are found. Any differences in research styles probably result from differences in institutional bases, stages of technology, or personalities of team leaders or participants.

Search for Applications

Important cultural differences, reflecting different national priorities, are likely to re-emerge at the applications stage. Although much U.S. technological research is of a very high quality, it has no immediate practical applications. SDI research, for example, may lead to weapons having important symbolic functions, but the sole purchaser of these systems would be the U.S. government. It is difficult to envision new industries arising on the basis of SDI technology to make consumer products for the world marketplace. In Japan, in contrast, one might expect diligent efforts to find new applications for the results of all technological research undertaken.

Because the exact course of a technology's development cannot be predicted, and because most of a technology's ultimate applications cannot be foreseen when development begins, the relentless search for applications--"solutions in search of problems"--takes on paramount importance and is subject to many levels of cultural influence.

Again, I stress that different mixes of institutional contexts in the U.S. and Japan for commercializing technological research may be the most visible effect of national differences. In both nations, there is an enormous capacity to bring new products quickly to market. However, products that initially fail in the marketplace may require several generations of redesign, sometimes over decades, until they succeed. Wealthy multinational firms (whatever their nationality) can indulge their faith that a particular product eventually will find a large market by investing in R & D heavily over the long haul. Obviously, an undercapitalized start-up company seldom has this luxury. In the U.S., many important products appear initially in firms that cannot sustain drawn-out periods of development; if the product fails at first, the company may give up, losing it to more able competitors (at home and abroad).

In Japan, it seems that much new product development takes place in larger firms capable of maintaining a long-term commitment. Institutional contexts patently affect invention and innovation at the application stage, and the mix of these contexts may reflect the influence of cultural factors at a national scale.

Concluding Remarks

Introduction of the anthropological concept of culture to discussions of international differences in technological invention and innovation can stimulate much fruitful discussion and debate. Technological research and development do not take place in a vacuum, propelled only by internal drives. Rather, most researchers work in institutional settings that, themselves, are nested within systems of overarching constraints and incentives that reflect national priorities and imperatives. These contexts are cultural, brought into existence by specific historical processes taking place in specific natural environments. Clearly, explaining international differences in invention and innovation requires us to take into account these differences in cultural contexts. Nevertheless, however attractive cultural explanations are, we must avoid falling into the trap of a naive cultural reductionism. There is simply too much cultural and behavioral variability in present-day nation states to permit such unicausal, unidimensional explanations. That is why a long-term perspective is so helpful; its lessons can help us to winnow out explanations based exclusively on shallow generalizations about contemporary phenomena.

References
Allen, Frederick Lewis
    1952   *The big change: America transforms itself, 1900-1950.* Harper and Brothers, New York.
Allen, G. C.
    1958   *Japan's economic recovery.* Oxford University Press, London.
Baranson, Jack
    1981   *The Japanese challenge to U.S. industry.* Lexington books, Lexington, Massachusetts.
Bolling, Richard and John Bowles
    1982   *America's competitive edge: how to get our country moving again.* McGraw-Hill, New York.
Davidson, William H.
    1984   *The amazing race: winning the technorivalry with Japan.* John Wiley and Sons, New York.
De Mente, Boye and Fred T. Perry
    1968   *The Japanese as consumers: Asia's first great mass market.* John Weatherhill, New York.
Fox, Richard Wightman and T. J. Jackson Lears
    1983   *The culture of consumption: critical essays in American history, 1880-1980.* Pantheon Books, New York.
Horowitz, Daniel
    1985   *The morality of spending: attitudes toward the Consumer Society in America, 1875-1940.* The Johns Hopkins University Press, Baltimore.
Johnson, Chalmers
    1982   *MITI and the Japanese miracle: the growth of industrial policy, 1925-1975.* Stanford University Press, Stanford, California.
Kotler, Philip, Liam Fahey, and Somkid Jatusripitak
    1985   *The New competition.* Prentice-Hall, Englewood Cliffs, New Jersey.
McElvaine, Robert S.
    1984   *The Great Depression: America, 1929-1941.* Times Books, New York.
Melman, Seymour
    1975   *Our depleted society.* Holt, Rinehart and Winston, New York.
    1974   *The permanent war economy: American capitalism in decline.* Simon and Schuster, New York.
Office of Technology Assessment
    1983   *International competitiveness in electronics.* U.S. Congress, Office of Technology Assessment, Washington, D.C.
Ozawa, Terutomo
    1974   *Japan's technological challenge to the West, 1950-1974: motivation and accomplishment.* MIT Press, Cambridge.
Plog, Fred
    1974   *The study of prehistoric change.* Academic Press, New York.
Rathje, William L.
    1979   Modern Material Culture Studies. In *Advances in Archaeological Method and Theory,* Vol. 2, edited by M. B. Schiffer, pp. 1-37. Academic Press, New York.
    1989   Rubbish! *The Atlantic,* December, pp. 1-10.
    1990   Archaeologists bust myths about solid waste and society. *Garbage: The Practical Journal for the Environment.* September/October, pp. 32-39.
Rathje, William L. and Michael B. Schiffer
    1982   *Archaeology.* Harcourt Brace Jovanovich, New York.
Reid, J. Jefferson, Michael B. Schiffer, and William L. Rathje
    1975   Behavioral archaeology: four strategies. *American Anthropologist* 77:864-869.
Reischauer, Edwin O. and Albert M. Craig
    1978   *Japan: tradition and transformation.* Houghton Mifflin, Boston.
Schiffer, Michael Brian
    1991   *The portable radio in American life.* University of Arizona Press, Tucson.

    n.d.   *Technological perspectives on behavioral change.* University of Arizona Press, Tucson (in press, ms. 1991).

Thomas, David H.
    1989   *Archaeology.* Holt, Rinehart and Winston, Fort Worth, Texas.

Tsurumi, Yoshi
    1976   *The Japanese are coming: a multinational interaction of firms and politics.* Ballinger, Cambridge, Massachusetts.

Vatter, Harold G.
    1963   *The U.S. economy in the 1950s: an economic history.* W. W. Norton, New York.

STRAINS IN THE AMERICAN INDUSTRY-SCIENCE-EDUCATION TRIAD

Robert McC. Adams, Secretary, Smithsonian Institution

In the face of intensifying competitive pressure from our international trading partners, basic features of the American industrial system are coming under increasingly critical scrutiny. I cannot independently confirm the respective strengths and weaknesses to which critics have alluded in this system, except to note that something approaching a consensus will be found in numerous publications -- and receives further amplification in the proceedings of this conference. But it may be useful to step back from the specificity of most of the present diagnoses of illness, and to sketch more broadly its historical context. To be sure, no easy solutions emerge from doing so. Instead, taking a longer view tends to magnify the challenges we face, by suggesting that problems approached in terms of their recent, sectoral impacts in fact have much deeper roots and wider institutional entanglements. But we ignore historical patterning at our ultimate cost, even where the chosen topic of our discussions has the immediacy of the creative act of innovation.

Salient features of our present condition provide a useful starting point. While specialists obviously differ on details, something like the following assessment seems to have wide support:

Granting that there is a great deal of firm-by-firm and industry-by-industry variation, this country has tended to rely on outdated industrial strategies that overemphasize mass production of standard commodities. Other obsolescent features include a slowness to adopt or adapt to such significant production innovations emerging abroad as "just-in-time production" and "quality circles"; an inattention to the need for more intensive and sustained workforce training; and a pronounced hesitancy about seeking greater workforce commitment through a wider delegation of production-line decisions. Driven by an external economic environment that imposes shrinking time-horizons and a growing preoccupation with short-term profits, we have been repeatedly outperformed in the design and manufacture of reliable, high-quality products (Cohen and Zysman 1988, Berger et al. 1989, MIT Commission on Industrial Productivity 1989). Most critically missing has been an emphasis on enhanced flexibility,

> a firm's ability to vary what it produces.... This notion is captured in the distinction between economies of scale and economies of scope.... Economies of scope are created by standardizing processes to manufacture a variety of products.... The capability to change quickly in response to product or production technology -- to put ideas into action quickly -- is the central notion (Cohen & Zysman, 112-113).

Granting the heroic overcompression that is unavoidable in this setting, some account of long-term trends in U.S. industrial history is a necessary point of departure: No less than with the recent rise of Japan as an industrial superpower, it is a gross distortion to view the transfer of England's Industrial Revolution to the United States as an act of passive reproduction by the recipient. This is not to deny the small but crucial part played by a direct movement across the Atlantic of trained mechanics, foremen, and entrepreneurs (Stapleton 1987: 29), which no doubt speeded the growth of an industrial foothold here and influenced the initial forms it took. But the initial stimulus of that successful prototype aside, industry in the New World quickly developed in different directions since it had to confront different challenges.

Dominant among them, in a well-endowed but sparsely populated new land, was the limited supply of local labor and the high costs of transport. Frontier living promoted familiarity with the use of tools and with principles of elementary machinery. A sustained emphasis on labor-saving agricultural machinery and on the production of serviceable personal weapons of high dependability were other predictable outcomes of the unique set of conditions encountered in the New World, as was the abundance of timber and hence a wider reliance on wood products. As exemplified particularly in the textile industry, high operating speeds and greater mechanization, and maximum use of inanimate power were other characteristically American features (Jeremy 1973). New, automatic milling devices, anticipating the subsequent emphasis on continuous flow technology, had already been introduced by the time of the Revolution. But perhaps the most persistent and widely celebrated theme was the development of machines and implements employing interchangeable parts.

Another influence of the risk, impermanence, and sparseness of settlement associated with the American frontier may well have been the emphasis it focused on rugged dependability and ease of maintenance in manufactured goods, at the sacrifice if necessary of variability of style or quality of finish. Standardization met this end. Probably supportive of the same trend was a "confident expectation of rapidly growing markets." That provided an "extremely powerful inducement" to a partricular form of industrial expansion:

> The use of highly specialized machinery, as opposed to machinery which has a greater general-purpose capability, is contingent upon expectations concerning the *composition* of demand -- specifically, that there will be strict and well-defined limits to permissible variations. Both of these conditions . . . were amply fulfilled in early-nineteenth century America (Rosenberg 1981: 52-54).

This "American System," as it came to be called, thus anticipated, and perhaps partly generated, a fairly tolerant form of consumer demand. American consumers were markedly more ready than their English counterparts to value function more highly than finish and to accept homogeneous final products. This made it possible for producers of capital goods to seize the initiative in setting the objective of standardization and suppressing variations in product design (Sawyer 1954: 369-71; Rosenberg 1970: 558-60). It also contributed to American leadership in pursuing the goal of standardized, interchangeable parts (however imperfectly it was met at the time), which was a special subject of admiration in the great Crystal Palace exposition in London in 1851 (Hindle 1981, Hindle and Lubar 1986).

> European comment often gave much attention to the crudity, the lack of finish, the use of wood in place of metal, or the light construction of various articles; but the admiration for their simplicity, originality, effectiveness, and above all their economy and volume of production, led not only to recognition at the world fairs but to increasing entrance of American products and methods into European markets as imports, or through licensing or notably less formal practices (Sawyer 1954: 371).

The truly sustained growth of American industry, and its mounting competitive success in world markets, developed on these foundations in the period following the Civil War (Hounshell 1984). Product and process innovation both played a substantial part, at first almost exclusively at the hands of managerial and shop-floor personnel with practical experience rather than in dedicated corporate research laboratories. Dependence on basic research in universities -- indeed, any effective linkage, save in the recruitment of college-trained personnel -- was of course virtually insignificant.

One of the roots of the prodigous industrial growth was, quite naturally, an unparalleled commitment to mass production. Another, more slowly developing as the ties of individual firms to the exploitation of particulars set of natural resources weakened (Wright 1990), was the hierarchical, multi-division corporations. But while initially they were sources of great competitive strength, these features contained seeds of weakness under a different set of world economic conditions that only began to materialize a century later. On the one hand, some industrial sectors such as automobiles and steel became state oligopolies with only marginally increasing demand and high barriers to entry. These structures diverted competition from production costs or basic technological development to marginal product, process, and style changes (Cohen and Zysman 111).

Another source of long-term structural weakness that began as a competitive advantage was the abundant supply of industrial labor that began to flood into America in the post-Civil War period. Large-scale immigration from southern and eastern Europe provided a huge but undisiciplined and ill-trained labor force, initially very limited in its familiarity with English. This could only reinforce the emphasis on highly standardized mass production as the dominant industrial outlook. In time, this came to be codified as Taylorism, involving managerial reliance on small cadres of educated planners and supervisors who could reduce complex assembly jobs into multiple, rote assignments that could be demonstrated rather than taught. Effective as it was in meeting the special conditions of the time, the inherent defect in this approach from the vantage point of the late twentieth century has become that

> what the world is prepared to pay high prices and high wages for now is quality, variety and responsiveness to changing consumer tastes, the very qualities that the new methods of organizing work make possible.... The new high performance forms of work organization,...rather than increasing bureaucracy,...reduce it by giving front-line workers more responsibility. Workers are asked to use judgment and make decisions. Management layers disappear as front-line workers assume responsibility for many of the tasks -- from quality control to production scheduling -- that others used to do....
>
> [But] because most American employers organize work in a way that does not require high skills, they report no shortage of people who have such skills and foresee no such shortage. With some exceptions the education and skill levels of American workers roughly match the demands of their jobs.... More than 70 percent of the jobs in America will not require a college education by the year 2000.... No nation has produced a highly qualified technical workforce without first providing its workers with a strong general education. But our children rank at the bottom on most international tests -- behind children in Europe and East Asia, even behind children in some newly industrialized countries.... Only eight percent of our front-line workers receive any formal training once on the job, and this is usually limited to orientation for new hires or short courses on team building or safety (Commission on the Skills of the American Workforce 1990: 2-4).

The U.S. educational system is often described as the best-funded in the world, but as this report implies the funding -- and performance -- of its primary and secondary sectors must be assigned to an entirely different category than its college and university sector. Our per capita expenditures rank rather low among those of industrialized countries (12th of 14 in OECD rankings) if we consider primary and secondary schooling only. Moreover, we need to look deeper than these aggregates. The gulf between per-pupil levels of spending in different states and localities is very large, and the dollar levels fail adequately to reflect differences in the availability of specialized equipment and advanced-level classes, and in the quality of teaching. Students in low-income and inner city neighborhoods, including a high proportion of minorities and of those destined to serve as the core of our industrial workforce, are particularly at risk. Possibilities for their upward mobility are seriously reduced by the barrier to college or university training that is imposed by these earlier educational impediments.

Abbreviated as it necessarily has been, the burden of this account is to outline how some original strengths of American industrial organization have gradually been transformed into weaknesses in a world order that has been still more profoundly transformed over the course of a century. Nothing in that will be found very surprising. But more disturbing and perhaps even counterintuitive are its implications for our own self-image.

We hear on all sides, including repeatedly at this conference, that self-reliance and openness, derived in some measure from the expansiveness of our western frontier in an earlier era, are fundamental features of the American character. Predisposing our society to an absence of rigid structures and hierarchical barriers to mobility, they are credited with playing an important part in our scientific preeminence. Yet the sweeping applicability of this stereotype to our society as a whole is obviously at variance with what we have seen of its mass production industries. While corrective steps are now underway, in many instances too large a proportion of our industrial workforce is still permanently consigned to a semi-skilled status, relatively well-paid but limited to highly routinized assignments. Typically lacking is any significant prospect for individual advancement, as well as opportunity for the mutual improvement and reinforcement -- not merely of the morale of the workforce but of the industrial product and production process -- that the Japanese have led the way in showing can result from closer group interaction.

Does it follow, however, that this self-image is no longer anything more than a memory? To the contrary, I would argue strongly that it still applies with considerable accuracy to circles of professional, intellectual, entrepreneurial, and in particular scientific life like those participating in this conference. Indeed, in these circles, there is very nearly the antithesis of Taylorism. Setting aside doubts about the capacity of the modern American research university itself to adapt to a changing, more stressful environment, this is perhaps best exemplified by the favorable internal environment for creativity that it has fostered.

While demographic, financial and other pressures have eroded the availability of tenure and other advantages, the research university continues to provide at least a relatively flexible, egalitarian, innovative setting for research. Disciplinary barriers to communication are often decried, but boundaries to association in furtherance of the research enterprise remain relatively negotiable both internally and externally. Virtually absent, particularly in the strongest institutions, are incentives for anything other than rampant, individualistic behavior -- if it is accompanied by high productivity.

There is a curious coupling, in other words, in the new version of the "American System." On the one hand, a complex of advanced educational institutions has developed (with substantial, if declining, federal support) that is exceptionally well attuned to its responsibility for encouraging the creativity essential for basic research. Clearly, its modern origins lie in the outstanding series of successful military applications of discoveries in university-based laboratories during and after World War II. Whether the post-Cold War transition to a market- or consumer-product-orientation will be met with comparable success is still in doubt. And on the other hand, on the shop-floors that must be the real front-line of any serious effort to attain greater productivity and competitiveness, no comparable support-system exists. A formidable set of barriers instead remains in place, severely retarding any comparable trend toward self-improvement and involvement in the innovation process by our industrial labor-force at large.

This is a strikingly bimodal system, in other words, although its two components are not equally visible. Those of us who usually represent the U.S. at international conferences like this one, and who in our publications articulate the understanding of basic American institutions and character on which colleagues in other countries largely depend, are drawn almost exclusively from the academic-professional-entrepreneurial component. The claims of freedom of action we make for ourselves, and perhaps even the preeminence we attach to individualist values, need to be taken with corresponding skepticism before they are extended to the whole of the body politic. To illustrate the extent of the resultant deviation, it may be useful briefly to consider the larger milieu of public acceptance within which the American scientific community has been able to operate freely for two generations.

Daniel Yankelovich (1984) has attributed -- correctly, in my view -- no small part of the success of American science to an 'social contract' with the larger society, more or less underwriting its own autonomy and creative separateness. Sheltered by this unwritten agreement, he argues, is considerable scientific exceptionalism with regard to prevailing norms and values. There is a strong popular predisposition, for example, to regard Truths as fixed and given, while the accepted and prevailing scientific posture is one that views essentially all findings are provisional and contingent. Similarly, it is the stance of science (though there are practical limits) to accept controversy and a lack of consensus not only as a tolerable but as a normal state. We think of solutions to problems as generating not truths but a cascading selection of new problems. This stance, too, is foreign and counter-intuitive to the general American populace. Yet its use by scientists as an operating credo is accepted even without being understood.

Another aspect of our remarkably extensive and loosely woven grant of autonomy applies not merely to science as a whole, nor only to groups of scientists, but to individuals. It is the freedom in principle, obviously qualified by considerations of funding and institutional setting, to work on 'discovered' as opposed to 'presented' problems. Basic research, which is what research universities profess to be all about, involves precisely this substantial degree of individualized control over the direction, scale, methodology, and pace of investigations.

Or further, consider science's freedom in drawing its boundaries. To be sure, there is some public fuzziness over who is and is not a scientist. But except in some domains of social sciences where relevant (if anecdotal) public experience is nearly universal, and in a very few other disputed areas like those involving creationism and opposition to abortion and the use of animals in biological experimentation, there has been little external disposition to question the line of separation between science and non-science that the scientific community thus has been able to draw essentially unilaterally.

In conclusion, these "contextual" and hence somewhat unfocused remarks need to be related more directly to the issues immediately before this conference. In my view, the American problem with lagging industrial competitiveness seems unlikely to be resolved if we regard it as a matter to be dealt with at the level of individual firms alone, whether by industrial financiers, managers, or research directors. Among the more general failings to which individual firms cannot easily frame an adequate response are our lack of flexibility in quickly developing commercially successful products out of basic discoveries, and our failure also to match our competitors in implementing innovations, improvements and cost reductions at the factory floor level. These difficulties reflect the growing inadequacy of the bifurcated approach to education which we have not yet found the means, or will, to abandon.

All of our labor force, not just our managers, scientists, and engineers, should be regarded as part of a single pool of human resources. Any real and permanent improvement in our competitive stance requires greater readiness to delegate decision-making authority to individuals and groups at the operational level, and greater *national* investment in education and training that targets the whole of our workforce and not just its managerial and scientific elite.

References

Berger, S. et al. 1989. Toward a new industrial America. *Scientific American* 260/6: 39-47.
Cohen, S. and J. Zysman 1988. Manufacturing innovation and American industrial competitiveness. *Science* 239: 110-15.

Commission on the Skills of the American Workforce 1990. Report of the National Center on Education and the Economy.

Hounshell, D.A. 1984. From the American System to mass production, 1800-1932. Baltimore: Johns Hopkins University Press.

Hindle, B. 1981. Emulation and invention. New York: New York University Press

____ and S. Lubar 1986. Engines of change: The American industrial revolution 1790-1860. Washington: Smithsonian Institution Press.

Jeremy, D.J. 1973. Innovation in American textile technology during the early 19th century. *Technology and Culture* 14: 40-76.

MIT Commission on Industrial Productivity 1989. Made in America: Regaining the productive edge.

Rosenberg, N. 1970. Economic development and the transfer of technology: Some historical perspectives. *Technology and Culture* 11: 550-75

____, 1981. Why in America? Pp. 49-61 in Mayr, O. and R.C. Post, eds., Yankee enterprise: The rise of the American System of manufacture. Washington: Smithsonian Institution Press.

Sawyer, J.E. 1954. The social basis of the 'American System' of manufacture. *Journal of Economic History* 14: 361-79.

Stapleton, D.H. 1987. The transfer of early industrial technologies to America. American Philosophical Society, *Memoir* 177. Philadelphia.

Wright, G. 1990. The origins of American industrial success, 1879-1940. *American Economic Review* 80: 651-68.

Yankelovich, D. 1984. Science and the public process: Why the gap must close. *Issues in Science and Technology* 1: 6-12.

THE DISTINCTION BETWEEN SCIENCE AND ENGINEERING

Bryan Pfaffenberger, University of Virginia

During the past week, participants in the Conference on the Influence of Japanese/American Cultures on Technical Innovation in Advanced Materials have explored a unique opportunity to look beyond our disciplinary concerns, and to try to interpret technical and scientific activities in a broad, cross-cultural framework. If the Conference has done its job correctly, the results should be both revealing and disturbing; they should highlight the unspoken, taken-for-granted assumptions that underlie our attitudes about scientific and technological activity. What is doubly fascinating about the Conference has been that we ourselves, through our speech and other actions, have perhaps illustrated some of the underlying themes that differentiate technological activities in the United States and Japan.

Of the many contrasts that seem to have emerged, I want to concentrate on what I think to be the most significant, the relationship between scientific and technological activities. What seems clear is that in the United States, scientific activities are differentiated from, and evaluated more highly, than technological activities. And equally clear is that in Japan, the distinction, and the differential evaluation, are by no means so clearly evident. In these comments, I want to explain briefly why this might be so, and what its implications might be for the management of technological innovation.

In Anglo-American engineering, the distinction between scientific and technological activities is closely related to the Victorian and post-Victorian professionalization of engineering (Bledstein 1975, Shapin 1989). To be a professional, and to lay claim to a social and economic status that is commensurate with professional training, one must have mastered an esoteric, abstract knowledge about the world--an explanatory and synthetic knowledge that allows the professional to peer beneath the buzzing confusion of reality to perceive the underlying truths and immutable laws. This knowledge is difficult to master, but it is objective and far superior to mere opinion or skill. It is the possession of this type of knowledge that justifies the high standing and compensation given to attorneys, physicians, and other professionals.

As engineers struggled to enhance their social standing in the late nineteenth century, they came to emphasize the application of science to engineering problems. They did so, to be sure, in large measure because scientific knowledge and mathematical reasoning were capable of increased control over the natural world, but at the same time this knowledge and reasoning style could be represented as fully akin to the kind of knowledge possessed by other, well-rewarded professionals. To be a professional engineer, in short, was to favor abstract knowledge and mathematical reasoning over the approach of the "mere technician," whose skills were tactile, visual, rooted in specific materials, and learned by experience. By the twentieth century, the distinction between "professional engineers" and "technicians" had been institutionalized in a pecking order of undergraduate engineering programs, with professional engineering taught at universities and technical skills taught at minor "tech and ag" schools.

The distinction between the high status, professional engineer and the low status technician was reinforced by the anti-industrial ethos common in universities by the late nineteenth century (Wiener 1981). In England, the universities became the last bastion of an aristocracy that had been completely marginalized by the industrial revolution, and in the aristocracy's defense the humanities departments of universities created and elaborated a mythos of the pre-industrial past, a past of Faustian depth and noble virtue that had been ruined by the crass commercialism and the environmental degradation wrought by the industrial revolution. In the face of this anti-industrial (and tacitly aristocratic) ethos, a person's interest in, and involvement with, technical matters and capitalism could only be seen as a sign of inferior breeding and taste. Although America lacks an aristocracy, its universities have nonetheless echoed the anti-industrial ethos of Oxbridge, introducing strong pressures for a concentration in engineering curricula on abstract knowledge and math (and away from experientially-learned, tactile and visual "technical" skills).

In the mid- to late-twentieth century, the distinction between professional engineering and technical skills has been deepened by the rise of the modern research university, in which promotion and status depend heavily on the conduct of scientific research, the publication of scientific journal articles, and the acquisition of scientific research grants. Funding agencies have often expressed a preference for basic research of a markedly scientific character, rather than "technical" projects aimed at resolving specific technical difficulties that would probably not have implications beyond a limited sphere. To succeed in the research and publication game, one had to prove one's mettle by demonstrating virtuosity in abstract knowledge and mathematical reasoning. These activities had come to define professional engineering and applied science as a distinct and justifiable area of activity within the modern research university, and one could de-emphasize them only at the peril of opening questions about the desirability of maintaining a "technical" program at the research university level.

At the Conference, the Anglo-American distinction between science and engineering manifested itself in several ways. Some U.S. members of the group, and in particular those with strong affiliations to the professional (scientific) model of engineering, were sharply critical of attempts to locate scientific and engineering practice in a broader, social and cultural context. That in itself is hardly surprising. The scientific ethos holds that, when science is done properly, social and cultural bias are eliminated. Indeed, when one scientist criticizes another, the criticism often focuses on the bias that explains why an inappropriate model or analytical approach was chosen. When the theme of one's training has been the acquisition of an esoteric, unbiased knowledge, it can be threatening to be informed--especially by someone doing "soft, qualitative work"--that these forms of knowledge might very well rest on tacit notions that are socially or culturally supplied.

When U.S. scientists and engineers described their work, the presentation style they chose seemed to echo the distinction between professional engineering and technical skill. They clearly delineated the boundaries of scientific disciplines, and their presentations tended to emphasize a verbal, linear history, with relatively few illustrations. When illustrations were used, they tended to show linear, cause-effect relationships that rarely involved more than two variables. In describing their activities, the U.S. researchers seemed less concerned with the materials or artifacts with which they were working than they were with the issue of credit--just who did what first. Emphasizing the theme of academic confrontation and argument, the U.S. participants--engineers and sociologists included--seemed more than willing to engage in conflict-oriented debate as a means of isolating and dealing with the significant issues.

I am not familiar with the literature on the history of science and technology in Japan, but it was obvious during the conference that Japanese culture (for reasons unknown to me) does not distinguish "professional engineering" and "technical skills" quite so sharply. No palpable distinction was drawn between university science professors and corporate engineers, who seemed to share much in common and indeed to perceive themselves to be involved in a common enterprise. There seemed little sign of the traditional academic disciplines in the Japanese researchers' talks; on the contrary, they showed an artifact or materials orientation, concerning themselves almost by definition with an interdisciplinary field of study that encompasses anything and everything that could conceivably be relevant. These fields, moreover, seemed to be defined from the beginning with an artifact-centered, problem orientation: we want to build this, so we've got to solve the following problems. In describing the evolution of these research areas, Japanese researchers seemed to place less emphasis on just who did what when, and much more emphasis on the growth of knowledge generally about a set of materials, and how the more vexing challenges have been overcome through patient, dedicated work. One gets a sense that the Japanese have little problem forming consensus communities around such problem areas. They didn't seem to have as much to lose if the research didn't work out--unlike their American colleagues, who often seemed to suggest that their scientific careers would be ruined if they made a disproportionate commitment to an unpromising area.

Other differences stood out during the week of presentations. The talks given by the Japanese researchers employed significantly more visual images, and these images differed in quality as well as quantity from their American counterparts. Visually attractive and complex, they often portrayed many variables in an interacting, organic system replete with feedback linkages. Judging from the illustrations, too, the Japanese seem to take more aesthetic pleasure in materials and artifacts; many slides depicted materials or machines lovingly, as if they were artistic as well as technical feats.

Judging from what I have seen this week, I would offer the following very provisional conclusions. I would suggest, and I hope I do not do so because of my cultural biases, that the American approach--conflict oriented, and placing science higher than technology-- would seem very productive of fundamental scientific insights, and rather less effective when it comes to creating an interdisciplinary consensus network oriented to the solution of an artifact-based problem (e.g, "How do we get this mag-lev train to work?") The American approach seems to underlie the evidence that Americans do better in science, but are less successful in devoting themselves to the start-to-finish construction of a high-quality, well-designed technological system or artifact. Precisely the opposite could probably be said of the Japanese; Japanese scientists sometimes complain that the environment for pure scientific research is better in the United States, and Japanese achievements in technology hardly need comment.

And what is to be learned from this comparison? When we think about the influence of culture on technological innovation, it is very easy to fall back into the trap of cultural determinism: we have to do it this way because we are American, or Japanese, or Sri Lankan, or whatever. What this week suggests is that we actively construct and reproduce our cultural worlds even as we do mundane things like sit around and talk about technological innovation! What underlies our actions are not inflexible cultural beliefs and values that have been forever stamped in our minds, but rather behaviors that, so long as we are unaware of them, continue to produce and reproduce the cultural world around us. Contemporary discourse theory holds that we actively produce the social and cultural world around us through the choices we make as we engage in discourse (Clark 1990), and this week's meeting would seem to supply ample evidence in support of this contention.

A significant corollary of discourse theory is that, to the extent that we structure our cultural world through our interactional choices, we are reponsible for them. But responsible action can follow only on awareness. What is so valuable about cross-cultural interaction is that it brings these culture-shaping behaviors to the fore, where we can finally examine them, and make rational choices and judgments about their desirability. The Japanese predilection for multi-dimensional illustrations, for instance, may very well stem from some facet of the Japanese world view that an American could not grasp without many years of residence in Japan. But it doesn't take too much work to learn how to make multi-dimensional graphs, and to start thinking about a field of expertise as an interdisciplinary research area, focused on the creation of a specific artifact, that is to be tackled by a solidary, noncompetitive team that isn't out to vanquish the others in the game of publication and grants. Precisely that kind of systems-level, group-oriented thinking, and a kindred goal of solving artifact-related problems in an interdisciplinary context, seems to be characteristic of some of the greatest feats of American technological innovation, such as the fabled Alto computer produced in the 1970s by Xerox Corporation's Palo Alto Research Center. Tellingly, it also seems to characterize the most successful American innovations discussed at the Conference, silicon nitride technology, which was from the beginning focused on the creation of a ceramic engine with considerable support from a government agency. World views aren't easy to change, but Americans have shown plenty of ability to learn and use new designs for the social organization of technical activity, and new ways of attacking pressing technological problems. Cross-cultural interaction and the sharing of perspectives, such as has been made possible by this Conference, are invaluable in providing rich new opportunities for this learning to occur.

References

Bledstein, Burton J. 1975. The Culture of Professionalism: The Middle Class and the Development of Higher Education in America. New York: W.W. Norton.
Clark, Gregory. 1990. Dialogue, Dialectic, and Conversation: A Social Perspective on the Function of Writing. Carbondale: Southern Illinois Univ. Press.
Shapin, Steven. 1989. "The Invisible Technician," American Scientist 77:554-563.
Wiener, Martin J. 1981. English Culture and the Decline of the Industrial Spirit, 1850-1980. Harmondsworth, U.K.: Penguin Books.

THE MANAGEMENT OF KNOWLEDGE AND THE ROLE OF CULTURE IN THE R/D ENTERPRISE

Rustum Roy, The Pennsylvania State University

Locally generated research is a rapidly decreasing factor in the useful management of knowledge which can lead to both invention and innovation. New roles for the university are seen in the gathering of information and knowledge by all means, negentropic organization of the same, and providing for systematic networking for groups of user industries.

Negative cultural factors operating in the U.S. R/D culture include: overemphasis on competition; inadequate cooperative networking with other groups within/outside the country; endemic neglect of the literature; gross, persistent structural bias against interdisciplinarity in universities; failure to understand the "system."

Introduction

In Figure 1, I reproduce from earlier work (1) the relationships among often confused terms ranging from data to wisdom. In the world of technology, the hierarchical level most involved is "knowledge." Most so-called university-industry technology-transfer is grossly misnamed as first pointed out by J. R. Johnson (2). At best, it is knowledge transfer. Figure 1 is my first attempt to chart knowledge flows in the R/D process. For the present purposes only a few brief relevant points are made from the flow chart.

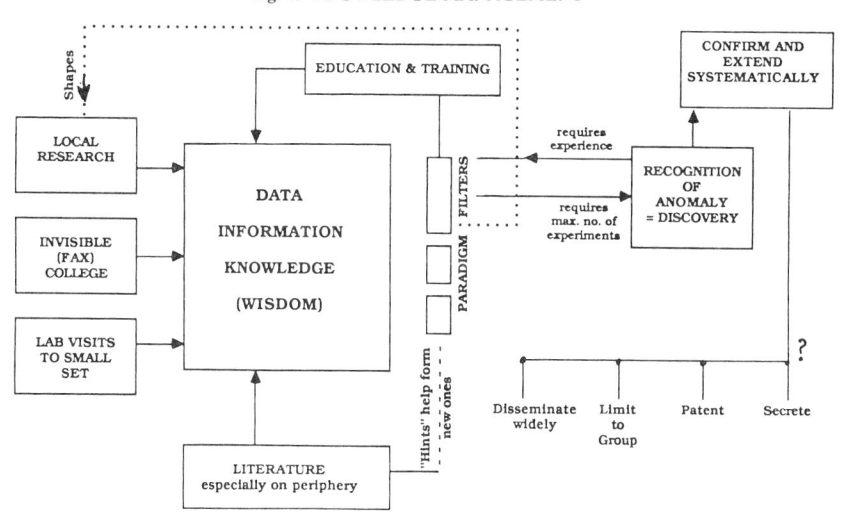

Fig. 1. KNOWLEDGE MANAGEMENT

1. A mixture of data, information and knowledge is constantly being scanned by our brains. In the R/D mode they are searching for inventions, i.e. anomalies or novelties or derivations from the previous *pattern*. Such patterns or paradigms constitute a high hierarchical assemblage of knowledge and experience from many sources and intuitive interconnects.

2. This search involves filtering the new knowledge (signals) + associated noise (most unorganized data) through sets of filters (patterns, paradigms, models) to see what does NOT fit. The probability of finding an anomaly or invention is directly proportional to the number of filters (models) one can insert.
3. The patterns themselves have been formed in part by education (K through graduate school) and are added to by life-long learning. Experience is essential in enriching our store of such models or paradigms.
4. The input to the filters comes mainly from four sources: (a) one's own (local) research; (ii) research results from the small set of colleagues who constitute our invisible college, i.e. leaders and labs we know and respect; (iii) research results reported from all over the world at meetings; (iv) the formal scientific literature.

We may compare, very generally, the R/D culture of the U.S. with those of Europe and Japan in the following comments.

	U.S. (1990)	Japan/Europe
Paradigm-richness	Usually poorer, due to lack of conceptual clarity at college level, and disciplinary narrowness of *all* universities. Experienced (>60) years of age) managers leaving industry will make situations much worse.	Team strategy helps to provide more models.
Literature	Almost totally ignored (except from invisible college, i.e., paradigm sharers). Most U.S. scientists are now proud to say they have *not* read the literature. "No time to read; busy writing proposals." Yet no effort is being made by government or professional societies to change this. No time to read; busy writing proposals.	Much more systematic use of literature. It shows in papers and every contact with Japanese or European science.
Local Research	Often narrow, instrument-intensive, discipline focussed, not problem-solving. Problem identification is often done by following fashions set in the media; or upgrading Ph.D. theses.	Problem-driven. Consensual, selection of targets by a group process inconceivable in the U.S.
Professional Meetings	Incredible, unsustainable increase over the last 10 years. Every paper is given and re-published 5-10 times. Very inefficient knowledge transfer.	Starting to follow U.S. models.
Lab Visits	After 25-year lag, starting to copy Japanese.	Since 1955 very systematic use of lab visits to garner new ideas and results.

Fig. 2. Three Discoveries (Past)
Comparative Research History & Strategy

	"Ferroelectricity" TiO_2-$MgTiO_3$ $BaTiO_3$	"Superconductivity" $LaBaCuO_4$ $YBa_2Cu_3O_7$	"Ferromagnetism" $Sm_3Ga_2Ga_2 3O_{12}$ $Y_3Fe_2Fe_3O_{12}$
U.S. Position at Time of Discovery	Industry in commanding position.	Industry v. weak universities v. weak in relevant fields.	Industry far ahead. Still small payoff.
Technological Prospects	Drop in replacement into existing products.	~No products. Short-term prospects unclear.	No impact on ferrites. Memory applications trivia. Microwave - minor.
Technological Reality	Enormous capacitor industry.	Major impact improbable in 10-15 year frame.	Impact minor.
U.S. Position at Time of Discovery	Industry in commanding position.	Industry v. weak universities v. weak in relevant fields.	Industry far ahead. Still small payoff.
Technological Prospects	Drop in replacement into existing products.	~No products. Short-term prospects unclear.	No impact on ferrites. Memory applications trivia. Microwave - minor.
Technological Reality	Enormous capacitor industry.	Major impact improbable in 10-15 year frame.	Impact minor.

To examine these claims it is valuable to compare the historical record of recent materials inventions. This I have done in Figure 2 following the history of ferroelectrics, ferromagnetics and superconductivity. *These charts show beyond any doubt that no rational, thoughtful processes were involved in the choices of problems by scientists or allocation of resources by governments.* Figure 3 shows the response of the materials innovation community to three major materials discoveries, two of them being discussed at this meeting—superconductors and diamond films. It is an interesting confirmation of the point made in Figure 3 (that policy attention is a function of hype) that Lanxide, the new material innovation furthest along the innovation ladder is not being discussed at this meeting.

Comments on Present Situation

The key difference between the two cultures—U.S. and Japan—has very little to do with science or engineering. It is to be found in the overall attitude to planning, and achieving

Fig. 3. Technology Opportunities Confronting U.S. Policy Makers Three Discoveries (Starting)

	MRS DEC. '85 Lanxide Announced	MRS DEC.'86 High Tc Superconductor Confirmed	MRS DEC. '87 Diamond Films Announced and Confirmed
Advertising Hype	Strictly avoided. Technical press missed it.	Enormous.	Very minor. Technical press picked it up.
Impact on Research	Not one proposal submitted to NSF, DOE in 2 yrs.	Cast of thousands working outside their field.	Considerable interest. Minor initiatives.
Impact on Agencies	Only DoD response. Ignored by ALL others.	Measured, studied response. Not overwhelmed.	DoD responds. Others ignore.
Impact on Industrial Research	Major potential. Replacement and new technologies.	Unknown. Most analyses dubious.	Major potential on wide variety of existing products.
U.S. Interest	100 % U.S.	U.S. role - minor	U.S. role - v. small
Technological Position	>1200 patents filed by one U.S. company.	Patent positions - appear irrelevant so far.	Japanese, far ahead in tech., USSR in science.
Attention from U.S. Policy Makers	1	1000++	10

harmony and consensus among all involved. When it comes to technology development and national policy the differences become even sharper. If we refer again to Figure 1 in connection with materials synthesis and processing one finds the following:

If one looks at the relative weights given to the *inputs* into the knowledge bank, there is very clear agreement that American labs put much more weight on local research than their Japanese counterparts. Indeed they do it almost to the exclusion of other inputs. The N.I.H. syndrome is rampant. Openness to other sources is minimum. The only exception is the invisible college or old boy network. Unfortunately even this valuable interacting group process is being poisoned by excessive competition for funds. The Japanese read the literature much more thoroughly and they become masters of "knowledge-scooping" by visits to labs all over the world. American researchers tend to stay narrowly focussed for decades. Good Japanese labs systematically arrange for many to move *contiguous* new problems every 5-10 years. The paradigm richness is much greater in Japan both because of the above factors and the team efforts. The Japanese respect for "sabi,"—wisdom and quality and value acquired by age—helps in the use of experience in invention recognition and problem selection. They also know the probability trees for payoff. Also the U.S. university world, which started with complete openness in disseminating its results, now is struggling with the bottom-right issue of "how much to divulge to whom." This will be a continuing problem.

Policy

We have read interminable reports on the Japanese R&D process and system. The U.S. knows the system well. We know full well that their carefully managed system is working and that it is beating the pants off our laissez-faire effort of absolutely no policy on technology. But the U.S. "fundamentalist" political attitudes from 1980 onwards have made any change impossible. Hence further detailed analysis is, in my view, a waste of time. We should only address the question of how the technical community can get political attitudes changed.

References
R. Roy, *Experimenting with Truth,* Pergamon Press, NY (1979).
J. R. Johnson, Science, _____, _____.

AN ETHNOGRAPHIC APPROACH TO THE CONFERENCE: AN ANTHROPOLOGICAL PERSPECTIVE

Eric Poncelet, Department of Anthropology, Univeristy of Arizona

There are a number of elements which distinguish this symposium from other technologically-oriented conferences. First and foremost, the focus of the symposium was primarily on cultural (rather than scientific, economic, or political) influences on technological innovation. Furthermore, the approach was cross-cultural, using two different countries (Japan and the U.S.) as the basis for comparison. Finally, in addition to the cohort of scientists, engineers, and managers of innovation in attendance, the conference was also attended by a number of anthropologists. The purpose behind inviting anthropologists to the conference was two-fold. First, as experts in the study of culture, they were uniquely qualified to present material concerning two major foci of the symposium: a description of the culture-technology nexus, and a comparison of Japanese and American cultures. Second, as practitioners of ethnography, they were asked to attend the meetings as outside observers. The purpose of this short essay is to briefly describe some of what transpired at the symposium from an anthropological point of view in the hope of further illuminating the influences of culture on the innovation process.

A few words describing who anthropologists are and what they do are appropriate at this point. Anthropologists are social scientists who share a common interest in culture. Culture refers to "the patterns of behavior and belief common to members of a society" (Spradley & Rynkiewich 1975:7). One of the primary techniques employed by anthropologists to study culture is "ethnography". Ethnography may be defined as "the firsthand, personal study of a group of people" (Kottack 1987:58). However, ethnography is more than simple, objective observations of activities and behaviors. Ethnography also involves interpreting the subjective meanings of these activities and behaviors (i.e. what they mean to the participants themselves). Consequently, ethnography has also been described as "thick description" (Geertz 1973:5) to distinguish it from the purely objective studies of the natural sciences. Two important ethnographic techniques used by anthropologists in the field include observation and participant-observation. Observation is performed when the anthropologist directly observes interpersonal interaction and records what is seen as it is seen. Participant-observation occurs when the anthropologist takes part in the events being observed and analyzed.

Both of these techniques were utilized by various anthropologists attending the symposium. Some of the resulting observations proved to be both interesting and pertinent to the goals of the symposium. It is a goal of this paper to describe some of these observations and their subsequent interpretations. As it would be beyond the scope of this paper to describe the entire conference, only a brief (though representative) sample will be discussed below. The discussion will proceed in three parts: a description of the conference setting, some ethnographic observations made by anthropologists, and discussions/interpretations of these observations.

Setting

The conference setting may be briefly described by examining who attended and how it was structured. The participants in the symposium came from wide cultural and occupational backgrounds. Both Japan and the U.S. were represented by representatives from industry (corporations), government (national labs), and universities, all with experience in the innovation of oxide superconductors, synthetic diamonds, or silicon nitride structural ceramics. Also in attendance at the symposium were experts in the management of innovations. Finally, social scientists were represented in the form of anthropologists and sociologists.

The structure of the symposium consisted primarily of a number of scheduled formal presentations followed by discussion and comments. The presentations and discussions were performed solely in English. The discussions were basically open-floor discussions characterized by voluntary participation and little solicitation of comments.

Observations

For the purposes of illustrating the types of observations made by anthropologists at the symposium, four specific observations will be discussed. These include observations of group-formation tendencies, general communication characteristics, the actual presentations, and the ensuing discussions.

There was a definite distinction between the Japanese and American participants concerning the manner in which they organized themselves into groups. Many of the U.S. participants tended to compartmentalize themselves into diametrically opposed groups. Thus, many of the Americans aligned themselves as scientists vs. engineers, natural scientists vs. social scientists, or academicians vs. business people. However, this compartmentalization did not restrict multiple group membership. It was common for U.S. participants to consider themselves as having ties to more than one group (e.g. ties to government and industry, or to the U.S. and Japan). Furthermore, among the Americans, there was often a sense of competition within and among groups. For instance, there was competition between the three material science subfields concerning their relative successes and importance. The Japanese participants, on the other hand, did not follow the group-formation tendencies pursued by the Americans. Instead, the Japanese maintained a more unitary appearance. They were less likely to compartmentalize themselves into groups and more likely to present a harmonious appearance.

There were also a number of differences between the Japanese and American participants concerning their communication styles. In general, the Americans tended to speak more frequently and at greater length than the Japanese participants. The U.S. style of communication tended to be distinguished by voluntary participation, rapid exchanges, and frequent interruptions. By contrast, the majority of the communication from the Japanese participants came when solicited from U.S. participants and in regard to specific issues. There were few if any Japanese participants interrupting other participants.

The third area of observations concerns the formal presentations made during the symposium. Many of the presentations made by the U.S. participants tended to be historically oriented. They focused on the historical development of the specific technologies with special attention paid to key individual actors and organizations. Credit was paid to those who made critical contributions. Relatively less attention, however, was paid to cultural influences. In the Japanese presentations, greater attention was placed on the innovation process and the key barriers which needed to be overcome. There was less separation of technological and social elements in the Japanese presentations, and they were more frequent in indicating cultural influences.

The final area of observations concerns the informal discussion sessions which ensued the presentations. There were relative differences between the Japanese and American participants in regards to the occurrences of disagreement and the use of criticism. In their comments to each other, it was not uncommon for the American participants to disagree with one another or to offer criticisms as part of the discussion. These types of comments were generally presented as individual opinions or based on individual experiences. There was a greater tendency for the American participants to focus on areas of disagreement rather than areas of agreement. Among the Japanese participants, there was less disagreement or contradictions offered as comments to the presented material. The Japanese participants presented descriptions of their systems or experiences (and even admitted problems) but never criticized each other or their institutions.

Discussion/Interpretation

Undoubtedly, much more took place at the symposium than has been described in the above, rather simplified observations. Furthermore, many influences besides cultural ones obviously played a role in determining the process of the conference.[1] Nevertheless, as Japanese and American cultural influences were the focus of the symposium, they will be my focus in discussing and interpreting the above observations. Three Japanese-American cultural distinctions will be discussed.

First, differences exist between Japanese and American styles of group formation. In general, Americans tend to be more individually-oriented and the Japanese more group-oriented. Furthermore, Japanese individuals typically restrict their membership to only one group, while in the U.S. multiple-group membership is both accepted and encouraged. Second, there are differences in communication styles. In the U.S., a relatively higher emphasis is placed on language as "the" means of communication. Furthermore, the basic American assumption concerning language is that "what you are is what you say." In Japan, the basic assumption is that "what you are is how much you say as well as when you say it" (Kii 1985). Thus, Americans and Japanese attach different levels of importance to what is said, how it is said, and when it is said. In regards to the directness of interpersonal communication styles, Americans value openness, egalitarianism, and "honesty". By contrast, the Japanese tend to see a too direct style of communication as being disrespectful and discourteous. Finally, there are differences in styles of interpersonal relations. American interpersonal relations tend to be dependent on individual principles and are hence more adversarial or confrontational. Japanese interpersonal relations, on the other hand, tend to be characterized by a general avoidance of confrontation and an emphasis on harmony and consensus.

Based on this brief discussion, it is apparent that an understanding of Japanese and American cultural differences is important in understanding the events which took place at the symposium. Indeed, many of the general cultural characteristics discussed as influential on technological innovation also played a role in the actual process of the symposium. If the symposium is itself considered as an innovation, then the form finally assumed by the symposium may be seen as both directly and indirectly affected by the cultural assumptions brought and held by the participants.

References

Geertz, Clifford
    1973 The Interpretation of Culture. New York: Basic Books, Inc., Publishers.
Kii, Toshi
    1985 Japanese and American Communication Styles. The Japan American Society of Georgia, News Letter, Vol. 5, No. 4, Atlanta, GA.
Kottack, Conrad Phillip
    1987 Cultural Anthropology. New York: Random House, Inc.
Spradley, James P. and Michael A. Rynkiewich
    1975 The Nacirema: Readings on American Culture, J. P. Spradley and M. A. Rynkiewich (eds.). Little, Brown and Company, Inc.

---

[1] For example, the fact that English was the primary language of the conference had a direct influence on communication style.

CONFERENCE DISCUSSIONS

W. D. Kingery, University of Arizona

Perhaps the most valuable aspect of the conference for the participants were the many discussions, both in session and extra-session, nucleated by the formal presentations. We do not have a complete transcript of these discussions and in any event, many were cumulative in their effect. It is not feasible to reconstruct the whole meeting. Separating out each strand of thought would leave us with none of the fabric. We have collated some flavor of the discussions in relation to ten frequently occurring topics, all of which are overlapping.

CHANGE

It might be inferred from some of the discussions that Japanese and American cultures and technology are stable islands which can be dissected and compared at one's leisure. The opposite view was the true sense of the conference. There was general accord that changes in culture and changes in technology are rapid and equally as significant as any stable differences. The development of global transportation and instantaneous worldwide communication provides a continuous series of new images, styles and impressions that sweep around the world almost instantly and have a strong impact on technological innovation. Professor Longacre described how changes in the social and economic context of the Kalinga, a rural tribe in the Philippines, have been a driving force for innovation in their ceramic technology. In Japan it was pointed out that the younger generation contains many *shinginrui* who are not accepting the classical culture of hard work and company family that is the standard for Japanese over thirty. Most young Japanese lie somewhere in a broad spectrum between the *shinginrui* and the classical *salary man*. In the United States there is a massive cultural transformation as Blacks, Hispanics and Asian immigrants become an increasing factor in our society. Half of the graduate students working on technical Ph.D.'s in the United States are foreign and it is problematic to talk of American graduate student behavior under these conditions.

There has been a globalization of manufacturing. For a longer time there has been a globalization of science and there is rapidly developing a globalization of engineering. On differing scales at different places in the world, there tends to be a globalization of culture. There was a consensus that the influence of cultural *change* on technological innovation is important and would be a worthy topic for some successor conference.

COMMUNICATION

In many of the discussions of innovation in relationship to the nature of feedback loops between designers, manufacturers, engineers, customers, the relationships between manufacturers and consumers and processes of technology transfer, the importance of communication was emphasized. Questions came up in various discussions about the effectiveness of communication within cultures and subcultures and also between cultures and subcultures. One problem in evaluating communication is the different concepts and images evoked in different cultures by the same terminology. During the conference it became clear that the Japanese involved in technological innovation think more in terms of artifacts than the Western participants who are more related to research, scientific findings and publication. A conceptual difference developed by Professor Netting is the idea of *efficiency*. Japanese agriculture is extremely efficient in terms of output per acre, but not particularly effective in terms of output per unit labor. In contrast U.S. agriculture is the most efficient in the world in terms of output per unit of labor but relatively low in terms of output per acre. In general, there is no Japanese equivalent to the American developments in efficiency engineering, making the worker's role specialized and automatic as a part of Frederick Winston Taylor's "scientific management." Workers at all levels in Japan are more generalists and less specialists with a concomitant decrease in the "efficiency" of labor utilization. Perhaps this is related to the very strong emphasis in Japan on robotics and the investment in automatic manufacturing machines.

Participants at technical meetings are well aware of differences between the American and Japanese presentations and discussions. American presenters tend to be full of confidence, emphasizing differences from previous speakers or results, explaining how their results are new and exhibiting an eagerness to expand the horizons of the audience. Americans are good at accepting criticism, at separating the self from the idea. They are quick to express opinions and eager to enter discussions and put forth new ideas, often not well considered, at meetings of all kinds. In contrast the Japanese tend to be much more self effacing in their technical presentations and hesitant about putting forth criticisms and offering new thoughts which have not been thoroughly considered. It has been suggested that the American approach is an exhibition of individualism and that the Japanese behavior is part of a *situational* approach to communication in which a harmonious relationship, the avoidance of conflict is seen as a measure of sincerity which should be admired. The Japanese are much more likely to identify self with the idea and consider that direct criticism is a form of personal rejection. Where Americans might see the opposite of honesty as dishonesty, Japanese might see it as harmony. Self oriented behavior and communication are seen by Americans as signifying positive attributes such as independence and individuality signifying control and rationality while the Japanese might see the same behavior as eccentric, selfish, disruptive or insensitive. Japanese prefer modes of communication which Americans may see as ambiguous, inclusive and irrational.

Much has been made of the fact that American communication is verbal in which being direct, clear, explicit and to-the-point is admirable. In Japan communication is much more ambiguous with an emphasis on gesturing, intonation and protocol that is difficult to translate and even difficult for foreigners to understand. In Japanese culture there is a major distinction between the inner group of an organization, *ouchi* and the outer world, *soto*. Most Japanese very much want to belong to, act like, and deal with the *ouchi* but communicate in a very different way and sometimes seem almost oblivious to the *soto* people, including *gaigin*. But we're also warned not to make too much of these distinctions in technological interactions.

Other deep differences in cultural conceptions can lead to difficulties in communication. As has been mentioned, *individualism* is valued in American eyes as signifying independence and self control, but may be more often equated with disruption, insensitivity and selfishness in a Japanese view. *Democracy* in American eyes is majority-rule, the result of confrontation and choice. America goes to war on the basis of a 52-48 vote. In contrast, Japanese see democracy as a process of obtaining consensus of as many people as possible, of trying to get the 48 to join the 52. For Americans, *freedom* is to be presented with options and selections and choices that one may choose from; in Japan it is more a matter of reaching harmony within one's self and one's environment.

The American emphasis on the straightforward spoken word may be related to the broader distribution of attributes, education, family, culture, language and national origins amongst its population. In Japan most traits have a much narrower distribution so that an ambiguous mode of communication can be effective.

JAPANESE AND AMERICAN COMMUNICATION STYLE[*]

Toshimasa Kii, Georgia State University

Personality differences among individuals aside, there appear to exist discernible differences in the communication styles of Japanese and Americans. Stereotypically or not, Japanese see Americans as open, frank and expressive and Americans often consider Japanese to be reserved, cautious and indirect. Some of the behavioral differences can be explained by cultural values.

Basic Difference in Styles

Self-assertiveness behavior: Americans consider a self-assertive communication style good because it symbolizes individuality, independence, autonomy, and even competency, which they find valuable assets in people. Japanese, on the other hand, often perceive a self-assertive communication style as arrogant, insensitive, egocentric, and even disruptive.

It is a matter of degree, of course. Too much self-assertive behavior may be considered obnoxious among Americans. And well-controlled self-assertive behavior yields a manner of comfortable self-confidence which is considered positive by Japanese. However, what is considered an acceptable self-assertive style of interpersonal communication among Americans often becomes an arrogant and insensitive manner among Japanese. Why?

It seems that there are two different ways of evaluating, examining, and sizing up the people with whom you interact. The American assumption is that what you are is what you say. Under this premise you must talk, interestingly enough so that others will judge you positively. If you are too reserved to speak out, you are often judged to be uninteresting or less sociable, or to have nothing to contribute.

The Japanese premise holds that what you are is how much you say as well as when you say it. In this case you must take into consideration the interacting others—their social positions and demeanors. It is not just what you have to say that is a criterion for judging you. In fact, when and how you say it is more important than what you say. Adjusting your interpersonal communication style to a given situation is of the utmost importance because it is important to show a concern for harmony with and sensitivity to others. Even if you have a definite opinion and idea about whatever topic is under discussion, you will be considered more sensitive and responsible if you are not forthright and eager to convey your distinctiveness.

Directness: Americans tend to be direct in their interpersonal communication styles. In the American value system directness symbolizes such positive traits as openness, egalitarianism, and even honesty. Japanese, on the other hand, tend to see direct style of interpersonal communication as disrespectful and discourteous. They often use an indirect style, as it signifies respect for and polite attitudes toward others. But most of all, it is highly valued among Japanese because it is nonconfrontational. Japanese are highly sensitive to confrontational situations, and they will avoid them at almost any cost. Indirectness certainly softens a potential confrontation which may lead to a conflict. Americans, however, see indirect styles of personal communication as inefficient, manipulative, and even cowardly.

---

[*]Reprinted from The Japan American Society of Georgia, Newsletter, Vol. 5, No. 4, Atlanta Georgia.

Openness: Americans value openness in interpersonal communication, even when the people involved are just getting acquainted. Openness symbolizes trust, kindness, and optimism about the relationship they might develop. This is one thing Japanese are often baffled about when interacting with Americans. Americans are not inhibited in talking about family life, personal matters, and likes and dislikes about work, even with people they do not know well. Japanese often consider this openness shallow, a sign that the speaker is insignificant and even insincere. Why are you opening up to me so fast? I don't even know you. Can I trust this person? Japanese are, thus, cautious in their interpersonal communication styles. Cautiousness signifies, in the Japanese culture, patience, dependability, and sincerity.

Understanding through logic: Japanese certainly aspire to understand through reasoning. But as a basis for interpersonal communication, too much logical reasoning is often considered threatening, confrontational, and argumentative. Japanese tend to base their understanding of people on intuition and a considerable amount of emotionality. They have a tendency to avoid logical argument to achieve a sense of understanding. Americans, on the other hand, try to use logic because it gives them a sense of control and rationality. Intuitive understanding of a situation without this analytical exercise would leave them with a sense of inconclusiveness, acquiescence, and irrationality.

Behavior and Language are Products of Culture

I would like to make a final comment on the utility of language. Language is a tool for communication, but it is also a product of culture. In this sense language delimits the range as well as the content of expression that is permissible in the culture in which the language is used. The English language is basically egalitarian. It does not differentiate among individuals in terms of the range of expressions they may use, because the language reflects the egalitarian value of the culture. The Japanese language, on the other hand, is hierarchical—reflecting the human relationships of a vertical society, and limits the permissible range of expression according to the status of the speaker. In English, what makes interpersonal communication tick is primarily the content of what is being communicated. The primary emphasis and concern in interpersonal communication in Japanese is the manner by which the content is communicated, taking into account the hierarchical order of the people involved.

It is fascinating to observe that Japanese who speak English fluently are quite open, straightforward, and frank when speaking English but reserved, formal and cautious when speaking in Japanese. The language they are using reflects the values of the culture in which it is spoken.

METAPHORS

Brian Pfaffenberger has suggested that the metaphors with which we communicate can tell us a good deal about the underlying cultural assumptions within which we function. During the conference discussions Americans tended to discuss group action in terms of a team of individuals whose object is simply to win. In contrast Japanese speakers use metaphors related to a family structure, even though the groups of men embedded in the company structure are not a family in the traditional sense. A number of the metaphors that have come up in the conference discussions are collected in Table 1.

Table 1. Root Metaphors

	U.S.	Japan
Group	*Team of Individuals: Object;* to win by any means	*Family;* Object is to win through technological superiority while keeping people happy
Motivation	Individual *ego-fulfillment*	Group spirit
Production System	A *machine* that must be controlled	A *social organization* that should grow
Technology	*Economics:* The measure of its success is profit and avoidance of liability	*Culture*: The measure of its success is profit and the enrichment of human experience
Company	*Stage* where the root conflicts are carried out	*Ritual* where conflict directly endangers spiritual well-being
Corporate Structure	*Physical system* that can changed only with great difficulty	*Social system* that should be sensitively and creatively adjusted to new situations
Communication	*Verbal mechanism* to be judged by its efficiency	*Expression* of group identity and solidarity
Anthropology	*Tool* that could prove useful for science and industry	*Map* that locates the human being in the universe

Source: Brian Pfaffenberger with acknowledgements to S. Kline and E. Poncelet

In Japan it's suggested that the development of group spirit, social organizations, the idea of company as ritual, corporate structures as social system are strengthened by Japanese participation in male bonding rituals which lead to work related cohesion and common goal setting not often seen in the American workplace. Similar activities in the United States occur in other aspects of the society. This suggests that when work groups involve many people from a variety of backgrounds and also both men and women, there may be new modes required for obtaining effective harmonious work groups.

THE CONCEPT OF *do*

Kazuo Kobayashi proposes that various aspects of Japanese culture such as religion, education system, family relationships, social system, etc. have an influence on the way of thinking and Japanese attitudes toward research, development and innovation. Among many factors the idea of *do* should be mentioned.

Ju-*do* is now a popular Olympic game. There are many traditional sports and arts which are named something-*do* in Japan. For example, there are Ken-*do* (Japanese swordsmanship or Japanese fencing), Kyu-*do* (Japanese archery), Iai-*do* (sword manner), Jyo-*do* (stick fighting), Karate-*do,* Sumo-*do,* etc. in traditional Japanese sports and Ka-*do* (flower arrangement), Sa-*do* (tea ceremony), Syo-*do* (Japanese calligraphy), etc. in traditional Japanese arts. *Do* means way or road in Japanese and the meaning extends in a wide sense to pursue a way of life or to cultivate one's spirit. *Do* seems to imply a sense of Zen.

Important points of modern sports are to win the game or achieve a goal which come from techniques and fighting spirit as seen in Olympic games. In addition to those points, Japanese traditional sports place great importance on manner, politeness, respect and spirit of harmony and cooperation. This seems similar to old western chivalry. So *do* in sports and arts is a way to cultivate personality, to learn manners, fellow-feeling, thoughtfulness, etc. Therefore, parents in Japan want their children to learn Ju-*do,* Ken-*do,* etc. We can think in a similar way of golf-do, business-do and research-do.

"According to my personal opinion (Dr. Kobayashi), Japanese generally have a tendency to tackle their jobs not only as work and not only for salary but also to pursue a way of life to approach a perfect person. The idea of *do* seems to be present even in the research works of Japanese."

THE COMPLEXITY OF SOCIO-TECHNICAL SYSTEMS

The scientist/engineer participants at the conference sometimes found it difficult to escape from traditional internalist linear histories of successive discoveries and innovation. The conference began with Stephen Kline's discussion of the complexity of socio-technical systems and the inadequacy of a linear research-development-manufacturing-marketing sequence to account for actual system behavior. Time and again participants reiterated the need for considering behavior and innovation in the context of complexity.

Stephen Kline described attending a seminar where it was suggested that since the principles of science were socially constructed, they are therefore relative. That doesn't follow because the principles of hard science only involve a small number of variables and are subject to strict reductionist mathematical tests of verification. That is not true of technological innovation and practice which are activities, not principles. People's behavior and perceptions are integral to the process and create a system of immense complexity in which multiple feedback loops and learning by doing require flexible non-dogmatic and adaptive approaches rather than dedication to fixed principles and procedures.

In general, participants thought that when Americans used charts to illustrate social relationships, they were simple and had definite linear lines of authority. In contrast, Japanese charts were complex and difficult to understand at first sight, because they showed multiple social relationships and many cross-linkages and feedback loops. Also, American descriptions of technical contributions tended to focus on individual achievements. After all, there is a deeply embedded scientific tradition of crediting one's predecessor. The Japanese were perceived as giving more organizitional credit and not mentioning disciplinary lines of development. From another viewpoint the Japanese were seen as more use oriented.

Somehow this developed into discussions of research funding. As discussed elsewhere, Japanese research funding has generally been for a group activity with a relatively long term commitment and is preceded by under the table testing of new ideas and then by extensive consensus building amongst a wide range or different viewpoints. American funding tends to be developed by an individual champion of himself/herself or a team preparing a proposal without much outside interaction. After all, it's their idea, their intellectual property. As a result, many of the Americans saw U.S. funding policies as erratic, unpredictable and ineffective in building the continuity and cooperation necessary for technological innovation (as opposed to basic science).

INNOVATION AS A PROCESS

The conference focused on advanced materials discovery and development prior to commercial innovation (high temperature oxide superconductors), on advanced materials inventions and innovations in nascent commercialization (diamond films) and in infant industries (silicon nitride structural ceramics). In all of these example, many of the conference discussions emphasized that we must keep in mind that the development of commercial products with advanced materials is a long-term process. Active development work with silicon nitride began twenty years before the first commercial application. Another decade passed before complex automotive parts entered the market place. Commercial markets for diamond films are just beginning some ten years after the critical innovation of growth on non-diamond substrates was achieved. Four years of intense research have not yet led to significant commercial applications of high temperature oxide superconductors.

Several lines of discussion arose from the idea of innovation as a process. One obvious topic was the Japanese commitment to long-term programs and their success in commercial production of silicon nitride auto parts. This stands in sharp contrast to the American experience. In order to have commercial development it is thought necessary to have a customer requirement that is matched by an appropriate level of manufacturing technology. Substantial government involvement occurred in the American DARPA/Ford/Westinghouse program that has been described. It was a successful demonstration with little prospect for early commercialization. Materials suppliers were subcontractors or vendors and corporate activities pretty much ended when government funding ended. In contrast, after the oil shock of 1973 a number of successive programs were initiated in Japan where a national consensus developed with regard to high technology fine ceramics of all kinds, including structural ceramics. The government (MITI), national laboratories and corporate research participated in what amounted to a coordinated but competitive program. One of the key technical innovations was pressure sintering which evolved in a national laboratory (NIRIM). Development of commercial equipment at NGK Spark Plug was supported by the government which subsequently was repaid through royalties. Kyocera with Isuzu and NGK Spark Plug with Nissan had long-term joint development programs that were ultimately successful. Kyocera developed a slip casting process using a liquid slurry in a mold to form complex shapes; NGK Spark Plug developed a very different injection molding process. The extensive networking and interactions between the private and public sector, the long-term commitment of all parties, long-term relationships between ceramic producer and engineer manufacturer cannot be explained by something so simple as cost of capital. Rather it seems that there is a national Japanese commitment to successful innovation, to learning by doing and perhaps to both national and corporate market share. This stands in sharp contrast to the American commitment to short-term profits and short-term stock holder interests. As part of the picture, some discussions emphasized the continuing series of corporate restructuring during the 80's. Others saw the low level of American employee loyalty to company and company to employee as favoring short-term projects in the U.S. environment.

Another aspect of innovation as process was seen as the danger of following the model of science history and identifying founding fathers and then tracing the record through their reminiscences (as we have largely done). An inevitable bias of linearity develops which tends to minimize the false starts, feedback loops, uneven progress and cumulative quality of technological history while over-emphasizing breakthroughs. It is thought we need more data about information transfer via graduate students and post-docs, more information about the vagaries of peer review and the personal nature of interactions between researchers and managers as well as researchers and sponsors. John Ogren mentioned the importance of a golf game in initiating TRW-Kobe Steel joint ventures. In a similar vein Robert Adams quoted John Reed, CEO of the Citicorp as saying "Bob, nobody knows the importance of golf games." David Larbalestier emphasized the importance of people in the history of superconductors. All of this emphasizes that we are just beginning to scratch the surface of the rich context of social organization, personal behavior and personal perceptions as they affect the on-going process of technological innovation.

With regard to process, Robert Adams suggested that by all means the most innovative approach to museum design in the United States is the one taken in San Francisco. Exhibitions are designed with what we are beginning to think of as "Japanese" process. People who want to develop an exhibit begin with a learning experience by bringing in specialists from a wide range of diverse fields. Then a multi-faceted design team including high school students (the ultimate customers) works as an unstratified group involving intense iteration and feedback from people coming through and using the exhibit. The result is often entirely unanticipated, and the methodology is becoming a shining example for the museum world.

GOALS, PLANS, AND OBSTACLES

National goals in the United States and Japan are set within a cultural framework of shared perceptions that pervade the corporate, political and legislative realms. In the United States a major share of research and development funding for technological innovation is focused in the Department of Defense and its support for the military-industrial complex. Goals are measured in terms of weapon systems which impact the advanced materials community in the form of material specifications set by weapons designers. The overall cost of defense systems is such that materials costs are but a drop in the ocean. The ultimate in performance is the objective rather than any concerns with economy. Other American goals for research, development and technological innovation are the health system in a country with a rising fraction of older citizens, energy conservation programs and environmental protection. With regard to commercial technological innovation the accepted myth in the United States is that the support of basic research will automatically lead to applied research, advanced technology, improved manufacturing and the development of marketable products and a strengthened economy along with an improved quality of life. There is an explicit faith that with this source of basic science discoveries the business of industrial innovation is best left to industry. The net result is that there is no American consensus or policy for purely technological goals outside the special fields of military, energy conservation, environmental protection and health research.

By contrast, in Japan there is an explicit national consensus and governmental policy aimed at expansion of the economy, improvement in the trade balance and placing manufacturing industry at the cutting of technological development in preparation for the future. Japan is a small island economy without many natural resources and there is strong national consensus among ordinary people, corporations, bureaucrats and legislatures to actively support what we might call the economic-trade-industry complex. This has led to extensive cooperation of government agencies and national laboratories with industry and the development of plans for technological development. It has led to a consensus that the consumer is the ultimate judge of innovative success rather than the designer. It introduces a much broader range of concerns. Within this national consensus the primary goal of manufacturing companies is focused on market share, as the governmental goal is based on national market share. Future position in the market is more important than current short-term profits. As a result, it is rational to have a much larger effort devoted to technological innovation than would otherwise be the case. The corporate goals in the United States have, during the 1980's, increasingly focused on stockholder return and short-term profits as opposed to long-term objectives. This has decreased the impetus for the development of long range technical innovation efforts.

Within these national goals developed on the basis of cultural perceptions of national interest, the American plan has been to focus on science as the source of new technology. In the Second World War the physics community contributed in a major way to the development of radar, weapons systems and the atom bomb. A myth was generated which is a strong American cultural commitment that basic science and America being Number One in basic science is the road toward successful technology and national welfare. In the light of the importance of science two foci have developed. One is the support of big science in the form of superconducting super colliders and space observatories. The second focuses on the importance of individual invention and the support of individual scientists. For this group the choice of problems is very individualistic and subject primarily to peer review of people in the same field. As a result, there is no significant overall plan with regard to the influence of science on technology innovation. NSF project managers have told me plainly that their directive is to select the best science independent of potential technical outcomes. This is different in the region of defense research in which the development of products has led to advanced materials studies focused on particular performance capability desired by weapons designers. Composites, for example, which have a combination of high strength and light weight, are particularly desirable for many weapons systems and have been the subject of much weapons-related materials research. Health research is another special field in which the clinical component and the development of applied medical technology and diagnostic systems have received a high level of support. Much of the basic research in the health field is identified with particular diseases and disease mechanisms.

The national consensus in Japan on developing manufacturing capability as a principle component of economic and trade objectives has led to a focus on technology and artifacts as the source of new innovation as opposed to basic science and military devices. The choice of problems has been much less individualistic than the United States and is coordinated with consensus-seeking interactions of government, industry, national laboratory and university participants. The development of a consensus for a plan of research has been followed by effective action in which there is also strong networking amongst the different constituencies. The overall plan involves a good deal of learning by doing. In advanced materials this can be seen in the production of trivial products such as zirconia scissors and knives and advanced silicon carbide and silicon nitride eyelets for fishing poles which are of little market value in themselves but provide experience in developing manufacturing equipment and manufacturing techniques. This consensus of purpose has also led to cooperative endeavors. We have mentioned the relationship between ceramic producer Kyocero and automobile manufacturer Isuzu and a similar arrangement between NGK Spark Plug and Nissan.

In the conference discussions a number of specific obstacles toward achieving planned objectives and goals were discussed. Since the more frustrated and more forthcoming discussants tended to be Americans, most of the discussions were about difficulties in American research. A number of people proposed that government regulations, particularly with regard to product liability and antitrust requirements, had a dampening effect on technological innovation. A number of people mentioned new products with a potential liability for which large corporations had formed new company entities divorced from the mother company so that the potential liability would be lessened. Other products in which liability problems seem significant were kept from the market. There was seen to be in the development of American research projects a focus on design and characterization of performance as opposed to processing. The general paradigm for Materials Science and Engineering is that processing and materials synthesis lead to properties, compositions and structures which in turn give rise to performace. In Japan there has been a much larger emphasis on processing and processing equipment which has led to a higher rate of technological innovation in the field of advanced materials. It is often thought that there is a rigidity of purpose in American development programs and an unwillingness to network with other researchers and change course in midstream. There is a tendency to suggest that one model fits all conditions while in reality technological innovation requires a flexibility of programs in order to adapt to the complexity of systems behavior. Some participants regarded the university tenure system which depends entirely on individual productivity as being harmful to interdisciplinary work at universities such as Materials Science and Engineering requires.

There's general agreement that a long time is required for the development of commercial innovation. This is seen in the materials discussed at the conference but it is also exemplified by transistor development, charge coupled devices development, superconductor wire technology and a host of other technological innovations. The short time frame of return for stockholders mentioned by a number of American participants was not repeated by Japanese who instead represented company interests as being more identified with the employees than with stockholders. The short range lead-time of American research created a difficulty in that there was a perception of continually changing targets that prevented technical innovations from ever reaching the manufacturing stage. Whatever the source, there clearly emerged feelings of frustration about the progress of American technological innovation that were not matched by our Japanese colleagues.

CORPORATE STRUCTURE AND CUSTOMER RELATIONSHIPS

Corporate structure has already been discussed to a great extent in the section on Management of Innovation. The participants present at the meeting represented Hitachi, Sony, DuPont, 3M, NGK Spark Plug and American Superconductor Co., all of which have been very successful at technological innovation. It seems that many individual companies in the United States and in Japan are successful in the initial stages of invention represented by high temperature oxide superconductors. There seems to be general equivalence in corporate results. In the diamond film nascent industry Japanese companies became involved earlier because the key technological innovation of diamond deposition on non-diamond films occurred in Japan and Japanese companies had a head start. In the infant industry of silicon nitride structural ceramics, the long-term process of development has led to the Japanese being clearly in the lead with their manufacturing capabilities. In a field with uncertain commercial prospects they have been willing to stay the course.

Many discussions focused on the long-term process of developing new technological innovations from a nascent to a commercial stage. This was thought to present problems for small venture units in that long-term support was required in advance of generating significant income, particularly for advanced materials developments. This has a number of consequences. In the United States long time support for venture efforts has been provided from mission-oriented units of the Department of Defense and also by companies like General Electric and IBM with large systems that justify substantial research on advanced materials. In Japan corporate structure has led to many large corporations, of which Hitachi is one example, to be involved in a variety of systems developments that make long-term research support appropriate. Also, Japanese companies are more concerned with market share and there is a corporate culture viewing the company as a family in which continuity and employee well-being are primary goals. In contrast American companies are required by financial markets to be much concerned with current profits and the 1980's binge of takeovers and job changes of corporate executives has further reduced any corporate feeling of family. There are exceptions on all sides, of course.

Another factor affecting innovation as a long-term process is the difference in employment practices. In the United States employment is by contract with the employer or employee terminating almost at will. As a result, employees tend to move from one company to another when better opportunities are presented. Under these circumstances it is rare for corporations to send their personnel to universities or national laboratories or other corporations for periods of outside training. In addition the team orientation with a team leader and the necessity to continue operations makes it uncertain whether a job will be available to such an employee coming back to the corporation. This decreases to a substantial extent any real person to person continued interaction with outside groups that might provide different points of view or new experiences. Employee specialists are optimum for both company and employee. In contrast, the long-term employment practice in Japan and also the organization of work in groups that have a variety of skills and a consensual approach toward their objectives makes it easy for Japanese employees to take a year or more off to attend graduate school, work as researchers at a university or national laboratory or even with a non-competitive company to increase skills and breadth of their experience. In the Japanese environment a generalist approach toward job responsibilities and activities is encouraged by the system. This differentiation between specialists and a generalists cuts both ways, of course.

An important factor in innovation has been shown to be the strength and extent of feedback from users and customers as to the performance of products and modifications in the use technology which should affect product design and manufacture. In Japan the long term vertical interaction and close relationships of suppliers and customers has provided a much more effective feedback system. Several times during the conference the relationship of NGK Spark Plug with Nissan and Kyocera Corporation with Isuzu was mentioned. These interactions developed and encouraged feelings of trust and commitment that are hard to imagine on the American scene where multiple suppliers has been a guiding principle. Dan Button of Dupont Japan pointed out that "In Japan our technical people are constantly with customers" and the technical service representative and researcher are the same person. The researcher visits customers to understand and help with their problems and then comes back to innovate. In the United States technical service and research are in separate compartments and the research people tend to look down on technical service. As a result, cycle times for Dupont are extraordinarily shorter in Japan.

Personal relationships developed through close interactions are important. In the United States these seem to develop most effectively in the military-industrial complex and in small venture companies that have not yet developed the bureaucratic structure of large corporations. In Japan these relationships extend throughout the culture. It has been suggested that this personal factor makes it important that many more engineers are represented in decision making positions in Japan than in the United States. Personal experiences at the highest level of Japanese corporate management give a bias toward technological innovations that is hard to quantify, but that many conference participants believed important.

Dr. Watanabe believed that for Sony the concerns of customers and feedback from customers require a constant attention. He indicated that in one aborted joint venture with an American company, the underlying cause for failure was Sony's perception that the American company was not sufficiently concerned with and listening to potential and current customers. He suggested that in Japan companies which serve only the government such as national railroad companies are inefficient as compared with private railroad companies. He lived in the United States for some years and developed the opinion, perhaps a bit excessive, that "basically American companies have no regard for customers".

IDEAS, INFORMATION AND TECHNOLOGY TRANSFER

The conference discussion on Japanese and American technology transfer essentially reinforced the discussion presented by Robert Cutler and presented earlier in this report. Technology transfer is a "people intensive" activity, a "contact sport". As is discussed in relationship to national laboratories, universities and corporations, Japanese researchers and technologists have a much more generalist role, while Americans tend to be much more specialists. This is important with regard to technological innovation in that the knowledge interface of technology and science discussed in the Innovation Model presented by Stephen Kline is only effective if there is substantial interfacing between the research side, scientific knowledge, and the engineering side, technological knowledge. All through the conference discussions of corporate scientists and engineers spending a year in national laboratories, the organization of the ISTEC laboratory for superconductivity being staffed by industrial researchers, the participation of national laboratory and university people with corporate researchers, the many meetings of industry, corporate and national laboratory representatives in establishing a consensus for technological programs (discussed by Kitazawa in the section on high temperature oxide superconductors) were indicators of an extensive networking of many different viewpoints in the Japanese technological culture. In contrast the more specialized orientation of American researchers tends to give rise to more intense and focused interactions. At Gordon Conferences and at professional society meetings contacts and information exchange are intensive between members within a specialty. Indeed, when members in this network change positions, moving from one university or corporation to another, they remain members of the intensive specialist network. These different cultural orientations are thought to give rise to more rapid process and product innovation in Japan and to more effective achievement of scientific innovations and breakthroughs in the United States.

An additional consequence of this mode of networking and science-engineering interaction in Japan as opposed to science-science interaction in the United States is a willingness and capability for Japanese to accept new ideas no matter what their source as opposed to what is described as the not-invented-here syndrome that has affected American innovation.

During the meeting a number of the American participants expressed the view that the Japanese were not only more willing to accept foreign ideas, but also that they spend more time in reviewing the literature and learning what had been done in a particular field of endeavor. Rustum Roy made strong comments about American researchers not being familiar with the literature in their field, an opinion which was supported by many participants. Several members of the conference recalled examples of early work which was unknown to current American researchers, but well known and referenced by their Japanese counterparts. It is well known that the ratio of Japanese visitors to the United States compared to American visitors to Japan means that the Japanese are much more familiar with American innovations than vice-versa. In addition though, several members of the conference who had spent many months in Japan reported that there was less interest or curiosity by their American colleagues than they had anticipated. No one had any deep interpretations of these observations. Perhaps they represent one aspect of the American focus on individual achievement as opposed to a Japanese appreciation of group performance.

THE SCIENCE-ENGINEERING NEXUS

During the Second World War physicists in the United States were drafted into the war effort, played an important role in the development of radar, proximity fuses and other developments including the atomic bomb. After the war, these successes led to the establishment of the National Science Foundation and the beginnings of a myth that basic science accomplishments inevitably lead to applications, technological innovations and an improved quality of life. As applied to the military, the application of basic science would lead to weapons technology that would protect that better life. That linear hypothesis that science leads to technology which leads to manufacturing and performance has permeated American thinking and remains a dominant theme of scientists, policy makers, corporate executives, legislatures and everyday people. This has led to a preeminence of science in the American technological culture that has no counterpart in Japan.

In discussions of science, there tended to be an invocation of a deterministic reductionist physics model. But a number of participants pointed out that in fact the nature of science has been changing appreciably with the development of quantum mechanics on a statistical basis and the development of research programs in medical science in which laboratory investigations and clinical investigations proceed simultaneously and a development of science evaluation of chaotic systems, particularly in applied sciences such as plate tectonics and meteorology. The very structure of science is being transformed in a way not dissimilar from previous discussions about cultural changes with time. That picture of science as a moving target is perhaps less applicable to technological innovations in advanced materials which depend in a large part or related in large measure to condensed phase physics and chemistry as their siblings.

The Japanese evidenced a different understanding of science than many of their American counterparts. Throughout the discussions it became clear that Japanese see both science and technology as a process of learning by doing as opposed to an American concept of the development of technology founded on principles. American participants in Japanese laboratories emphasized the Japanese concern with an experimental approach whereas American laboratories were more concerned with theoretical questions. In one laboratory almost half of the Japanese Ph.D's. received their doctorate while working at the Institute. They were hired with a bachelor's degree and received the Ph.D. by reporting their accomplishments and publications to a university program. American participant observers saw the Japanese as being more attentive to detail and data, being more deliberate in evaluations and avoiding jumping to premature conclusions. The difference in focus became clear during a discussion of a program at Tokyo University aimed at developing a "science" of manufacturino. An American participant suggested that a science of manufacturing was impossible. There are so many variables involved in the socio-technological manufacturing system that describing this process in terms of "science" was just inconceivable. In reply it was argued that a science of manufacturing was very much like computer science which is a recognized field in many American universities. (However, computer science is almost always coupled with Electrical Engineering and exists in Colleges of Engineering). This also engendered the comment by Japanese participants that in general university professors in Japan are more involved with and concerned about manufacturing. In the United States university professors generally consider manufacturing too complex a problem to fit into an academic environment.

Many discussions turned on the idea that the relatively sharp distinction between science and engineering in the United States did not have a comparable counterpart in Japan. Brian Pfaffenberger proposed a illustration of a Japanese culture of science and engineering as a uniform subculture in Japan as compared to separate subcultures of science and engineering in the United States, Fig. 1. Science in the United States is emphasized as a more praiseworthy activity with substantially more prestige attached to it than is engineering. Science allows the practice of individualism and individual accomplishment to be particularly rewarded as compared with engineering which tends to be a group activity. Since in Japan groups function both in science and engineering, this distinction is not generally made. The diffezence in status or pecking order between scientists and engineers in the United States does not have a counterpart in the Japanese technological environment. It is thought that this separation also tends for Americans to be more concerned with disciplinary boundaries, a differentiation between Ph.D.'s, researchers, and technicians and between science and engineering cultures.

Some anthropologists suggest that it may be difficult to develop an anthropological approach to American science because scientists are trained to eliminate any indication of bias from their work in the effort to be objective. *Ex post facto* scientists portray their work as if it conforms to a model of the scientific method and the suggestion that there may be cultural factors at work can be threatening, perhaps even frightening. An indication of cultural bias in hard science might be difficult to sustain a participant observer basis. Technological innovation is essentially an heuristic process which may or may not have serious theoretical foundations. As a result, the culture of different research and technology laboratories was perceived as contributing to innovative success.

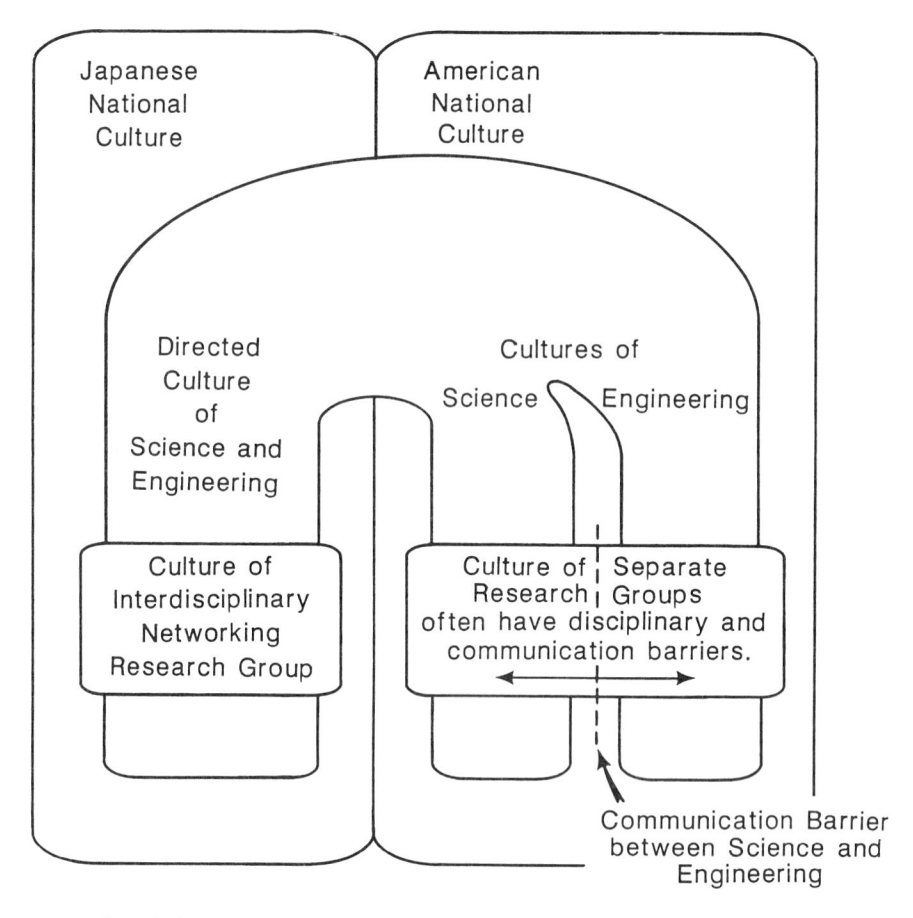

Fig. 1. Schematic view of the way in which Japanese highly directed culture of science and engineering contrasts with the separate American cultures of science and engineering. Each is embedded within its own national culture and both act as bridges between the national culture.

UNIVERSITIES

First of all, we need to introduce the caveat that the discussion was specific for advanced materials and applies pretty much to colleges of engineering but not to the totality of universities. A number of participants suggested that medical and biomedical fields which involve practical work in a clinical setting at universities may be quite different. With regard to advanced materials, both in Japan and the United States, there was general agreement that universities are not contributing much in the way of technological innovation. It was suggested that in large part this results from the fact that the substantial capital investment necessary for modern manufacturing facilities is not available at the university; the university role has been to educate students in fundamentals rather than with actual experience in manufacturing. It also seems that faculty members are rarely well informed about production activities. The interaction that faculty members have with industry is largely research university people interacting with industrial research people or government laboratory research people, all of whom are relatively naive about actual manufacturing processes.

In Japan the most popular way that industry supports universities is to contribute research funds identified with a particular professor. These funds are donated to the university for the use of the professor and the university has no obligation toward the company but it is often agreed between the individual professor and the industrial donor to work in a particular field. Industrial participants from Japan indicated that these amounts are not very large, usually about 1 million yen ($7,000), and the major industrial motive is to develop relationships with professors in order to be able to obtain the best students for employment. Exchanges of information and interaction between university and industry mostly occur at academic meetings of one sort of another where industrial people are always looking out for new ideas, new techniques. Many students who work in a company keep in a close relationship with their professors. Participants who have experience both in the American and Japanese context commented on the fact that networking with former professors, interaction of students with professors and the esteem with which students regard former professors is markedly higher in Japan than in the United States. In Japan it is mostly through contacts with former students that university faculty continue to be aware of technological problems.

As in Japan, industry is a minor source of support for universities and university research in the United States. By far the largest support of American university research is the government. Industrial support has increased to a certain extent and one of the common forms is consortia arranged around a particular area of technology such as the industrial consortia on diamond research at the Pennsylvania State University. Industrial participants at the conference indicated these are not particularly expensive and the main objective of industry is a defensive one to insure that they are not missing something which may turn out to be important. University participants agreed that industry, while willing to contribute a limited financial support, was generally unwilling to send their best people (or any people) to actually participate in these programs at the university. There was a general feeling amongst American university faculty at the meeting that industry was not really serious about their support for joint programs with university researchers. For an industrial participant to take time to seriously interact with a university represents a diversion from a career path. At the same time not very many university faculty spend time in engineering activities of industry as opposed to interacting with research departments.

During the last few years, beginning in 1987, there have been established in Japan centers for cooperative research at national universities. At present there are total of 18 national universities in different areas of Japan which have established this sort of center in order to promote closer relationships between university and industry. In Fig. 2 there is illustrated the organization of this type of center at Nagasaki University. One element of this

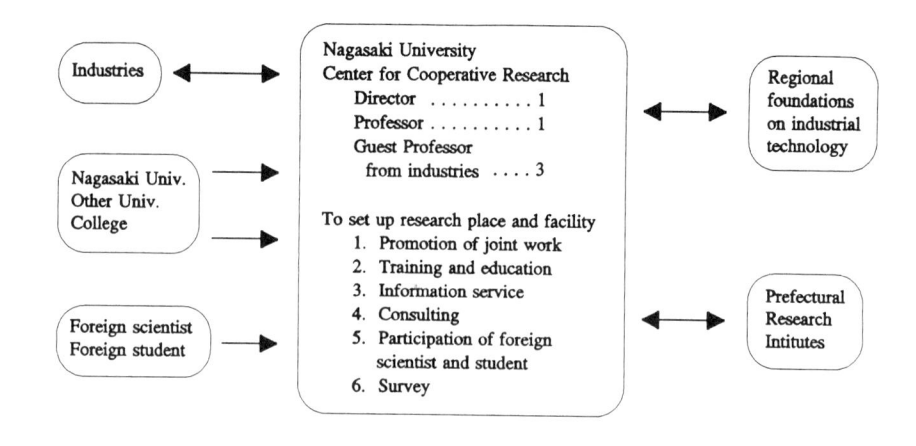

```
┌──────────────┐         ┌─────────────────────────────┐
│  Industries  │ ◄─────► │ Nagasaki University         │         ┌──────────────┐
└──────────────┘         │ Center for Cooperative Research │ ◄───► │ Regional     │
                         │    Director .......... 1    │         │ foundations  │
                         │    Professor ......... 1    │         │ on industrial│
┌──────────────┐         │    Guest Professor          │         │ technology   │
│ Nagasaki Univ.│ ─────► │      from industries .... 3 │         └──────────────┘
│ Other Univ.  │         │                             │
│ College      │ ─────► │ To set up research place and facility │
└──────────────┘         │  1. Promotion of joint work │
                         │  2. Training and education  │
                         │  3. Information service     │         ┌──────────────┐
┌──────────────┐         │  4. Consulting              │ ◄───►   │ Prefectural  │
│Foreign scientist│ ───► │  5. Participation of foreign│         │ Research     │
│Foreign student │       │     scientist and student   │         │ Intitutes    │
└──────────────┘         │  6. Survey                  │         └──────────────┘
                         └─────────────────────────────┘
```

Fig. 2 Organization of the Nagasaki University Center for Cooperative Research

is that there are guest professors from industry, resident at the university, in an effort to develop closer relationships of university-industry programs.

It was generally agreed that the tone of university research is largely set by funding patterns in which the industrial contribution is not generally significant. In the United States, the largest funding for university research comes from the National Science Foundation and from mission-oriented arms of the Department of Energy and Department of Defense. The National Science Foundation generally supports science research independent of technological objectives. The mission-oriented arms of the Department of Energy and Department of Defense support both basic research and also research which is aimed at more technological objectives. It is the efforts of these organizations, such as the Office of Naval Research support of diamond film studies at Pennsylvania State University and the Air Force Office of Science and Research which has supported research on the sol gel processing of ceramic materials and the introduction of more chemical applications to ceramics processing that have had the most influence on technological innovation at universities. In Japan the principle support for university faculty and programs comes from the Department of Education and influence on technological innovation generally comes when new professorships are established.

In the United States there was a period when new ventures originating in university developments were supported by venture capital companies with strongly industrial and technological innovation impetus. Increasingly venture capital companies in the United States have been concerned with other than start-up activities and the main support for start-up venture research has been contract research from mission-oriented Department of Defense units. As a result, venture capital start-up research in the United States is increasingly oriented toward technological innovations related to military applications rather than consumer markets.

One further topic that came up is the effectiveness of the materials research laboratories set up at American universities beginning some twenty years ago. They were founded because it was felt advanced materials research was interdisciplinary in industry and that universities' departmental structure and disciplinary orientation made(it desirable to have block funding. Several hundred million dollars have now been invested in these material research centers. The general opinion of participants was that this has not much changed materials research at the universities. Funds have normally been divided between Departments of Physics, Materials Science and Engineering, Electrical Engineering, Mechanical Engineering and Chemistry in a way that has retained disciplinary emphasis. There has been little significant difference between levels of technological innovation and behavior patterns between the haves and have nots. There was also a good deal of feeling that the funding pattern for most advanced materials research required faculty to spend a substantial part of their time, estimates varying from 20% to 40%, on writing proposals and pursuing funding. Faculty are not normally provided with funds for research assistants or assistant professors or students as is done by the Ministry of Education in Japan. As a result, faculty interests and research orientations are largely dictated by funding agencies, principally NSF, uninterested in technological innovation and contributions to economic competitiveness.

NATIONAL LABORATORIES

In the United States the major national laboratories developed in the course of weapons development, particularly the Manhattan Project, in World War II. Some have continued as primarily weapons laboratories, mostly focused on nuclear weapons, but most have searched for a rationale for continuing existence and development. They have become part of the national science infrastructure built around the support of large-scale machines such as nuclear fission reactors, nuclear fusion experiments, synchrotrons and particle accelerators. The latest in the series of development is the superconducting super colliding facility now being built at a cost of some ten billion dollars. In the last two years, the mission of supporting technological innovation and carrying joint research with industrial forms has been added to their role. In contrast, the national laboratories in Japan are mostly much smaller facilities, managed by the Agency of Industrial Science and Technology (AIST) under MITI. Under the AIST program there are sixteen national laboratories dispersed throughout Japan and about 2600 researchers. The basic objective of the AIST laboratories is to strengthen government-industry-university interactions and foster technological innovation. In the last few years, beginning in March 1986, the fundamental direction of the Japanese science and technology program included the promotion of creative science and technology, strengthening fundamental and basic research, a harmonization between technology and human beings and international cooperation. That is, changes are underway in which it is proposed that the American national laboratories contribute more to the country's technological base and Japanese national laboratories contribute more to that country's science base.

Argonne National Laboratory which has an active program in place attempting to develop high temperature oxide superconductor technology in cooperation with industrial partners is more or less typical of American national laboratories. The laboratory has three components, one concerned with breeder reactors and breeder reactor development, an application of large scale science that would not be possible without a national laboratory role. The second component of the laboratory is basic science and a third componmnt concerned with energy and environmental materials is associated with applied science and technological development. During the history of the laboratory, more or less continuing managerial efforts to encourage interdisciplinary activity and interactions between these separate components have not been very successful. The laboratory personnel have been very interactive with university researchers but there has not been much interaction with industry. One of the results of the discovery of high temperature oxide superconductors was to establish a common set of objectives between basic scientists eager to characterize and study the new materials and people in the applied group who had the capability of processing and preparing samples.

Until recently the national laboratories had no mandate to become interested in technological innovation or industrial competitiveness. It is too early to tell how effective the national laboratories will be in these new responsibilities. Laboratory participants indicated that there were serious problems related to changing a culture which had been devoted to science and the objectives of scientists who have no particular special knowledge with regard to engineering or technological innovation. A second difficulty seems to be in relationship to the funding method in which there is a competition between basic science and industrial cooperation whereby funds are awarded on a yearly basis by bureaucrats in Washington who some participants thought were not particularly skillful in separating hype from solid content in the proposals they receive. The conference heard reports that continuity of funding is a continuing concern of staff members who might better be focusing on their technical activities. A third difficulty is initiating and finding collaboration between the highly scientific Ph.D. on the Argonne staff and industrial representatives, the most needful of whom have but little technical background. There was a feeling that technology transfer is only feasible between almost equal partners and that is difficult to achieve for companies not having research laboratories of their own. The upper levels of the Department of Energy are supportive of this new national laboratory-industrial cooperative program but the difficulties in changing the national laboratory science subculture should probably not be underestimated.

Most of the discussion of the operation of Japanese national laboratories focused on the specific example of the Government Industrial Research Institute of Kyushu (GIRI-Kyushu) sponsored by AIST under MITI because Dr. Kazuo Kobayashi, former director of GIRI-Kyushu was a conference participant. At GIRI-Kyushu there are three types of research projects undertaken. One is a large national project, a second are special regional technology projects and a third is ordinary basic research undertaken by staff members.

For large national projects, there is extensive interaction between AIST headquarters and advisory academic committee between the Government Industrial Research Institutes and industrial consultants as to the needs and plannings and benefits of a proposal as a national project. Examples of these have been the Sunshine Project, the Moonlight Project discussed under silicon nitride research and basic technology for future industries. AIST headquarters works as a coordinator of the project and the term of this kind of project is usually five to ten years. Different government research institutes jointly participate in a project in which industrial research laboratories are also associated. The interaction between government and industry continues throughout the program with member companies and national laboratories working together to achieve the project goals which are in the field of "precompetitive" technological development. The schematic organization is illustrated in Fig. 3.

A second type of technological research program carried out by the regional GIRI laboratories are special regional technology projects. AIST initiated this program in 1982 in order to promote technologies in regional industries. The regional GIRI prepares a plan in association with a number of regional private companies and some prefectural government research institutes. The lifetime of the project is usually four or five years. An example of this in the GIRI-Kyushu laboratory was a program on "Utilization of Lime and Lime-Based Compounds"; another was "Development of New Pottery Clay Using Low Refractory Pottery Stone". This is coordinated by the GIRI laboratory with a structural organization such as illustrated in Fig. 4. Another type of special project is the joint government-private sector research program in which half of the budget comes from GIRI and the other half comes from the private sector. Physically the research is carried out in

Big National Project
(ex. Sun-shine, Moon-light, basic technology for future industries)

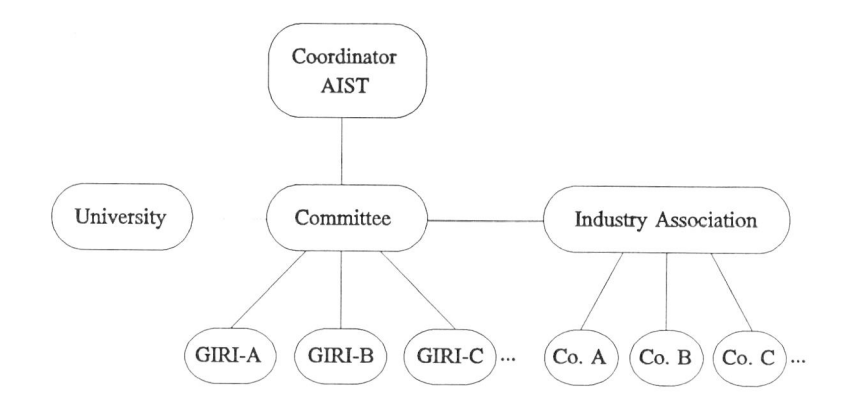

Fig. 3 Joint research system of the big national project.

Regional Technology Joint System

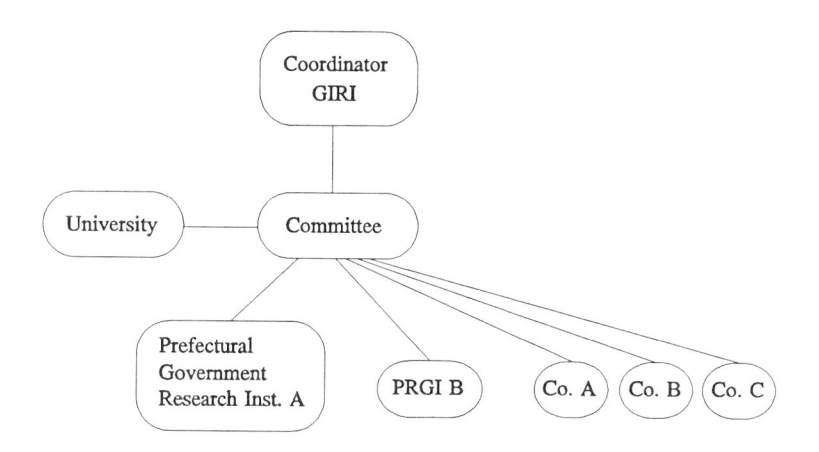

Fig. 4. Joint system of the project on regional technology.

the GIRI laboratory to which researchers are sent from the companies. Guest researchers do not necessarily have to be associated with a joint government private sector research. GIRI-Kyushu generally accepts about thirty researchers each year in which the company does not provide funding except for the researchers salary and costs of materials. This is usually done in order to take advantage of GIRI capability in a special area while for GIRI guest researchers are a source of additional manpower.

Another form of research carried out at the Government Industrial Research Institute is basic research built on proposals prepared by individual researchers. These proposals are evaluated by the director of GIRI on the basis of originality, research potential and influence on industry through discussions with individual researchers. It often happens that this basic research grows into becoming special projects.

Other activities of the GIRI regional laboratories are to serve as technical information centers. At GIRI-Kyushu the number of requests for technical information is about 300 each year. The GIRI laboratories also sponsor symposia on particular subject matters such as new materials, the development of natural resources and so forth at which 100-150 participants from regional industries participate. GIRI and the other national laboratories also have an invitation program for university professors. At the GIRI-Kyushu each year three or four professors are invited to stay three or four days for discussions with research groups at the Institute. Finally, GIRI has a system of sending its researchers to universities for study and research for a period of one year each. In the case of GIRI-Kyushu there are one or two researchers sent to universities each year.

The other Japanese national laboratory which was discussed was the National Institute for Research on Inorganic Materials (NIRIM) which has been described by its director general Dr. Setaka under a previous discussion of diamond films. At NIRIM there are about one hundred ten staff of whom about forty at any one time are visiting researchers from industry. From both the GIRI Research Institutes and the national laboratories such as NIRIM, there is an active program encouraging the application and licensing of technology by industry. For promising technologies funds are provided for the necessary development of equipment for industrial processes which is paid back by industry if and when processes become commercially successful.

An impressive characteristic of the national laboratory program in Japan is first of all a dedication to process studies, new material synthesis and technological innovation. A second characteristic is the extensive networking of the national laboratories with university faculty, who often serve as chairman of planning committees, with industry and with government coordinators. Programs are funded for a period of four to ten years so there is a continuity absent from most of the American national laboratory programs. In addition there is a strong consensus building prior to the establishment of programs which has been documented above for the case of ISTEC in Dr. Kitazawa's discussion under high temperature oxide superconductors. The efforts since 1986 to develop a larger science component in the national laboratories has probably not been so successful. Some Japanese university scientists suggest that the government research projects are not considered very successful when compared with university programs from a science point of view or with industry programs from a technological point of view.

SUMMARY

As discussed in the opening of this volume, the interactions of culture and technical innovation constitute a multiple variable, complex social-culture-technological process that is not subject to a facile reductionist one-line sound bite conclusion. That's in the nature of the questions we have posed and all the discussion at the meeting argues against giving too much credence to simplistic statements based on anecdotal evidence. That said, the conference has made some progress in initiating a process leading to better understanding of these issues. There was unanimity amongst those present that each participant took away with them a solid sense of benefit.

The idea that Americans are highly creative and that the Japanese have taken our breakthroughs and then transformed them into commercial success has come to be accepted wisdom in many circles. Our discussion of the discoveries associated with high temperature superconductors, with low pressure diamond synthesis provide no support for this view. In high temperature superconductors where Japanese and American researchers have left the starting gate together, the rate of new discoveries has been pretty much equal. Japanese research has been as creative as their American counterparts and vice versa. Similarly, in the discovery phase of low pressure diamond synthesis the Japanese have taken the lead. Dr. Robert DeVries, a distinguished General Electric Company researcher in the field and author of some histories has said: "Most of the creative advances in CVD diamond are of Japanese origin." In consequence it seems that many analyses of American vs. Japanese creativity is resulting from different educational systems, for example, are based on a faulty premise and largely without merit.

It is clear that the number of Nobel prizes in the basic sciences that have been garnered is overwhelmingly in favor of U.S. scientists. The conference discussions suggest that this may result from a profound difference in the Japanese and American concepts of "science" and of learning. In the United States the essence of science is taken as the development of principles and laws that can serve as fundamental concepts underlying future experiments and providing a basis for technological development. In Japan science is rather taken as a process of methodical experimental approach toward learning. Experimental results are important because an experimental fact is "truth". Explanations and interpretations are less important. In the examples of advanced materials we have discussed, this Japanese concept of science has led them to focus more strongly on processing of materials and synthesis. In contrast American research has been much more dominated by characterization of materials based on condensed state physics. Japanese participants suggest it is possible to develop a science of manufacturing analogous to computer science. In contrast most American participants suggested that manufacturing was a socio-technological activity depending heavily on human behavior, human perceptions and social organization. As a result, manufacturing analysis involve thousands of variables, perhaps millions, that can never be reduced to a "science". There was the view that a reductionist Taylorism approach to time and motion studies or any more comprehensive models that developed hardly warranted the application of the word "science".

This fundamental conceptual difference is deeply embedded in the American and Japanese cultures. It was influential in many aspects of our discussions. On the American side there is a national consensus that science is the goose that lays the golden eggs which inevitably lead to applied science, technology, innovative manufacturing and a better quality of life. This has been shown to be myth by many historians of technology, but is a widely and deeply held view of most scientists, policy makers, legislators and ordinary people. A successful moon landing is widely perceived as a scientific success while the Challenger disaster was viewed as an engineering failure. There is a current 1991 proposal by AAAS President Leon Lederman that we should increase science funding so that every American physicist, chemist and biological scientist is fully supported. This has also led to bifurcation of the science and engineering communities in the United States. It is exemplified by our separate National Academy of Science and National Academy of Engineering. By and large the intellectual pecking order places scientists above engineers. Many in the advanced materials community feel that the ten billion dollars or so funding the superconducting super collider program would be better devoted toward technology which would enhance American competitiveness. But there exists no national American consensus as to the importance, perhaps even the desirability, of technological innovation.

In contrast, in Japan science and engineering are seen as constituents of a single community. There is, in Japan, an air of technology enthusiasm, which has passed into history in the United States. Thomas P. Hughes, in his book *American Genesis* quotes Perry Miller who describes the American exhilaration with technology at the turn at the century. Americans "flung themselves into the technological torrent, how they shouted with glee in the midst of the cataract, and cried to each other as they went headlong down the chute that here was their destiny...." This same spirit seems to exist now in Japan and is exemplified by what has been described as a "ceramic fever" engaging all segments of the population during the 1970's. There is in Japan a national consensus shared by scientists, engineers, corporate CEO's, politicians, legislatures and ordinary people that Japan has a national imperative to place itself in the forefront of evolving technologies. These different cultural visions of science and technology in Japan and the United States have a profound influence on the course of technological innovation in the field of advanced materials. In America our national laboratories are mainly centers of science. In Japan the clear objective of the national laboratories is technological advancement.

Having seen a rough equality in the creativity and inventiveness in Japan and the United States, we turn to the question of transferring inventiveness into commercial innovation. The Japanese cultural mindset of "learning by doing" as compared to the American mindset of "learning from principles" has proved to have a profound influence on technological innovation. It has led to the Japanese accepting, even enthusing about, incremental innovations which have a large cumulative effect. This is contrasted with an American predilection for home grown breakthroughs. It has also led to a Japanese willingness to undertake the manufacture of advanced materials for insignificant markets as a way of learning, with confidence that as learning proceeds and costs and quality improve, markets will develop.

A second profound cultural difference that permeates the mode of technological innovation is the strong American emphasis on individual achievement vs. the strong Japanese emphasis on group achievement. Japanese groups are interdisciplinary and consist of Ph.D.'s mixed together with technicians to provide a broad array of interactive experience and approaches toward problem solutions. This has been commented on by many observers, as has the difference in Japanese system of lifetime employment and promotion based mostly on seniority as compared with the American system of contract employment, frequent job changes and promotion based on individual accomplishments. With regard to innovation an important consequence of these national characteristics is immensely more effective interdisciplinary, intracompany and interorganizational networking by the Japanese. The number of visiting industry researchers working alongside the regular staff at national laboratories in Japan, for example, is astounding by American standards. Americans leaving their home organization find themselves departing from established career tracks and, in any event, their companies can never be sure that they will return. These same characteristics of individuality give rise to more intense specialization, stronger disciplinary boundaries and effective networking *within a scientific field* in the United States, reinforcing and strengthening the effectiveness of the American science establishment, but isolating scientists from technological innovation. In Japan a group member in an industrial (or university or national laboratory) organization is encouraged to spend a year or more in a different organization, learning new skills and returning to a secure position able to enhance group performance. The Japanese social structure makes individual stars and nobel prizes rare, but much enhances technological and commercial innovation.

The strong "vertical" social organization in Japan that Chie Nakane identified as central to the family, village and company in her classic 1970 book *Japanese Society* continues to exert a strong influence in many ways. It is a factor in the effort to develop consensus and harmony rather than confrontational decision-making. One result is that there is a strong input to decision-making in Japan coming up through the hierarchical organization, bringing more contributions of potentially creative and innovative ideas. The emphasis on consensus has led a number of observers to quote a Japanese saying, "The nail standing out from the rest is the one hammered down," suggesting that creativity is hampered. In fact, Japanese managers appreciate this and have generally assured that there is time reserved for researchers' risky ideas to be accomplished as under-the-table activities that only surface when it becomes appropriate to solicit group consensus. Perhaps the most important aspect of vertical organization is the deep trusting relationships between suppliers and manufacturers and between manufacturers and customers. The most important feedback loops for technological innovation are from customers to manufacturers and manufacturers to suppliers. The strength of these feedback loops in Japan is a strong factor in the rapidity and effectiveness of Japanese technological innovation. Similar close interactions seem to be a characteristic of American military technological innovation.

The examples of invention, nascent industry and infant industry that were the focus of our discussions all converged on the realization that technological innovation is not an event, but rather a long term process. Japanese corporations have a focus on market share rather than short time profits that gives them a greater continuity of effort which is reinforced by employment practices and multidisciplinary group efforts. Some of this has been attributed to lower costs of capital, but that can hardly explain the Japanese predilection for four-five year funding of development programs as compared to the normal one year period (subject to renewal) common in the United States. Single year funding emphasizes rapid results independent of any costs of capital (it even leads to proposals being written for work already done so as to guarantee results).

Policy implications have been mentioned by some participants but were not the subject of much discussion or debate. However, it must be clear that the effects of an American culture so strongly ingrained with the importance of science that "scientific literacy" has become a buzzword for our educational requirements with little or no consideration of engineering and technology are substantial. Similarly, the Japanese immersion in the faith that technological innovation is the decisive factor shaping the future can hardly be ignored.

Finally, these experimental discussions made clear that there is a strong opportunity for a new anthropology of technology. Technological innovation is an activity, a process, involving social organization and human behavior that are equally or more important than technical considerations. This conference and this approach toward thinking about technological innovation must be seen as the initial often uncertain beginnings of a process. Several caveats must be heeded before making too much of any conclusions. First, all the discussions were between and focused on a particular age group. Second, technological innovation in advanced materials seems to be quite different than innovations in medicine or biotechnology and some other fields. Third we have not had the depth of discussions of technological innovation by the American military that their success would seem to warrant. Most important, the development of rapid transportation, international television, telephone and fax machines, the globalization of science, industry and technology combine to produce a breathtaking rate of change in all aspects of the industrialized world.

Author Index

A
Adams, R. McC., 233
Angus, J. C., 136

B
Bradt, R. C., 91, 198
Buckett, M. I., 81

C
Cutler, R. S., 67

E
Elder, T., 224

K
Kamo, M., 143
Katz, R. N., 166
Kii, T., 93
Kingery, W. D., 59, 263
Kitazawa, K., 119
Kline, S. J., 43
Kobayashi, K., 184, 192

M
Maquet, J., 35
Matsuda, S.-P., 105, 131
McLean, A. F., 211

Moriyoshi, Y., 147

N
Netting, R. McC., 99

O
Ogren, J., 157

P
Pfaffenberger, B., 249
Poncelet, E., 3, 23, 259

R
Richerson, D. W., 176, 201
Roy, R., 115, 151, 253

S
Sato, Y., 143
Schiffer, M. B., 235
Setaka, N., 142

T
Tajima, Y., 194

W
Watanabe, S., 216

Subject Index

C
ceramics
  silicon carbide
    research in Japan, 184
  silicon nitride, 161
    automobile engine components, 194
    gas turbine components, 166, 184
    innovations, 201
    research in Japan, 184
  structural
    research in Japan, 192
  synthesis, 143
cultural influences
  in Japan, 3, 23, 35
    and technological innovation, 105
  in Japan versus in the United States, 259
    and ceramics research, 184, 192
    and innovation, 23, 35
    and management behavior, 93
    and superconductor research, 119
    and technological development, 235
  and management, 235
  and managing knowledge, 253

D
diamond film technology, 151, 157
  policy and cultural influences in, 151
diamond synthesis
  chemical vapor deposition, 136
  high-rate synthesis, 147
  Japan–United States joint ventures, 157
  low-pressure, 135, 147, 151

E
engineering versus science, 249

I
innovation
  in advanced material technologies, 59
  and high-technology transfer practices, 59
  Japan versus the United States, 23, 35
  and management policies, 207
  and public policy, 43

J
Japan
  ceramics research, 184, 192
  cultural effects on innovation, 3, 23
  culture, 1
  diamond synthesis research, 147
  Japanese research students in the United States, 91
  Ministry of International Trade and Industry policies, 119, 192
  research environment in, 67
  superconductor research, 119, 131

M
management
  cultural influences on, 235
  effects on innovation, 207, 211
  in Japan, 99
  Japan versus the United States, 216
  and small venture groups, 224
  in the United States, 243
Ministry of International Trade and Industry (MITI). See Japan

O
Onnes, Heike Kammerlingh, 111

S
science versus engineering, 249
silicon nitride ceramics. See ceramics
superconductors
  history of, 111, 115
  research in Japan, 119, 131

International Superconductor Technology
 Center (ISTEC), 131
Superconducting Generator and Materials
 Program (SUPER-GM), 131
research in the United States, 115, 119

T

technology transfer policies, 59

U

United States
 cultural effects on innovation, 3
 culture
  and innovation, 23
 industry–science–education triad, 243
 Japanese research students in the United States, 91
 superconductor research, 119